Finches, Bowerbirds & Other Passerines

OF AUSTRALIA

Finches, Bowerbirds & Other Passerines

OF AUSTRALIA

Ronald Strahan, AM

Editor

National Photographic Index of Australian Wildlife

Angus&Robertson

An imprint of HarperCollins*Publishers*

PRODUCTION TEAM

EDITOR-IN-CHIEF

Ronald Strahan

SCIENTIFIC CONSULTANTS

Walter E. Boles

John H. de S. Disney

COLLECTION MANAGER

Denise Greig

SPECIAL PHOTOGRAPHY

Maurice Ortega

Angus&Robertson

An imprint of HarperCollinsPublishers, Australia

First published in Australia in 1996

HarperCollinsPublishers

25 Ryde Road, Pymble, Sydney NSW 2073, Australia
31 View Road, Glenfield, Auckland 10, New Zealand
77–85 Fulham Palace Road, London W6 8JB, United Kingdom
Hazelton Lanes, 55 Avenue Road, Suite 2900, Toronto, Ontario M5R 3L2 and 1995 Markham Road, Scarborough, Ontario M1B 5M8, Canada
10 East 53rd Street, New York NY 10032, USA

National Library of Australia Cataloguing-in-Publication data:

Finches, bowerbirds and other passerines.
Bibliography.
Includes index.
ISBN 0 207 18880 7.
1. Australian finches. 2. Bowerbirds – Australia.
3. Passeriformes – Australia. I. Strahan, Ronald, 1922 – .
II. National Photographic Index of Australian Wildlife.
598.8

Printed in Hong Kong

Cover photograph: Golden Bowerbird; H. & J. Beste

9 8 7 6 5 4 3 2 1

99 98 97 96

Foreword

This tenth volume concludes the work of rather more than a decade. It was conceived by Donald Trounson, founder and first Executive Officer of the Australian Museum's National Photographic Index of Australian Birds, which was subsequently expanded to include all Australian mammals, reptiles and frogs, and renamed the National Photographic Index of Australian Wildlife. This photographic collection now resides with the Museum's commercial photo-agency, Nature Focus.

In addition to creating an administrative basis for the project, Trounson's main contribution was to recruit a uniquely comprehensive photographic coverage of the birds of Australia, calling upon the resources of virtually every wildlife photographer – professional or amateur – in the nation. Although he came to know a great deal about birds, he made no claim to be an ornithologist, leaving biological matters to a number of professional colleagues, particularly Terence Lindsey, who mapped out the contents of the 10 volumes and played a leading role in compiling the first four of these.

The authorship of the first two volumes – *Wrens and Warblers* (1982) and *Waterbirds* (1985) – involved a unique partnership. Lindsey compiled copies of all relevant scientific references and passed these to the eminent writer, John Douglas Pringle, who used them as the basis of the text, subsequently checked for accuracy by Lindsey and other scientific advisers. The third volume, *Seabirds*, was written entirely by Lindsey. The fourth, *Shorebirds*, reverted to the original collaboration, but this was a field in which Pringle had considerable knowledge so it was appropriate to credit him as author of the text, Lindsey being responsible for the technical synopses.

No books were produced in 1989 or 1990, a period during which the future of the program was uncertain. Production was resumed with the publication in 1991 of *Robins and Flycatchers* by Walter Boles and *Honeyeaters* by Wayne Longmore in 1992.

Moving thereafter into more heterogenous groupings of birds, it became impracticable to rely on single authors. *Parrots and Pigeons* (1992) was the joint work of Francis Crome and James Shields. *Birds of Prey and Game Birds* (1993) involved Penny Olsen, Francis Crome and Jerry Olsen. *Cuckoos, Nightbirds and Kingfishers* (1994), which dealt with 17 groups of birds, called upon the expertise of 13 ornithologists. *Finches, Bowerbirds and Other Passerines* is the work of seven ornithologists, all but one of whom were contributors to the previous volume. Altogether, some 40 people have been responsible for the creation of the series, which comprises almost 4000 pages. Vincent Serventy edited the first two volumes, while I have had the honour, and travail, of supervising the remainder.

The editorial process has been significantly assisted by such expert consultants as Walter Boles, Stephen Davies, John Disney, the late Shane Parker and Ederic Slater. Documentation and care of the tens of thousands of photographs that have been considered for publication have been in the capable hands of successive archivists: Mary Rose Gordon, Heather Lawrence, Sally Bird and Denise Greig. Special photography, mainly of eggs, has been carried out by Anthony Farr, John Fields, Kate Lowe and Maurice Ortega.

Donald Trounson was highly skilled in raising funds from government and corporate sources. He also obtained support from individual donors by introducing an arrangement whereby illustrations could be sponsored in return for tax-exempt donations. The project remains indebted to these sponsors, many of whom expressed their faith more than a decade ago and have had the patience to see it gradually come to fruition.

With the conclusion of this enterprise, it can be said with some confidence that nothing of its size and scope, based on photographs, is likely to be produced for a long while.

Ronald Strahan
The Australian Museum

PHOTOGRAPHERS

H. & J. Beste
R. Brown
M. Carter
N. Chaffer
G. Chapman
B. Chudleigh
F. Dowling
A. Foster
C. & D. Frith
T. & P. Gardner
C. Gill
P. Green
K. Griffiths
D. Hadden
C. Andrew Henley
T. Howard
K. Ireland
F. Lewitzka
J. Lochman
M. Lochman
A.R. McGill
M. & I. Morcombe
P. Munchenberg
A.Y. Pepper
J. Purnell
D.R. Rowland
V. Serventy
M. Seyfort
J. Sharland
B. Shepherd
R. Slater
D. & M. Trounson
R.E. Viljoen
T.A. Waite
G. Weber
C. Webster
B. & B. Wells
D. Whitford

Contents

FOREWORD v

INTRODUCTION ix

BUNTINGS 1
- Yellowhammer 3

FINCHES 7
- Goldfinch 10
- Greenfinch 13
- Chaffinch 16
- Redpoll 19

SPARROWS 22
- House Sparrow 24
- Tree Sparrow 27

WEAVERS 31
- White-winged Wydah 33
- Red Bishop 36

GRASS-FINCHES 39
- Red-browed Finch 42
- Star Finch 45
- Crimson Finch 49
- Beautiful Firetail 52
- Diamond Firetail 55
- Red-eared Firetail 59
- Painted Firetail 62
- Zebra Finch 65
- Double-barred Finch 69
- Masked Finch 73
- Long-tailed Finch 76
- Black-throated Finch 79
- Plum-headed Finch 82
- Pictorella Mannikin 85
- Chestnut-breasted Mannikin 88
- Yellow-rumped Mannikin 91
- Black-headed Mannikin 94
- Nutmeg Mannikin 97
- Blue-faced Finch 100
- Gouldian Finch 103

STARLINGS AND MYNAS 107
- Metallic Starling 110
- Tasman Starling 112
- Common Starling 115
- Common Myna 118

ORIOLES AND FIGBIRDS 121
- Yellow Oriole 123
- Olive-backed Oriole 126
- Figbird 129

DRONGOS 132
- Spangled Drongo 135

BOWERBIRDS 138
- Golden Bowerbird 142
- Satin Bowerbird 146
- Regent Bowerbird 151
- Spotted Bowerbird 157
- Western Bowerbird 161
- Great Bowerbird 165
- Fawn-breasted Bowerbird 171
- Tooth-billed Bowerbird 175
- Spotted Catbird 180
- Green Catbird 185

BIRDS OF PARADISE 189
- Paradise Riflebird 191
- Victoria's Riflebird 195
- Magnificent Riflebird 199
- Trumpet Manucode 203

MAGPIE-LARKS 206
- Australian Magpie-lark 208

MUD-NEST BUILDERS 211
- White-winged Chough 213
- Apostlebird 217

WOODSWALLOWS 221
- White-breasted Woodswallow 223
- Masked Woodswallow 225
- White-browed Woodswallow 227
- Black-faced Woodswallow 230
- Dusky Woodswallow 232
- Little Woodswallow 234

MAGPIES, CURRAWONGS AND BUTCHERBIRDS 236
- Black Butcherbird 238
- Grey Butcherbird 241
- Black-backed Butcherbird 244
- Pied Butcherbird 246
- Australian Magpie 249
- Pied Currawong 254
- Black Currawong 258
- Grey Currawong 260

CROWS AND RAVENS 263
- Australian Raven 266
- Forest Raven 269
- Little Raven 272
- Little Crow 275
- Torresian Crow 278
- House Crow 280

EGGS 283

BIBLIOGRAPHY 294

INDEX 301

SPONSORS

E.W. & M.J. Abrahams
J.H. Allison
Amatil Limited
R. Arnheim
Douglas B. Ash
A.C. Atwill
Alex W. Auldist
J.O. Baldock
K.A. Blofeld
Sir Ron Brierley
A.B. & C.D. Britton
J. & J. Broinowski
J. Browne
P.J. Buckley
N. Burton Taylor
R.F. Butler
A. Byrne
N. Chaffer
R.J.C. Christian
Clampett Properties Pty Ltd
M.D. Cobcroft
K.H. Cousins
M.J. Crouch
F.D. & J.N. Cumbrae-Stewart
P.J. Davidson
R. Davidson
E.S. Dean
G. Dechert
R.E. Dickinson
J.H. Dutton
Sir Vincent and Lady Fairfax
C.A. Fay
Field Naturalists Society of South Australia Inc.
N. Filewood
B.C. France
M.D. Frecker
R. Gibson
Sir Archibald Glen
Gould League of N.S.W.
K. & H. Grant
Clive M. Hall
J.D. Hamann
Hunter Wetlands Trust
L.J. Hartigan
G.P. Jackson
R.C. James
A.C. Keating
H.W. & M.A. Kinnersly
L.B. Kirk
L. Lemke
T. Lewis
B.W. Lithgow
A.F. Little
J. & A. Lucas
K. Lyall
M.J. Mashford
P. Maund
P. Miles
S. & R. Miles
J. Minter
H.G. & R.J. Mondel
B.N. Morrison
R.L. Morrow
J.S. Newlinds
N.S.W. Field Ornithologists Club
B. & C. Payne
D.H. & E.A.S. Pearse
Peddle Thorp & Walker
Sir John Proud
Reader's Digest Services Pty Ltd
Renison Goldfields Consolidated Limited
R.S.P.C.A., N.S.W.
M.J. Ryman
P.G. Sayers
D. Solomon
South Australian Museum
S. Stevens
B. Stevenson
Strathlea Pty Ltd
Sydney Hospital Medical Staff Council
B.I. Walker
D.N.O. Walker
C.H. Warman
Westpac Banking Corporation
N. Wettenhall
M. White
J.H. Wootten

INTRODUCTION

This, the final volume in our series, completes the coverage of the birds of Australia by presenting the remainder of the passerines. None of the buntings, "true" finches, sparrows or their kin are native to the continent. Some are vagrants, some are introduced species which survive in limited areas and a few, such as the House Sparrow, are widespread and firmly established. The colourful grass-finches, which are widely distributed over much of the Old World, are well represented in Australia and the native Zebra Finch is familiar internationally as a domesticated cagebird.

Starlings and mynas occur in Africa, Europe and through Asia to the western Pacific. Two introduced species, the Common Starling and Common Myna, have been so successful that they are now significant pests. Australia has some native orioles and a single species of figbird. Of the drongos, which extend from Africa to Asia, one species is native to Australia.

Bowerbirds, noted for the elaborately decorated structures built by males to entice females to a rendezvous, include some strikingly coloured male birds. This group is essentially Australian but extends into Melanesia. Primarily Melanesian but with four species in Australia, the birds of paradise are characterised by the most fantastic male decoration of any of the world's birds.

The last four families to be dealt with have sombre colouration, often black and white. The magpie-lark is the Australian representative of a family with only two species, the other occurring in New Guinea. It builds a nest of mud but is unrelated to the solely Australian mud-nest builders, comprising only the White-winged Chough and Apostlebird.

Woodswallows, which feed on flying insects and nectar, are a predominantly Australian group. Currawongs, butcherbirds and the Australian Magpie (unrelated to the magpie of Europe) comprise an Australian family, Cracticidae, members of which have a superficial resemblance to the crows, jackdaws and shrikes of Europe. Nevertheless, Australia has its share of the crow family, Corvidae, with five species of the genus *Corvus*, divided on rather minor characters into crows and ravens. Because of their highly adaptable behaviour, corvids are regarded by some ornithologists as representing the acme of avian evolution. This may mean no more than that we see them as possessing such human characteristics as inquisitiveness and problem-solving, while overlooking the likelihood that the House Fly, House Mouse and House Sparrow may have a greater chance of long-term survival than our species. It may be appropriate to complete this survey of the Australian avifauna with recollection of the mournful cry of the Australian Raven. Perhaps it has a message for us.

Male Yellowhammer; Rotorua, New Zealand; September 1988 (B. Chudleigh)

Sponsors: Arthur F. & Mollie N. Little

Buntings

Generally sombre in appearance, and in most cases lacking the beautiful songs that we usually associate with other birds of their size, buntings and New World sparrows are members of the Emberizidae, one of the most widespread bird families in the world. Emberizids have successfully spread to most regions of the world, whether as residents, migrants, introductions or – as in Australia – as vagrants.

In addition to the buntings and New World sparrows, the Emberizidae also includes the tanagers and cardinal-grosbeaks (given family status in some earlier classifications). Other members include the South American Swallow-Tanager, *Tersina viridis*, and the Plush-capped Finch, *Catamblyrhynchus diadema*, which is found in the Andes of South America.

Most emberizids are essentially New World in origin and, in that region, are referred to as sparrows. This confuses them with sparrows of the family Passeridae, such as the House Sparrow *Passer domesticus* and its allies. Although the taxonomy of the Emberizidae is fairly straightforward, the relationship between the buntings and the sparrows, finches, weavers, grassfinches, Galapagos finches, and other similar groups is very complex, and some ornithologists have suggested that they should all be placed in a single family, the Fringillidae. Taking account of recent morphological studies of the palate and gut structure of the emberizids, they are treated here as a distinct family.

The subfamily Emberizinae (buntings, cardinal-grosbeaks and tanagers) includes approximately 554 species in about 133 genera. These occur naturally throughout Europe, Africa, Asia, the Arabian peninsula, Japan and the Philippines. Several species have been successfully introduced into New Zealand, Peru, the Hawaiian Islands, Bermuda and the United States. In fact, 34 species have been introduced to new regions and 11 of these attempts have been successful. Most introductions have been deliberate but, due to their popularity as aviary stock, at least five of the 11 successful introductions were the result of escapes or inadvertent releases during transit accidents. Only one introduced species, the Yellowhammer, *Emberiza citrinella*, has been recorded in Australian territory. It is a member of the tribe Emberizini (buntings and New World sparrows).

Members of the Emberizini are small to medium sparrow-like birds with short, conical bills and sharply angled mandibles, specialised for seed crushing. Several species have been responsible for major crop damage, especially the Common Cardinal, *Pyrrhuloxia cardinalis*, in the Hawaiian Islands. Its destructive behaviour occurs during harsh conditions. When food is scarce, large flocks of birds descend on food-rich paddocks. Even though insects make up approximately 30% of the adults' diet and young birds are fed almost exclusively on insects for the first months of their life, the benefit in the area of insect control is far outweighed by the destruction of crops.

The Yellowhammer also poses some problems. This beautiful yellow bird, typical of the Old World buntings, was successfully introduced into New Zealand and started a chain reaction of crop destruction that is still worrying farmers today. Its entry onto the list of Australia's rare visitors on the basis of a single confirmed sighting in 1949 might have been perceived by Australian cereal growers as the beginning of trouble, but 1400 kilometres is a vast distance for major migration from the New Zealand mainland, and the fact that at least three attempts to introduce the Yellowhammer into Australia have been unsuccessful might set farmers' minds at ease.

Although not brightly coloured, some members of the tribe exhibit a striking appearance. The Snow Bunting, *Plectrophenax nivalis*, the reputed passerine king of the Arctic, shows beautiful tone contrast in its summer plumage which is pure white except for black on the back and in the wings.

The majority of the Emberizini are sexually dimorphic and all exhibit a characteristic form of sexual behaviour in the period after pairing and before breeding. This generally involves sexual chasing, an energetic acrobatic pursuit of the female by the male, normally ending in a raucous tumbling of birds from branch to branch or to the ground. Nest sites vary from species to species, and are usually in low hedges or bushes or simply on the ground. Three to five eggs are laid in a normal clutch, though this tends to be higher in the colder Arctic areas. Incubation is mostly the responsibility of the female, but in some species, the male assists her. Most species, except the Arctic breeders, rear two to three broods per season and both adults feed and care for the young. A few species of *Emberiza* have been recorded as being polygamous, including the Corn Bunting, *E. calandra* in which the male has up to seven hens; obviously this does not leave much time for assisting the female in feeding the young.

P. Rowland

YELLOWHAMMER

Emberiza citrinella

Following its deliberate introduction into New Zealand between 1862 and 1871, the Yellowhammer was added to the Australian list of rare visitors in August 1949 when Keith Hindwood and J.M. Cunningham reported:

> A male Yellow-hammer was seen by Cunningham near the summit of Mount Eliza [Lord Howe Island] on August 18, 1949. This is an interesting record as presumably the bird had flown from New Zealand where it is common, having been introduced into that country in 1862. New Zealand is some 700 miles south-east of Lord Howe Island, without any intervening islands.

Other sightings of "finch-like" and "sparrow-like" birds resembling Yellowhammers have been reported over the years, but these are unsubstantiated and must remain in doubt. However, the presence of an individual bird on Lord Howe Island indicates what is possible. Several unsuccessful attempts were made by the Victorian Acclimatisation Society to introduce the Yellowhammer into Australia. The birds were released in Royal Park near Melbourne in 1863 and 1864. Records also exist of the release of a number of Yellowhammers in New South Wales in 1880 but fortunately, these also

Adult female at nest; Katikati, New Zealand; December 1983 (B. Chudleigh)

Sponsor: D.N.O. Walker

failed to become established. In Great Britain records exist of large Yellowhammer flocks invadingnewly sown fields, causing considerable losses. Since its introduction to New Zealand, there have been similar reports of crop losses. At first these were few and far between, but Yellowhammer numbers increased so rapidly that a bounty was put on their heads and birds were slaughtered as pests. Subsequently, the Yellowhammer was banned from importation to the United States, but records exist of an unsuccessful introduction attempt between 1972 and 1974.

Superstition and folklore has the Yellowhammer closely linked with the devil. The name "devil's bird" was bestowed on it because its song could be interpreted as "de'il, de'il, de'il take ye", and in Scotland the eggs are said to be "gouted with the taint of the de'il's blood". Even more strangely, the Yellowhammer was believed to drink one to three drops of the devil's blood every May morning, leading to a popular rhyme:

Half a paddock, half a toad,
Half a yellow yorling;
Drink a drop o' the de'ils bluid,
Every May morning.

Because of this belief, Scottish children used to kill every Yellowhammer (yellow yorling) they could get their hands on. Jamison recorded that:

> They often take the bare "gorbals", or unfledged young, of this bird, and suspend them by a thread tied round the neck to one end of a crossbeam, which has a small stone hung from the other. They then suddenly strike down the stone end and drive the poor bird into the air. This operation they call spangie-hewit.

Until recently, there was some doubt regarding the subspecific identity of the New Zealand population, no records having been kept of the source of the original birds. It was generally accepted, following the work done by Dr James M. Harrison in 1954 and 1955, that the birds are a geographically adapted form of Emberiza citrinella sylvestris. However, studies by G. Niethammer in 1971 indicate that they are E. c. caliginosa. (E. c. sylvestris is now considered to be a synonym of the nominate subspecies, E. c. citrinella.) The vagrant on Lord Howe Island was presumably self-transported from New Zealand and thus it is provisionally considered to be of the same subspecies as the New Zealand population.

The Yellowhammer is gregarious, forming fairly large winter flocks, but is fiercely territorial during the breeding season. Its song is not very extensive; a single "tink" or "twick" can be heard when in flight, occasionally a "twitup" or "twitic". The more common song is a "chitty, chitty, chitty...swee" popularly worded as "A little bit of bread and no cheese". Courtship display is energetic: the male chases the female early in the breeding season and, later parades around the female with crest raised, tail spread and open bill pointing upwards. Flight is erratic, often undulating with periodic closing of wings.

P. Rowland

Yellowhammer

Emberiza citrinella

Male at nest; Rotorua, New Zealand; September 1988 (B. Chudleigh)

OTHER NAMES

Yellow Bunting, Yellow-hammer, Skite, Devil's bird, Yellow Yorling, Goldie

MEASUREMENTS

LENGTH: 150 – 180 mm
WING: 80 – 93 mm
TAIL: 66 – 76 mm
BILL: 9.5 – 14 mm
TARSUS: 16.9 – 21 mm
WEIGHT: 17 – 26 g

DESCRIPTION

ADULT

Male: Nape, crown and forehead yellow, streaked with green-black across forehead and on top and sides of head. Underparts yellow, except for indistinct cinnamon band across chest, becoming broader towards sides. Upperparts chestnut, streaked black; rump and uppertail coverts chestnut; tail feathers brown-black, central pair browner, two outer pairs with conspicuous wedges of white on terminal part of inner web. Iris dark brown; bill blue-horn, upper mandible with tinge of brown and dull black ridge. Legs, toes and claws light yellow-brown with tinge of red

Female: Similar to male, but yellow parts more obscured; head darker and lower neck has greenish tint. Breast streaked with yellowish-red. General plumage duller; white spots on tail are smaller

IMMATURE

Male: Similar to adult female; yellow parts brighter; breast and flanks more chestnut

Female: Similar to adult female; virtually no yellow on crown; general plumage dull yellowish-brown above, yellowish-grey below; breast and sides streaked with brown

JUVENAL

Male: Generally browner and more streaked than immature female; underparts pale yellow; less chestnut on rump

Female: Greyer than juvenal male; less yellow on crown and underparts

IDENTIFICATION

Medium-sized bunting; slighter build and longer tail than typical finches. Back and wings chestnut, central part of each feather brown-black. Male: Head and throat bright yellow; breast, brownish-red. Female: Yellow of head obscured by dusky and brown; breast pale greyish-brown

VARIATION

Negligible. Three subspecies; Australian record considered as E. citrinella caliginosa, *same as New Zealand subspecies.* E.c. caliginosa *is slightly darker and more richly coloured than the nominate subspecies* E.c. citrinella

VOICE

Song, monotonous "chitty, chitty, chitty......swee", losing "swee" late in the season. Call, a single "tink", "twink" or "tweek" and a more liquid "twitup" when in flocks

HABITAT

Generally open country, especially farmland; wintering in urban parks and gardens

FOOD

Mostly seeds, also insects and spiders; young birds fed exclusively on insects

HABITS

Sedentary, but some migration evident; territorial in breeding season but gregarious in autumn, winter and early spring. Favours lower levels but has been found at up to 1600 metres. Crepuscular and diurnal. Perches freely on low bushes and trees, normally toward top limbs; feeds almost exclusively on the ground, adopting a crouched hopping motion. Shy around humans, generally quiet and unobtrusive

BREEDING

Late November to early January, normally two broods per season. Female carries out the bulk of incubation, though male helps occasionally. Both parents feed and care for young. Injury feigning has been recorded at the nest

NEST: Shallow cup of grass, plant stems, roots and moss, lined with hair and feathers; normally on or close to the ground in thick vegetation

EGGS: 2 – 5, normally 3 – 4; pinkish-white, scribbled with dark brown lines, some pale spotting; 14.3 – 17.8 mm x 18 – 25.9 mm

INCUBATION: 12 – 24 days; fledging 12 – 13 days

DISTRIBUTION

Western Palaearctic; Europe, Great Britain and Asia. Introduced to New Zealand between 1862 and 1871; vagrant to Australia in 1949

P. Rowland

Adult male Goldfinch; Katikati, New Zealand; June 1988 (B. Chudleigh)

Sponsor: Lillian B. Kirk

Finches

In general usage, "finch" implies a small, stout bird with a strong conical bill adapted for husking and crushing seeds. This is a fair description of a finch but around the world there are at least 10 families of small to medium-sized seed-eating birds with stout bills, all commonly referred to as "finches". Despite their similar adaptations, differences in details of their anatomy and behaviour leave no doubt that they are derived from several different ancestral stocks.

The so-called "true" finches comprise the family Fringillidae, in its narrow sense consisting of the subfamilies Fringellinae (chaffinches or Old World seed eaters) and Carduelinae (goldfinches and their allies). These are distinguished from other seed-eating birds by skull structure and in having nine (instead of the usual 10) primary wing feathers, and 12 (instead of the usual 10) tail feathers. They also are distinctive in behaviour.

There is considerable controversy about the boundaries of the Fringillidae and its relationships within the nine-primaried passerines as a whole. On the basis of various biochemical studies, it is now suggested that the Fringillinae, Carduelinae, Emberizinae (buntings) and Cardinalinae (cardinals) are part of one family, comprising two subfamilies: Fringillinae tribes (Fringillini, Carduelini, Drepanidini); and Emberizinae tribes (Emberizini, Cardinalini, Thraupini, Parulini, Icterini). Here, we retain the traditional classification, regarding the Fringillidae as restricted to about three species in the subfamily Fringillinae (chaffinches) and about 122 in the Carduelinae (goldfinches and their allies). The cardueline finches are distributed widely in the Americas, Africa, and Eurasia, most species occurring in subarctic, temperate, or desert climates.

As a whole, the Fringillidae is one of the more widespread and common groups of passerines. Its distribution has been greatly assisted by acclimatisation societies but only four have made an appearance in Australia. The two most familiar species are the Goldfinch, *Carduelis carduelis*, and the Greenfinch, *Carduelis chloris*, both having been successfully introduced into Australia. Two species recorded in Australia a number of times over the years as vagrants to Lord Howe Island, are the Redpoll, *Carduelis flammea*, and the Chaffinch, *Fringilla coelebs*, which were successfully introduced into New Zealand in 1862 along with four other species; five of the six are now among the most common passerines in the North and South Islands.

The Greenfinch and the Goldfinch were first released in Australia in the 1860s near Melbourne.

They rapidly spread throughout the south-eastern coastal areas. Since then, both species have self-colonised Norfolk and Lord Howe Islands.

Perhaps the best known of all finches is the Canary, *Serinus canaria*. Endemic to the Azores, Canary and Madeira Islands, it has become one of the most widespread of cagebirds. Its popularity is largely due to its beautiful songs and to its ease of maintenance in cramped conditions. Selective breeding has produced a variety of interesting colour forms but its native plumage is green above and greenish yellow, streaked with darker green below. Brought to Europe as a cagebird in the sixteenth century, it has been cross-bred in captivity with other carduelines such as the bullfinches (genus *Pyrrhula*) and the hybrids often have exceptional singing abilities.

Fringilline species possess strong melodious songs that are short, loud and clear. Each cock may have two to six different songs, used to attract females and warn off rival males. In contrast, the song of the cardueline finches is generally quieter, but can be extremely melodious, as exemplified by the Canary.

The subfamily Fringillinae comprises three species which, unlike the Carduelinae, lack a crop: the Common Chaffinch, *F. coelebs*; the Brambling, *F. monterfringilla*; and the Blue Chaffinch, *F. teydea*. The first two are primarily Eurasian, inhabiting open woods, gardens, and farmlands; the males are boldly patterned. The Blue Chaffinch is confined to the Canary Islands, where it inhabits mountain pine forests; the males, as the name suggests, have an overall blue colouration.

The beaks of finches vary in size and shape in their adaptations for holding and shelling different types of seeds. An extreme specialisation is seen in the crossbills of the genus *Loxia*. As the name implies, the mandibles are crossed, making the bird an expert at extracting seeds from pine cones in particular, though other seeds are well handled if the need arises. This unusual feature makes the crossbills one of the few markedly asymmetrical vertebrates.

In Scotland, one of the more familiar of the finches is the Twite, *Acanthis flavirostris*. Essentially an inhabitant of open moorlands, it has replaced the House Sparrow, *Passer domesticus* in the more isolated areas. Also found in the Himalayas, the Twite has been recorded breeding up to an altitude of 4580 metres.

Members of the Fringillidae are found in a variety of environments: closed forest, scrubland or savanna, to open tundra, steppe or desert, and exhibit both terrestrial and arboreal tendencies. In Australia, the distribution of the Goldfinch and Greenfinch are closely linked with human activity, being confined to areas of cultivated and cleared land that carry traditional food plants and exotic trees for nesting and roosting.

Cardueline finches feed mainly on seeds, buds and fruits. Unlike other grain and fruit-eating songbirds, they do not take insects to any great degree, even during the nesting season, preferring to feed their young on a mix of predigested plant material.

Pair formation in carduelines is a prolonged process with the male courting a female within a flock. Courtship is an elaborate combination of various wing, tail and body movements, as well as courtship feeding of the female by the male. Once pair formation is complete, a nest site is selected and a territory established. Fringillines differ from the carduelines in that the male first establishes a territory, and then attracts a mate to his territory by singing, leaving the selection of a nest site until the pair has been established.

The nest of fringillid finches is a tightly woven cup of twigs, rootlets and soft fibres, lined with plant down or feathers. It is built solely by the female. The male accompanies this activity with an occasional overhead display.

Most carduelines usually lay four or five eggs in a clutch, producing one per day. If conditions are favourable, a female may produce up to three broods in a season, but this is an exception rather than the rule for the Goldfinch in Australia. Incubation is carried out solely by the female and usually begins with the last or penultimate egg, ensuring that the eggs will hatch over one or two days. In cardueline finches, the female rarely leaves the nest during incubation, being fed by the male. Female fringilline finches feed themselves.

Young fringillines develop very rapidly and are fully fledged at 11 to 14 days. They can fly within a few days of fledging but full independence is not reached for another two or three weeks. The young are fed on insects, usually about every five to ten minutes. Parents defend large feeding territories while nesting.

In contrast, the cardueline finches feed their young on seeds (although some supplement the diet with a few insects in the early stages), and fly outside their territories to forage in flocks wherever seeds happen to be plentiful. They can pack large quantities of seeds into their crops for later regurgitation to the young, thus enabling them to forage at long distances from the nest. The young are fed at intervals of 20 to 60 minutes. To date, there is little evidence that the Goldfinch or the Greenfinch have caused any major economic damage to fruit and grain crops in Australia, but some damage to strawberry crops has been recorded. Since they inhabit gardens and farmlands and prefer exotic vegetation, they do not appear to compete directly with native birds.

P. Rowland

GOLDFINCH
Carduelis carduelis

The European Goldfinch, native to Eurasia and northern Africa, was introduced near Melbourne in 1863 or perhaps earlier. Other introductions followed at several centres in Victoria, and in Adelaide in 1879. Although no introduction attempts have been recorded for Queensland or New South Wales the species had become established in Sydney by 1886 and probably spread into Queensland from there. No records exist for deliberate introductions to Tasmania, although some arrived there in 1827 aboard a ship, and were well established in that state by 1902. Liberation of the Goldfinch in Perth occurred at some time between 1899 and 1912 and it was common around Perth and Albany between 1950 and 1970. It has since declined and persists only in isolated pockets.

In New Zealand, the Nelson Acclimatisation Society liberated 10 birds in 1862. During the following 20 years, the Goldfinch was released in several other places and soon became established. It has become common throughout the country and winter flocks usually number 50 – 500 birds, although flocks of 1000 – 2500 and even one of 15,000 have been recorded.

The range of the Goldfinch in Australia is largely restricted to the temperate parts of the country where there is farming, and is generally restricted to areas where exotic plants have largely replaced the native vegetation. Its range may still be expanding but it is unlikely to change markedly unless irrigation should permit agriculture in the present arid interior.

The Goldfinch is gregarious by nature. In Australia, it is found throughout the year in groups or "charms" which are small in the breeding season but, at other times, may number several hundreds. The birds move between feeding places with light and undulating flight. On the ground, a flock usually advances in a rolling fashion, with birds at the back continually fluttering over the flock to the front. A soft twittering comes from feeding flocks with occasional harsh notes and threat displays from upset individuals. When disturbed, the flock rises and flies off to nearby trees, drifting back to the feeding area when the disturbance has passed. When in a tree, the Goldfinch tends to perch on uppermost or outer twigs rather than among foliage, in small parties, unmixed with other species. It feeds mainly on weed seeds but occasionally catches small insects, especially to feed the young. When feeding on vegetation, it uses a foot to pull seed heads to the bill. It flutters lightly from one plant to another or hangs upside-down as it extracts seeds from thistles and other tall weeds. It also feeds on short pasture. Strawberry growers dislike Goldfinches because they peck the seeds from the ripening berries, leaving minute holes that permit the entry of fungi and bacteria and spoil the fruit.

The onset of the breeding season leads to an upsurge in sexual activity, increased singing, greater aggressiveness between birds within the flock, and the frequent sexual chasing of females by males. Pairing is completed by late August and the large winter flocks break up into small groups heading towards the breeding areas.

A courting male sways its body from side to side and quickly turns its slightly expanded wings backwards and forwards, giving a golden flashing effect. The nest is built solely by the female and usually takes a week to complete, the male giving an occasional overhead display in which he circles above the nesting area. Single eggs are laid daily, for three to six days, until the clutch is complete.

The female incubates the eggs alone for about 13 days and seldom leaves the nest, being fed by her mate. When the eggs have hatched, the male brings food to the nest, where it is regurgitated to the female and then fed to the hungry chicks. Twelve days after hatching the young fledge and are ready to leave the nest, although they remain dependent for a further three weeks.

In the early part of the nesting season the male vigorously defends his territory against intruders. As incubation begins, he becomes more occupied with finding food for the female. Although the Goldfinch is capable of producing at least three broods per season, this is unlikely to be the case in Australia. The prolonged breeding season probably results from repeated nesting attempts by unsuccessful pairs, or from birds reaching breeding condition at different times.

P. Rowland

Female at nest; Sisters Beach, Tas.; December 1980 (T.A. Waite)

Goldfinch

Carduelis carduelis

Female at nest; Christchurch, New Zealand; January 1987 (D. Hadden)

OTHER NAMES

European Goldfinch, Goldie, Thistle-Finch

MEASUREMENTS

LENGTH: *117 – 141 mm*
WING: *73 – 81 mm*
TAIL: *45 – 53 mm*
BILL: *9.9 – 13 mm*
TARSUS: *13 – 15 mm*
WEIGHT: *15 – 21 g*

DESCRIPTION

ADULT

Male: Crimson facial disc covering front of crown, forehead, cheeks – extending just beyond eye, chin and upper throat. Cheeks, area behind eye, and lower throat white. Black cap from centre of crown, behind ear coverts and down side of neck behind white area. Nape and upper mantle greyish-white; rest of upperparts pale chocolate brown, lighter on rump. Wings black with white tips to flight feathers; conspicuous golden-yellow on basal half of outer webs of primaries and secondaries, except outer primary which is completely black. Tail black with white tips. Underparts predominantly white, except for breast and flanks which have varying amounts of fulvous brown. Eye dark brown. Legs and feet buff to reddish brown. Bill pinkish with small black tip

Female: Similar to male, but slightly buff on wings. Crimson facial disc bisects eye. Golden-yellow on wings not as bright

IMMATURE:

Similar to adults; lacking red facial disc; plumage more buff and finely striped

JUVENAL:

Similar to adults, except plumage finely striped and head greyish-buff

IDENTIFICATION

Small, brightly coloured finch with striking red facial disc (absent in young birds). Predominantly buff upperparts with contrasting white rump and black flight feathers. Underparts white with varying amounts of buff. Crown to nape black. Conspicuous bright golden-yellow bar on upper wing

VARIATION

Twelve subspecies, one present in Australia, C. c. britannica, *introduced from Sussex, England*

VOICE

Call, a repeated "swilt-witt-witt"; Occasional prolonged "tu-leep". When alarmed, a regular harsh "geez". Song: short liquid warbling, composed of similar notes to call

HABITAT

Agricultural areas, urban parks and gardens, orchards and areas sparsely scattered with trees and bushes

FOOD

Mainly seeds; occasionally small insects, especially while rearing young. When regular food is scarce birds have been observed eating grass leaves

HABITS

Locally dispersive; diurnal. Generally arboreal, favouring lower levels, but feeds predominantly on the ground, hopping over other members of the group in order to feed at the front of the flock. Gregarious, found throughout the year in small groups called "charms", but occasionally flocks in excess of 500 birds are seen. Shy and non-confiding in humans

BREEDING

Late September to March. Can produce up to three clutches per season, but generally only one or two. Nest losses are high and long breeding season is probably due to repeated attempts by unsuccessful pairs, or because birds come into breeding condition at different times. Female incubates eggs and is fed at the nest by the male. For first few days of brooding, male feeds the female which in turn feeds the young; later, both adults feed the young. Injury feigning has been recorded at the nest

NEST:

Delicate cup of fine twigs, grasses, moss, wool and spider webs, lined with down and other soft material; approximately 60 mm in diameter and 30 mm in depth. Usually placed in fork of thin branches towards extremities, 2 – 12 m above ground

EGGS:

3 – 7, blue-white with delicate brown spotting, concentrated toward larger end. Laid one day apart. 15.6 – 20 mm x 12.3 x 14.3 mm

INCUBATION:

12 – 13 days; fledging period 13 – 14 (occasionally 15) days

DISTRIBUTION

British Isles, most of Europe, South-west Asia, Middle East and North Africa; introduced and established in New Zealand and parts of South America. Predominantly eastern to south-eastern Australia, from Bundaberg, Queensland through to Port Lincoln, South Australia; favouring coastal fringes, but range extends to the western slopes of Dividing Range and throughout temperate regions; Tasmania, except for undisturbed south-western areas; few isolated populations around Perth, due to the escape of aviary birds

STATUS

Introduced and established; common throughout range

P. Rowland

GREENFINCH

Carduelis chloris

Native to Europe, western Asia and northern Africa, the Greenfinch was successfully introduced into Australia, New Zealand and Uruguay in the mid-1800s. First brought to Melbourne in 1863, it quickly became established, but failed to spread as widely through temperate Australia as the related European Goldfinch, *C. carduelis*. It is now essentially confined to areas of conifer plantations, tall hedges, orchards and garden shrubs, although in the winter, it ventures into woodland and open farmland to forage for food. Other introductions were made near Adelaide (1879 or earlier) and on Norfolk Island. No deliberate introductions have been recorded for Sydney but a population was established there by the 1890s, possibly derived from escaped aviary stock. In Tasmania, the Greenfinch was first reported in 1941 and has since spread throughout the eastern areas. In western Tasmania it is confined to the coast.

On 18 March 1912, an English newspaper reported the story of a gentleman who in January 1911 enticed a Greenfinch into a cage. It wore a leg ring marked "Aberdeen University, 7185". The bird was released the following March but, when winter came round once more, it returned and "on its cage being presented to it the Finch hopped contentedly into it and settled comfortably down for the severe season".

Adult male at nest; Wynyard, Tas.; October 1977 (T.A. Waite)

Sponsors: Sir Archibald & Lady Glen

The Greenfinch usually perches in trees, but it also hops around on the ground. In the air, it exhibits the undulating flight that is common in varying degrees to most finches: a few rapid wing beats alternated with momentary closures and brief glides on extended wings when about to perch. In song flight, the male circles around in a peculiar, erratic, bat-like flight with slowly flapping wings.

The onset of breeding in September is signalled by the return of scattered pairs to their nesting areas. A long drawn out "zeep" call announces their return, the birds having paired while still in their winter flocks. In winter, males dominate the females but once pairs are formed, this order is gradually reversed. Courtship usually consists of a series of chasings of the female by the male; in a less aggressive display, the male points his bill upwards and lowers his wings, displaying yellow feathers on the wings, rump and tail. The male also feeds the female. Acceptance of the male by the female is signified by a crouching stance and shuffling wings, to which the male responds by gently pecking her bill. A pair begins nesting in early October within a small colony of two to six pairs. As with other members of the family, the female builds the nest, but the male occasionally gathers material for the female to use. Incubation is carried out solely by the female, which is fed on the nest by the male. During nest construction and incubation, the male makes repeated overhead display flights, in which he circles the nest area with a slow flapping flight, and repeats his "zeep" call.

In Europe, nest construction is undertaken in an erratic fashion. Nest material is added for a period of 10 to 40 minutes, followed by hours or even days when nothing is added. The time taken to complete the nest varies from a few days to two weeks or more. In the same fashion as the Chaffinch, *F. coelebs*, completed nests may remain empty for up to three weeks before eggs are laid.

At the onset of breeding a Greenfinch becomes very aggressive to other species, such as the Goldfinch, that stray into its territory. If their breeding habitats happen to overlap, the Greenfinch tries to remove the smaller Goldfinch from its territory, either by destroying its nest or by puncturing any unattended eggs.

Outside the breeding season it is sociable, feeding in large flocks and often using a communal roost. In New Zealand, winter feeding flocks can sometimes grow to enormous size, occasionally exceeding 10,000 birds.

The Greenfinch has a powerful bill for dealing with a wide variety of seeds. In its native European habitat, it feeds mainly on the seeds of elm, yew and hornbeam, but as these are rare in Australia, it is often found feeding on the seeds of weeds in open wasteland. When food is scarce it enters farm areas where it eats fruit buds and ripening grain. Greenfinches studied in New Zealand tended to eat mainly seeds and a few insects such as weevils and sawfly larvae, but not in sufficient quantity to control these pests. The most common seeds taken were of maize, redroot, chickweed, boxthorn, wild turnip and peas. The seeds of the Scotch Thistle are the major summer food in coastal areas, as well as those of small fruits such as rosehips, crab-apples and cotoneasters. A preference for woodland habitats has restricted its range in Australia, so much so that it has had little opportunity to interact with native species. Where it has established itself, it has now become an integral part of the avifauna but is unlikely to achieve the success of its relative, the Goldfinch.

P. Rowland

GREENFINCH
Carduelis chloris

Adult female at nest; Katikati, New Zealand; December 1984 (B. Chudleigh)

OTHER NAMES
European Greenfinch, Green Linnet

MEASUREMENTS
LENGTH: 145 – 158 mm
WING: 83 – 90 mm
TAIL: 54 – 60 mm
BILL: 13.7 – 15 mm
TARSUS: 17 – 19 mm
WEIGHT: 26 – 36 g

DESCRIPTION
ADULT
Male: Back of head and most of upperparts dull olive-green, feathers edged with grey. Ear coverts brownish-grey, edged with green. Uppertail coverts brownish-grey, edged with green. Tail feathers brownish-black, edged with grey; outer tail feathers edged with yellow. Wings grey-brown, tinged with green; flight feathers brownish-black, edged with yellow-green. Underparts yellow-green, except flanks which are olive-grey. Under wing dark grey with bright yellow leading edge, forming distinct yellow wing bar. Lower abdomen and undertail greyish-white. Eye brown; bill pale brown above, white-horn below; Legs and feet pale pinkish-brown. In winter, becomes browner and duller, similar to female

Female: Similar to male, but generally duller and browner. Less yellow on wings and especially on tail

IMMATURE
Similar to adult female and wintering male

JUVENAL
Generally streaky-brown. Yellow on outer edges of primaries and outer tail feathers, more noticeable in males

IDENTIFICATION
Thick-set, stout-billed finch. Predominantly olive-green with conspicuous bright yellow bands on wings and sides of tail

VARIATION
Negligible. Four subspecies recognised: Australian subspecies is C. chloris chloris. *Lacks brighter yellow and grey washes of other subspecies*

VOICE
Song consists of various ringing notes; most common is a monotonous drawn out "dzwee", rising in pitch at end. Call is a loud, rapid "chichichichichi-teu-teu-teu"

HABITAT
Gardens and parks of urban areas; farmyards and orchards near cities and towns; exotic tree plantations

FOOD
Mostly seeds, occasionally insects. Small fruits often pulled apart to obtain seeds

HABITS
Mostly sedentary, although some winter dispersal. Diurnal, forming large winter night roosts, varying in numbers from night to night. Gregarious, usually travelling in small groups. Largely arboreal but frequently feeds on or near to ground. Inconspicuous even where common. Often with groups of goldfinches and sparrows

BREEDING
Mainly October – January, sometimes later; usually only one clutch per season. Both adults build nest and care for young. Female incubates eggs alone, being fed regularly by male

EGGS: 4 – 6; pale bluish-white with scattered brown spots and blotches, slightly more concentrated toward larger end; 17.2 – 24.1 mm x 12.2 – 16.1 mm

INCUBATION: 12 – 14 days; fledging period 13 – 14 days

DISTRIBUTION
Coastal areas of south-eastern Australia, from Lismore, New South Wales to Port Pirie, South Australia; extending only a short distance inland. Throughout Tasmania, except undisturbed south-western areas. North-west Africa, Europe, New Zealand and south-west Asia. Isolated population in Turkestan

STATUS
Introduced and established in Australia. Common throughout range

P. Rowland

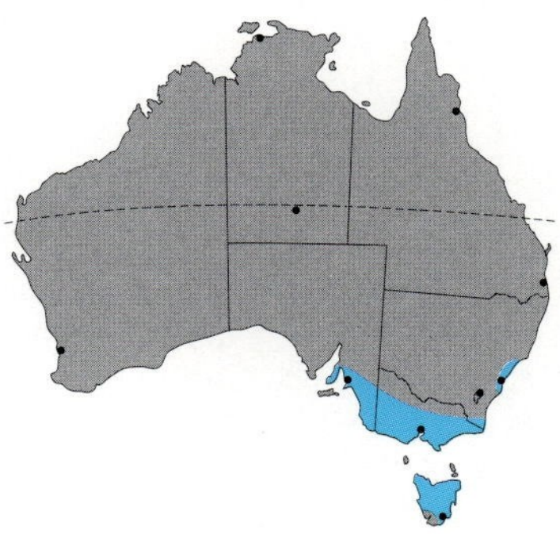

Chaffinch

Fringilla coelebs

Unsuccessfully introduced into many parts of eastern Australia between 1863 and 1872, the Chaffinch is now a rare vagrant to Lord Howe Island. Occasional sightings reported on the Australian mainland are probably due to the escape of aviary stock. Although the sightings on Lord Howe Island have been poorly documented, it is likely that, like the Redpoll, *Carduelis flammea*, these birds were self-transported from New Zealand, being initially introduced there at Nelson in 1862, followed by numerous releases in other parts of the country by both acclimatisation societies and private individuals.

The Chaffinch, the meat of which was once a part of a supposed cure for epilepsy, could be regarded as one of the few useful introduced species since it feeds on insects such as aphids and the caterpillars of White Butterflies, as well as the seeds of weeds. In summer, however, it takes a certain amount of grain and, in New Zealand, it has become a pest in orchards where, like the Redpoll, *C. flammea*, it damages the buds of apricots, peaches and other cultivated fruits. For the past two centuries or more, the Chaffinch has attracted much attention because of large flocks, often up to 600 birds, which seem to be composed almost entirely of females. This habit prompted Linnaeus to give the species its scientific name, which translates as "unmarried finch". In Gilbert White's *Natural History of Selborne* there is a well-known passage in Letter XIII which was written to Thomas Pennant and dated 22 January 1768:

Adult male at nest; Katikati, New Zealand; November 1981 (B. Chudleigh)

Sponsors: Bruce & Norma Lithgow

> For many years I have observed that towards Christmas vast flocks of chaffinches have appeared in the fields; many more, I used to think, than can be hatched in any one neighbourhood. But when I came to observe them more narrowly, I was amazed to find that they seemed to me to be almost all hens. I communicated my suspicions to some intelligent neighbours, who after taking pains about the matter, declared that they also thought them mostly females – at least fifty to one.

The flight of the Chaffinch is similar to that of other fringilline finches, being rapid and undulating, due to quickly repeated flapping, with short intervals of cessation. It roosts at night in thick hedgerows or among evergreens in plantations. On the ground it proceeds by a succession of hops, or short, quick steps, in a "tripping" fashion.

Chaffinches have been observed "anting", a process whereby the bird rests upon an active ant nest and lets the ants crawl all over it, or actively rubs the ants through its feathers. This procedure is still not fully understood, but its main function appears to be to remove external parasites from the skin and feathers.

According to Chambers' *Popular Rhymes*, the note of the Chaffinch is taken in the north of England and in Scotland as a sign of rain, the notes being interpreted by the children as "Weet, weet!". Chaffinches also act as sentinels for other birds. When danger approaches, they utter a short alarm call, giving other birds warning. They also readily join in with other species to mob the unwelcome intruder.

In contrast to other cardueline finches, the male Chaffinch first establishes his territory, then advertises possession by singing vigorously. He is soon joined by a female and a nest site is selected. During this time, dominance switches from the male to the female. Mating generally takes place over a period of about 10 days, usually before the completion of the nest. The male's courtship display consists of short display flights which involve a swerving, headlong chase of the female by the male, followed by the male drooping and fluttering his wings. The female accepts the male by shivering her wings and raising her tail. Other courtship displays can consist of the male strutting towards the female with body upright, chest out, and head back, in a similar style to a walking pigeon. Sometimes he simply perches stiffly with crest elevated, turning slowly from side to side to display his white wing patches.

The female usually builds the nest on her own, taking from three to 18 days to complete it; older birds tend to build more quickly than young ones, especially when replacing a previously destroyed nest. Curiously, nests completed early in the season usually remain empty for up to four weeks before egg laying commences.

Eggs are laid a day apart, incubation commencing with the final or penultimate egg, thus ensuring that the three to five eggs hatch within two days of each other. Parents eat the faeces of the young during the first two or three days but carry the wastes away after this. The young are largely fed on insects, the bulk of these being caterpillars. Territorial defence becomes less intense after the young become independent. If both parents are fortunate enough to survive until the following year, they probably pair again, although during the winter they do not seem to maintain an obvious pair bond.

P. Rowland

Chaffinch

Fringilla coelebs

Adult female at nest; Katikati, New Zealand; November 1981 (B. Chudleigh)

OTHER NAMES
Beech-finch, Pink, Shell-apple, Scobby

MEASUREMENTS
LENGTH: *146 – 159 mm*
WING: *78 – 90 mm*
TAIL: *58 – 67 mm*
BILL: *13 – 14 mm*
TARSUS: *17 – 19 mm*
WEIGHT: *19 – 26 g*

DESCRIPTION
ADULT
Male: Forehead black; crown, nape and sides of neck slate-blue. Upper back and scapulars chestnut; lower back and rump yellow-olive. Primary wing coverts black; greater wing coverts black with broad white tips; median and lesser wing coverts white. Primary and secondary flight feathers dull black; all but outer primaries and inner secondaries with white bases and broad white edges to outer webs. Uppertail coverts grey. Sides of head and underparts rich pinkish-brown, becoming paler on breast and flanks. Vent and undertail coverts white. Two central tail feathers grey, remainder black; outer two feathers on each side have conspicuous white wedges. Eye brown; bill slate-blue with dark tip; legs and feet brown

Female: Similar to male, though slightly smaller and duller. Forehead, crown, nape and mantle ashy-brown. Rump pale olive. Underparts pale ashy-brown, breast darker than belly. Bill brown

IMMATURE
Similar to adult, but browner. Rump and uppertail coverts olive-brown

JUVENAL
Similar to adult female, but rump brownish-green. Whitish patch on nape. Mantle of male tinged with chestnut

IDENTIFICATION
Sparrow-sized finch; brown above with black and white wings. Breast pink. Slate-blue cap. Female more brown, distinguished from female House Sparrow by obvious white wing bar. Undulating and bouncing flight

VARIATION
Slight. Fourteen subspecies; Australian subspecies same as New Zealand, F. c. gengleri

VOICE
Song is a rollicking cadence of several short notes, ending in a flourish. Call is a soft "pink-pink" and a questioning "wheet"; in flight a subdued "tseep"

HABITAT
Variety of habitats: exotic pine plantations, bush, orchards, farmland, etc. Especially at home in urban parks and gardens

FOOD
Insects, grubs, fruit and seeds of weeds; largely grains in summer

HABITS
Dispersive, diurnal, largely arboreal, but often feeds on ground. Gregarious in winter; no pair bond exhibited during winter but couples usually pair up again to breed. Generally unobtrusive, being more common to the ear than to the eye

BREEDING
Does not breed in Australia. In New Zealand, September to January (mainly October to December). Unlike British forms, normally only one clutch per season, but will replace failed nest. Female incubates eggs and cares for young, male occasionally helps in food gathering

NEST: Elaborate deep cup of moss, dry grass, wool, rootlets and cobwebs. Outside decorated with lichens, inside lined with hair or feathers. Usually situated in the fork of main trunk and branch of a tree, up to 18 m above ground (usually within first 6 m)

EGGS: 3 – 6, greenish-blue, spotted and streaked with dark purplish-brown. 17 – 22 mm x 13.2 – 15.8 mm

INCUBATION: 11 – 15 days; fledging period 10 – 16 days

DISTRIBUTION
British Isles, Europe and central Asia; Wintering in North Africa and Middle East. Introduced and established in South Africa and New Zealand. Single males recorded three times on Lord Howe Island in 1982; regarded as vagrants from New Zealand

STATUS
Rare vagrant to outlying islands

R. Rowland

REDPOLL

Carduelis flammea

The Redpoll *Carduelis flammea* was one of the species included in a shipment of various kinds of birds imported to Nelson, New Zealand, in 1862. It was subsequently released in other parts of the country and soon became well established. Although Lord Howe Island is some 1100 kilometres north-west of New Zealand, Roy Bell collected two female Redpolls there on 26 August 1913; others had also been observed on the island. Apparently these birds had recently flown from New Zealand and were all exhibiting very worn plumage. Almost exactly 70 years later, on 9 June 1983, another confirmed sighting of Redpolls was reported by G.J. Ingram and G.J. Roberts on Blackburn Island, some 600 metres off the west coast of Lord Howe Island:

> Soon after landing, our attention was drawn to a small charm of finch-like birds which flew into the top of one of the pines. For the next half-an-hour we closely studied these birds as they filtered and fed amongst the trees and herbs. The party of birds was comprised of two males and three females. The males were brightly coloured with conspicuous red crowns and a crimson wash extending from the throat to the lower breast ... We believe that these birds were Lesser Redpolls *Carduelis flammea cabaret.*

There is still much debate over the subspecific identity of the Redpolls observed on Lord Howe Island. While there is little doubt that they were self-transported from New Zealand, some authors believe that just before the first introduction of Redpolls to New Zealand, large numbers of Mealy Redpolls, *C. f. flammea,* descended upon Britain from Europe and that many of these became trapped

Pair at nest; Christchurch, New Zealand; February 1985 (D. Hadden)

Sponsor: Betty N. Morrison

in bird catchers' nets and were transported to New Zealand along with the Lesser Redpoll, where they have been interbreeding ever since. It is widely agreed that the bulk of the New Zealand population consists of the Lesser Redpoll, only 10 – 15% being Mealy Redpolls (slightly larger and lighter in colour, but otherwise very similar).

Although Redpolls have been recorded breeding on Macquarie Island, some 1000 kilometres south-east of Tasmania, it is doubtful that the species will become established on mainland Australia. This is fortunate, since it has become a severe pest to fruit growers in New Zealand by destroying the buds of apricots and peaches. A Redpoll moves along the branches and systematically pecks every bud, thus destroying all or part of the bud's protective sheath. This pecking behaviour is not readily explained, since the bird's main food in orchards is weed seeds. It also removes seeds from ripening strawberries, leaving minute holes which allow bacteria or fungi to enter the berries, causing them to rot. This is particularly the case for birds that winter in the high country, which apparently run short of food early in September, and descend upon orchards and other cultivated areas. The Redpoll is of some use in eating turnip greenfly and similar insects, but this benefit is outweighed by the harm it does to other crops.

Redpolls are sociable birds roosting and feeding together through winter in flocks of up to 250. Coastal wastelands are a favourite feeding ground. Even in the breeding season, several adults may form small feeding flocks away from the immediate area of their nests. In Britain, the roost changes as one feeding area is exhausted and birds move to a richer one.

In New Zealand the Redpoll is usually more common at higher altitudes (up to 1500 metres), inhabiting hill scrub, pine plantations, forest edges and sub-alpine scrub, as well as coastal sand dunes. When feeding and flying with others it keeps up a continuous twittering "chuch-uch-uch" or "tiu-tiu-tiu". The alarm note is a plaintive "tsooeet" and a grating "geez" is uttered when one bird attacks another. Most song is delivered in a display flight, but sometimes from perch or in ordinary flight, commonly at a good height and often by several males together.

Normal flight is undulating, as with other members of the group, whereas the display flight is more often in loops and circles, with slow, hesitant wing beats and sometimes a glide, six metres or more above the tree tops.

Pair formation begins in late winter while birds are still in flocks. Courting males can be observed hopping backwards and forwards from one branch to another with tail spread, wings raised above the back and sometimes slowly fluttering, while uttering low hissing notes. Frequently, several pairs may join together to form a large nesting colony, each pair defending a small territory around its nest. However, many nests are isolated. The nest for a second brood may be far from that in which the first brood is reared.

Young are fed insects and insect larvae by both parents, which continue to feed them for a further 10 to 14 days after they have fledged. Redpolls live up to eight years. They are sometimes kept in aviaries but lose their red colour after the first moult.

P. Rowland

Redpoll

Carduelis flammea

Nest and eggs; Christchurch, New Zealand (D. Hadden)

OTHER NAMES
Lesser Redpoll, Lesser Read-headed Linnet

MEASUREMENTS
LENGTH: *115 – 125 mm*
WING: *63 – 79 mm*
TAIL: *47 – 55 mm*
BILL: *7.7 – 10 mm*
TARSUS: *13 – 15 mm*
WEIGHT: *10 – 15 g*

DESCRIPTION
ADULT
Male: Crown and forehead dark crimson except blackish area above bill; upperparts predominantly light brown, streaked with dark brown; head, sides of neck and lower back paler; rump pink. Greater and median wing coverts blackish, edged with buff and tipped with sandy buff, making indistinct pale wing bars; primaries dark brown; secondaries brown, edged with buff; uppertail coverts and tail feathers dark brown edged with lighter brown, tail feathers white on inner webs. Chin black; throat, cheeks and breast washed with pinkish-red. Belly, flanks, underwing coverts and undertail coverts greyish-white. Eye dark brown; bill dark brown with yellow sides. Legs and feet dark brown

Female: Similar to male but lacking redness on breast and rump; black on chin more extensive; crimson on crown duller. Slighter in build

IMMATURE
Similar to adult female; less black on chin; darker streaking on throat and breast. Males have pink rump

JUVENAL
Similar to immature except no crimson on crown (this appears after first autumnal moult; males also develop pink rump at this time)

IDENTIFICATION
Small finch; predominantly streaky-brown; crimson crown, more noticeable in males; males exhibit rose-pink wash to breast and rump, varying in degrees of intensity

VARIATION
Four subspecies recognised. Lord Howe Island specimens believed to be C. f. cabaret, *self-transported from New Zealand. Colour varies within* C. f. cabaret; *depth and distribution of red or pink plumage of both sexes ranges from orange, yellow to fairly pure red and in some cases an almost purple bluish-red. Variation possibly geographic or the result of breeding from escaped aviary birds which lose red colouration in captivity*

VOICE
In flight, call note is characteristic high-pitched and metallic "chich-ich-ich-ich"; a plaintive "tsooeet" when alarmed. Song a short, sustained series of brief rippling trills, often preceded by flight call; and a softly modulated sub-song audible at close range

HABITAT
Common in wide variety of habitats, favouring country areas and coastlines to urban sites. Commonly found in orchards, exotic forests, sub-alpine scrub and coastal sand dunes, but successful inhabitant of many other environments

FOOD
Seeds, small insects and buds of fruit trees

HABITS
In New Zealand, mostly sedentary, with winter dispersal to lower altitude. Diurnal, mostly terrestrial, feeding on ground and low vegetation. Gregarious, forming large winter feeding flocks and roosts. Does not readily mix with other finch flocks. Quiet and unobtrusive. Near-colonial nesting habits

BREEDING
Non-breeder in Australia. In New Zealand, pairs form and nests are built late in winter but eggs are not laid until September – January. Female constructs nest and incubates eggs, while fed at the nest by male; both parents feed young. Sometimes double-brooded

NEST: Small compact structure of grass, twigs, rootlets and wool, decorated with moss and lined with feathers or hair, placed in low fork of small tree or bush, sometimes near trunk but usually in outer branches; generally within 3 m from ground

EGGS: 3 – 6; bluish-green, non-glossy; blotched and spotted with pale purplish-brown (occasionally streaked with dark reddish-brown), roughly concentrated toward larger end; sometimes unmarked; 14.3 – 15.2 mm x 10 – 11.6 mm

INCUBATION: 10 – 14 (usually 10 – 12) days; fledging period 11 – 15 days

DISTRIBUTION
Europe, northern Asia, Greenland, Canada and, in non-breeding season, mid-Asia and North America. Several confirmed sightings and captures on Lord Howe Island, regarded as vagrants from New Zealand

STATUS
Occasional vagrant to oceanic islands of Australia

P. Rowland

Sparrows

The family Passeridae is interpreted in several different ways. In the narrowest sense, currently used in Australia, it comprises only three genera and about 30 species of sparrows from Afro-Eurasia. They are generally birds of more or less extreme environments: deserts and mountains to drier open savanna. Except by colonisation, they are absent from primarily forested South-east Asia. Several species have become domestic over thousands of years, living in and around the biologically extreme environments on farms and in villages, towns and cities. The House Sparrow now occupies these habitats over much of the earth's surface.

Some authorities unite sparrows with the weavers, which occupy the same general area, but mainly in less harsh environments; there cannot be much doubt that they are closely related. This composite family has often been called Ploceidae, but Passeridae is the older name and is coming back into use. Many ornithologists nowadays expand the family still further to include the Grassfinches (Estrildidae) and there is a movement to put all or most of the seed eaters and their supposed relatives (here separated at Emberizidae and Fringillidae, among others) back into a single family, where they were a century ago.

The largest group of "true" sparrows, in the strict sense, is the genus *Passer*, including about two dozen species, of which the House Sparrow and Tree Sparrow have been introduced into Australia and elsewhere. They are all essentially similar in breeding habits, building closed nests usually of dried grass and other vegetation, often in close proximity as "colonies" or "neighbourhoods". Males are monogamous and contribute, more or less equally, to nest building, incubation and care of young. They frequently initiate nest building, but do not use the warp-and-weft technique of the weavers. Pairs often remain mated for an entire season, and sometimes for life.

Most species are more or less sedentary, but populations that nest in the coldest areas tend to move to warmer ones in the non-breeding season, and some undertake considerable migrations from higher latitudes or altitudes. All are primarily seed eaters, but opportunistically eat all sorts of vegetable matter and small animals. The beak is more or less conical, strong and capable of a painful pinch. They have no true song, and the males advertise ownership of a nest site by a monotonous repetitive chirruping.

It is interesting to compare and contrast the two species that have been introduced into Australia. The House Sparrow is larger, heavier, and – in this region – overwhelmingly more successful. Yet size is not all-important: the native Cape Sparrow of South Africa is bigger still, and also likes domestic situations, but the House Sparrow can displace it. The Tree Sparrow is so called because, in Europe, it is a bird of woodland, and normally nests in trees; in eastern Asia it is fully domestic, living in and around houses and other buildings. As a dweller in many airports, it is often the first species that a visiting birdwatcher sees. In many countries to which it has been introduced, it remains confined to the immediate vicinity of urban areas. The Australian population is at least mainly of this urban east Asian strain, and remains close to human development, playing second fiddle to the House Sparrow in most areas. It has died out in most of New South Wales, although for reasons not understood it is the common sparrow of many Murray Valley towns, and has spread considerably within a few decades since the 1950s.

While all Tree Sparrows look alike, the House Sparrow is strongly dimorphic: females and young are quite different from adult males. Most other members of the genus and family tend to be monomorphic or nearly so, but there are a few other dimorphic species; the significance of this is unknown. Almost all sparrows are more or less cryptically coloured. Certainly the male House Sparrow is a handsome bird when one bothers to look at it but the various colours blend into a whole that is probably inconspicuous to a potential predator. It is perhaps useful in such a pugnacious species for the males to be able to recognise possible rivals from possible mates but this fails to explain why other pugnacious sparrows are monomorphic.

Not all sparrows are "little brown birds": most of the plumage of the male African Golden Sparrow is bright yellow, a common hue in the numerous African and few Asian Weavers. It is proper to call these shades "canary yellow" since the feather pigments are the same as in domestic canaries.

W. Filewood

Juvenile House Sparrow moulting to female plumage; Buderim, Qld; February 1995 (R.E. Viljoen)

House Sparrow

Passer domesticus

The House Sparrow was living beside humans in the Middle East when Western agriculture began and it still nests in cliffs around the oldest known granary caves. It has remained an urban and grainland bird for some nine millenia, following humans almost wherever they have gone, spreading quickly along highways and particularly favouring stables where there is plenty of food both before and after it is given to the horses.

Introduced to Victoria in the 1860s, it became well distributed from Atherton to Tasmania, and west as far as Tarcoola in less than a century. It now breeds from Torres Strait to southern Tasmania. The founding stock was mainly British, belonging to the widespread subspecies *P. d. domesticus*; sparrows from China and Java, released at the same times and places, were presumably Tree Sparrows, *P. montanus*. Although a small distribution map shows the House Sparrow to be spread over all of the eastern third of Australia, it is a readily observable fact that only stray birds are found at any distance from human settlement.

In 1865 two boys were fined for destroying a House Sparrow's nest in a tree in a street of Warrnambool but by 1868 the House Sparrow was considered the criminal, and accused of eating fruit in Collins Street in the heart of Melbourne. Within 20 years of the first release of the species in Australia, the South Australian Government had paid out bonuses on almost 40,000 heads and nearly a quarter of a million eggs. Early in the twentieth century, the New South Wales Department of Agriculture compiled an annual Sparrow Atlas, charting its expansion, but it now occurs virtually everywhere near habitation. Dubbed an "avian rat" in 1899, it is successful for many of the same reasons as domestic rodents: it tolerates humans and eats a similar mix of food. Perhaps equally important, it has the potential for rapidly increasing in numbers.

The House Sparrow has been prevented from establishing itself in Western Australia by deliberate destruction of every individual seen to land by boat or arrive overland from the east. However, it has reached the Northern Territory from Queensland, and will no doubt soon be reaching the Kimberley Division, which will be harder to defend.

The House Sparrow arrived in Norfolk Island around 1939 and has become one of the two commonest birds there. It may have arrived by ship, or possibly by flying 720 kilometres over the sea from New Zealand, as it apparently did to reach Campbell Island, 560 kilometres from the South Island of New Zealand. There have been several reports from Port Moresby, where it may have bred at least once before dying out. These birds almost certainly arrived by ship, as the first pair was seen at the main wharf, and the last was reported over a decade later at Paga Hill. It is now a common breeding bird on Thursday Island, only one and half degrees south of Port Moresby.

Although often seen in numbers, the House Sparrow is not really social. It seems that individuals land and feed near others of their kind because there is a better chance of finding food there than elsewhere or because, at least for the moment, there are no predators nearby. Birds are quite nervous when feeding alone, and spend more time scanning for danger than for food. Nests are built quite close together but nest entrances are separated by a minimum distance, below which aggression becomes savage. They fight frequently – the more birds the more fights per bird – and this is probably an important population control.

At the beginning of the breeding season, several males of all ages may court any unmated female which does not flee the first encounter; they hop around with tail half-cocked and wings drooped, continuously chirruping. The female is torn between stretching up tall to indicate that she does not wish to mate, and trying not to appear too aggressive, although she will peck and wrestle males that are too insistent. Males too may spar and battle briefly among themselves; such "sparrow parties" usually ending with the flustered female dashing off with the gang in pursuit. A receptive female actively visits nest sites advertised by the bachelor males, eventually emerging to solicit mating on the threshold, or as near as practical. Copulation normally takes place in the open, near the nest, and is frequent from shortly before the eggs are laid until incubation begins. The male treads his mate as many as 30 times in succession whenever they are reunited, even after a brief separation. There is apparently a good reason for this: a female is not at all discriminating, and neighbouring males often court her successfully during their rival's absences. Repeated copulation, it is supposed, ensures that at least most of her eggs are fertilised by her mate. A long-mated pair may occasionally indulge in copulation well outside the egg fertilisation period.

Since nest sites are at a premium in many areas, successful breeders guard their property for a large part of the year. After the last brood has become independent there are usually several months during which both parents join the autumn grain harvest, sometimes in large numbers. At least some older males frequent their old nest sites from early June, thereby discouraging first-year males from taking these over. The surplus of potential breeders ensures that vacated sites are occupied and that new ones further afield are discovered. Sparrows are notoriously clever at finding suitable places, following roads and railways to homesteads quite remote from the nearest colony. In many parts of the world, the spread has exceeded 100 kilometres a year in suitable country, although even a relatively narrow band of hostile habitat may slow the rate of colonisation.

In central Europe, and much of Australia, the House Sparrow produces two, three, four or even more broods in quick succession; in the ancient home range, and in nearby Turkestan and Afghanistan, where the Sparrow is migratory, there is usually only one brood of seven eggs. Both sexes incubate about 14 days and both feed the young, which remain in the nest for about 17 days and for a short time after fledging. Newly independent young gather in often large and noisy flocks at roosting places.

W. Filewood

HOUSE SPARROW

Passer domesticus

Juvenile moulting to male plumage; Christchurch, New Zealand; March 1992 (D. Hadden)

OTHER NAMES
English Sparrow, European Sparrow

MEASUREMENTS
LENGTH: 143 – 167 mm
WING: 72 – 78 mm
TAIL: 49 – 60 mm
BILL: 13.4 – 18.7 mm
TARSUS: 18 – 19.6 mm
WEIGHT: 20 – 30 g

DESCRIPTION
ADULT
Male: Predominantly grey below, with black face and bib (more pronounced during breeding). Upperparts tawny, streaked with buff and black; conspicuous white wing bar. Crown grey; nape chestnut. Rump grey; tail brown. Eye dark brown

Female: Head, neck and underparts pale brown, with paler cream-buff eyebrow. Upperparts similar to male, but duller

IMMATURE
Similar to adult female, but duller with some mottling on the crown

JUVENAL
Similar to immatures, but with darker bill, with creamy-yellow gape. Lacks distinct colour pattern

IDENTIFICATION
Unmistakable. Small to medium finch with streaked dark brown, buff and black plumage (male); female paler with more brown below and brown head. Both sexes of P. montanus *resemble the male Tree Sparrow, but* P. montanus, *in southern New South Wales and central Victoria, has a rufous crown and black cheek patch*

VARIATION
Many subspecies, the exact number still under debate. Separated by amount of rufous plumage on back and depth of general colouring. Australian subspecies nominate P. d. domesticus, *introduced from England in 1860s*

VOICE
Variety of chirruping and twittering. Most typical call a harsh double-noted "chiisck" or "cherrup"

HABITAT
In and around human habitation; also in cultivated areas and some wooded country

FOOD
Seeds, insects, spiders, flower buds, fruits, berries and food scraps

HABITS
Monogamous, forming permanent pair bonds. Gregarious in small to medium groups, though often forming vast flocks. Diurnal. Mainly sedentary, though believed by many to be partially migratory

BREEDING
All year round, though more concentrated in spring and summer. Sometimes two to three, or more, broods per year. Both sexes build the nest and care for the young, though the female alone incubates the eggs

NEST: Large, untidy sphere of grass, wool and feathers, lined with feathers and finer plant material; placed in crevices in buildings, under bridges, in thick bushes or in tree hollows. (If placed in bush the nest is generally dome-shaped, with side entrance.)

EGGS: 3 – 6; grey-white, blotched with brown and grey; oval-shaped; 24 x 16 mm

INCUBATION: 11 – 14 days; fledging period 14 – 15 days

DISTRIBUTION
Throughout the entire eastern half of the Australian mainland and Tasmania. Particularly associated with human settlement. Probably absent from mid-western Cape York Peninsula

STATUS
Introduced. Very common and abundant around human habitation

P. Rowland

Tree Sparrow

Passer montanus

The Tree Sparrow so resembles the familiar House Sparrow that many people do not realise there are two species of sparrow in Australia. Indeed, there was no hint of two kinds of sparrow in European literature until 1713, when a different bird was described from the Italian Alps. In this sense, it is the last widespread European species to have been "discovered" scientifically. This could be taken as the fledging date of real ornithology. Linnaeus wrote the first version of the *Systema Naturae* shortly after this discovery, and called the Tree Sparrow *Fringilla montana*, since it was not understood that it is also common at sea level. In fact, its natural range (to the extent that this can be determined) is at least as great as the House Sparrow's: it is common in China and Japan, where, in the absence of the larger and bolder species, it is *the* domestic bird, nesting in houses, temples and other buildings. It is absent from peninsular India (living up to the name *montanus*) and it is also absent from most of intertropical Asia, except by introduction. However, it is often very common in urban areas where it has established itself, as in Singapore and parts of Malaysia, Borneo and the Phillipines.

It is uncertain whence the Australian Tree Sparrow populations were recruited. Some were surely from Britain – and perhaps elsewhere in Europe – but apparently some came from the Orient, since "Chinese Sparrows" were imported by at least one acclimatisation society. The current Australian population appears superficially more like the Japanese subspecies *P. m. saturatus* than the European *P. m. montanus*, and seems to be largely confined to environments close to human settlement. The name was coined by Stejneger to indicate that his new race was "saturated" with pigment, but the Australian Tree Sparrow is said to be slightly darker still, and more brightly coloured, as well as slightly smaller.

Adult at nest; France; 1975 (V. Serventy)

It is common in middle Victoria, from Geelong to about Sale, and perhaps more so (or more obviously so) in country towns towards the Murray than in Melbourne and the south. In New South Wales it has increased in some Murray Valley towns since the middle of the century, and in some towns of the southern slopes, north to Cowra. It used to occur around Sydney and other localities in the central east, but records are very few for most of this century.

The subspecies recently established on Christmas Island is Oriental, supposedly from a vessel from Hong Kong (where *P. m. saturatus* is native) but the birds could have joined the ship elsewhere. Whereas the temperate form has not succeeded as well in Australia as the House Sparrow, it is quite possible that a tropical form such as *P. m. malaccensis* could become quite numerous if it reached suitable areas of Australia. This possibility diminishes somewhat as the House Sparrow becomes established ever more firmly throughout tropical Australia, now breeding north to Thursday Island and west to Camooweal, if not beyond.

Considering how similar the two species are in most respects, it is interesting to contrast their success, and also to puzzle over certain differences. For example, it is hard to explain why the sexes of Tree Sparrow are indistinguishable externally, while those of the House Sparrow are very different. What bearing does this have, if any, on their relative success? Is the Tree Sparrow less successful simply because it is smaller, and unable to better the heavier and decidedly quarrelsome House Sparrow in direct conflict? In Europe there is evidence that the numbers of Tree Sparrows wax and wane independently of the House Sparrow, at least locally. At Oxford, the nesting success of Tree Sparrows is apparently related as much as anything to the supply of caterpillars for the hatchlings, but even this does not explain its arrival and quick increase in suitable areas from which it has been absent for many years. In Australia, the Tree Sparrow has not been found anywhere that the House Sparrow is not also established; in most areas the latter is the more obvious and often very much commoner, but the Tree Sparrow apparently prevails in a few small towns. It is absent from both western Victoria and eastern Victoria, and failed to survive in Tasmania. It mixes with the House Sparrow: each species attracts the other, in ones and twos, when they are feeding; but at places like Shepparton, where they come around picnickers for scraps, I have often seen a dozen Tree Sparrows defer to a single cock House Sparrow. It may simply be that the latter is bolder and less fearful of people.

Both species eat a similar mix of seeds, with some insects, eked out with bread crusts and pie flakes. Both obtain much of their food on the ground, but also forage in trees: there is no indication however that the Tree Sparrow spends more time in trees. The House Sparrow is quite at home in such all-Australian trees as eucalypts and casuarinas; the Tree Sparrow may be slightly less opportunistic.

A peculiar difference has also been described in the behaviour of nestlings, at least in England. Young Tree Sparrows come right to the entrance of the nest from an early age to beg energetically as their parents approach; House Sparrow nestlings do not generally make an appearance at the entrance until they are ready to fly. Perhaps nestling Tree Sparrows are more visible to predators than those of the House Sparrow.

Tree Sparrow

Passer montanus

Juvenile (D. & M. Trounson)

OTHER NAMES
Mountain Sparrow

MEASUREMENTS
LENGTH: *132 – 140 mm*
WING: *65 – 71 mm*
TAIL: *53 – 58 mm*
BILL: *14 mm*
Female slightly smaller on average but ranges overlap

DESCRIPTION
ADULT
Sexes similar. Back and wings brown, streaked with black and buff; rump grey; crown chocolate or chestnut; cheeks whitish with black crescent; two distinct white wing bars; underparts grey with black bib; iris dark brown to red-brown; bill black; legs and feet light to dirty brown

JUVENAL
Like adult, slightly smaller

IDENTIFICATION
Distinguished from House Sparrow by chestnut crown, black crescent on cheek

VARIATION
7 – 8 subspecies in Eurasia, Indonesia and western Pacific. Australian population founded with European P. m. montanus and east Asian P. m. saturatus

VOICE
Variety of chirps. Higher pitched twitter than House Sparrow

HABITAT
Sea level to mountains, open woodland, cultivated country, hillsides, villages, towns and cities; rarely far from human habitation and cultivation over most of its range

FOOD
Seeds, food scraps and small invertebrates, mostly taken on the ground

HABITS
Mainly sedentary like the House Sparrow, but less bold. Mingles freely with House Sparrow but defers in competition for food. Male's courtship display includes bowing, spreading the wings and raising the crown feathers

BREEDING
Mainly spring to summer, often several broods per year

NEST: a domed structure of grass, usually in a crevice

EGGS: 3 – 6; grey-white to brown-white; slightly glossy, with fine brown dots

INCUBATION: about 14 days; fledging period about 14 days

DISTRIBUTION
Most of Eurasia from Britain to Japan and Taiwan; South-east Asia to Malaya and western Sundas. Introduced to North America, Australia, Christmas Island, Borneo, Philippines, Pescadores, Sulawesi. In Australia, from Sydney to Melbourne, mostly west of Great Dividing Range

STATUS
Generally sparse but locally common, particularly in some settlements in the Riverina

W. Filewood

Adult male Red Bishop at nest; Taronga Zoo, N.S.W.; March 1977 (D. & M. Trounson)

Sponsor: Claude A. Fay

Weavers

The family *Ploceidae* is a predominantly African group of small to medium-sized seed-eating passerines, with only a few species in Asia and islands of the Indian Ocean. It comprises approximately 126 species of buffalo-weavers, sparrow-weavers, "true" weavers (which include bishops and widowbirds) and wydahs. Though there is still some debate as to the true boundaries of the group, and recent biochemical studies suggest that weavers may be more closely related to the estrildid finches, fringillid finches and sparrows (Passeridae) than previously reported, the traditional classification is retained here.

Buffalo-weavers (subfamily Bubalornithinae) are noisy birds which live throughout the year in flocks and colonies in African savannas. The male Black Buffalo-weaver, *Bubalornis albirostris*, is unique among songbirds in having a phallus and, in the breeding season, it grows a protuberance on the bill.

The African sparrow-weavers of the subfamily Plocepasserinae, (also including the sociable weavers) have gained the somewhat dubious title of "thatchers" rather than "weavers", due to their habit of inserting ends of sticks or grass into the nest structure, somewhat resembling the roof of a thatched cottage. The Sociable Weaver, *Philetarius socius*, found in the Kalahari Desert, builds a huge structure of sticks, sometimes weighing up to a tonne with a common roof and separate nest holes below. This compound nest protects the birds from the vast temperature ranges of the desert as well as minimising the effects of predators.

Wydahs (subfamily Vidualinae) are brood parasites which, like cuckoos, lay their eggs in other birds' nests – usually those of estrildid finches – where they are incubated and reared by the foster parents, at the expense of their own young. The vidualine wydahs do not kill or evict the chicks of foster parents, but merely mimic the brightly coloured and patterned gapes of their chicks and, in so doing, compete with them for food. These parasitic weavers, along with the true weavers, have caused great confusion over the years, in that both subfamilies contain birds called, with varying degrees of accuracy, wydahs and widowbirds. Both also contain bishops, weavers and indigo birds.

Conversely, nearly all of the bishops and wydahs, of the genus *Euplectes*, rear their own young. One exception is the African Cuckoo-weaver, *Anomalospiza imberbis*, which lays its eggs in the nests of grass-warblers, which rear the intruders at the cost of their own young.

Bishops are named for the hood of red or yellow in the male's breeding plumage. Widows have long tails (resembling the long train of a widow's veil of mourning). Wydahs are named after a Portuguese word which also implies mourning. In the Long-tailed Widow, *Euplectes progne*, found in Kenya, the male has a long flowing tail used in displays over his grassland territory. In an experiment, the length of the tail was artificially increased in several birds by adding parts of tails from other males, and in all instances these birds were found to be more successful than normal males in attracting females to nest.

All of the Ploceinae are gregarious but, at the commencement of the breeding season, males break away from the flocks to establish their territories, which are vigorously defended against intruders. Once a territory is established, the task of attracting a mate is undertaken with great fervour. In polygamous species, this involves great effort.

Two weavers have been introduced into Australia, both belonging to the genus *Euplectes*. Small colonies of Red Bishops Birds, *Euplectes orix*, were temporarily established in South Australia between the 1920s and 1950s. The White-winged Wydah, *Euplectes albonotatus,* survived on the Hawkesbury River, New South Wales, until the late 1960s. No recent records exist for either of these species, and both are now considered to have been extirpated within Australia.

Probably one of the best known and most feared of the Ploceinae is the Red-billed Quelea, *Quelea quelea*, which inhabits the more arid areas of Africa. It forms flocks of up to millions, which feed on crops of rice and cereals: in breeding times a single tree may hold 6000 nests and a single colony may boast as many as 10 million nests. The destruction caused by these flocks is incalculable, and it is not surprising that people have tried everything from rattles and drums to dynamite and flame-throwers in combat against the species. In South Africa over 400 million birds were destroyed over four years by aerial spraying without visible effect on their numbers, probably eradicating no more than the surplus population that would have otherwise died during the dry season. Control is made harder by the rapid rate at which the Quelea breeds, taking only two weeks to incubate its eggs and another two weeks for the young to fledge, with almost 90% success. On the other hand, humans benefit from the success of this species, which provides a high protein food in the poorer areas of Africa.

Some members of the Ploceinae are forest-dwellers, have only one mate, and eat insects. Most inhabit savanna areas or marshes, are polygamous and feed mainly on grass seeds, sometimes supplemented with insects in order to feed the chicks.

The nest of a weaver is truly woven, starting as a single loop suspended from an outer branch of a bush or tree, and is constructed by the male. This then serves as a display perch for attracting a mate. Once a willing female is found the male continues constructing the nest, and as soon as it is finished he begins another to attract another mate.

P. Rowland

White-winged Wydah

Euplectes albonatatus

Native to Africa, the White-winged Wydah, *Euplectes albonotatus*, was accidentally introduced to New South Wales in 1931, following the wreckage of a ship, the *Malabar*. The birds quickly became established in an area of rank vegetation along the Hawkesbury River, near Windsor. Flocks of as many as 80 – 100 individuals were soon reported in the area, and it was feared that it would become a major pest. Large flocks are capable of causing considerable damage to grain crops, although this is uncommon. Numbers along the Hawkesbury fluctuated over the years, but the numbers of males observed in breeding plumage appeared to diminish and the species has not been seen in Australia since 1975.

Male in breeding plumage (A.R. McGill)

Sponsor: Westpac Banking Corporation

The White-winged Wydah is polygamous, the male having several females in a territory of about 2500 square metres. When grasses have begun to seed, the male commences nest building by constructing a skeleton nest between two grass stalks, normally up to two metres above the ground. Once this is finished, he uses it as a stage on which to perform his display to a prospective mate. During the display the male points his bill upwards and quivers his wings, similar to the begging posture of a young bird, and, with the tail spread, utters a continuous series of "zicker-zicker-zicker-zicker" calls. When the female approaches, he turns his head around to face her and opens his mouth. Then, as she reaches him, he flies away, leaving the female to finish the nest.

Once the nest has been completed, courtship begins. The male drops down into the grass just below the female and depresses his beak into his chest while erecting the feathers of his nape and breast. After a short period, he jumps up from the ground to a point directly in front of the female or directly below her on the same stem. He then begins to bob up and down, with his body held erect and his beak still depressed into his chest, and utters a harsh "squi-squi-squeege-squeege". Copulation soon follows and the male flies away to repeat this behaviour with another of his several mates. The female is left to incubate the eggs and care for the young but the male remains in the area to defend his territory and keep a watchful eye on any unattended nests. He advertises his territory with a display in which he flies quickly from seed head to seed head with his tail spread, uttering a series of repetitive raspy notes.

Outside the breeding season a male White-winged Wydah loses its striking black plumage, and takes on a rather drab brown plumage like that of the female or young. Such "sparrow-like" birds form large gregarious flocks that roam the countryside in search of suitable feeding sites, mainly agricultural land where small grain can be found.

P. Rowland

White-winged Wydah

Euplectes albonotatus

OTHER NAMES
White-winged Widow-bird, White-winged Wydah

MEASUREMENTS
LENGTH: *150 – 180 mm*
WING: *(male) 72 – 81 mm; (female) 62 – 72 mm*
TAIL: *82 – 90 mm*
BILL: *c 14 – 16 mm*
TARSUS: *c 21 – 23 mm*

DESCRIPTION
ADULT
Male: (Breeding plumage) Predominantly black; bases of flight feathers, underwing coverts and the greater part of the primary and secondary wing coverts white. Shoulders have conspicuous yellow patch (in some populations overlaid with cinnamon and appearing only at the bases of the feathers). Tail black; bill blue-grey. Eye dark brown. Legs and feet dark grey

(Non-breeding plumage) Similar to female but larger and has wings and shoulder patch of breeding dress

Female: Generally brown. Crown, nape and mantle streaked with dark brown and black. Wings dark brown, edged with buff; bases of flight feathers pale buff. Underwing coverts pale buff-white. Sides of head and underparts pale yellow-brown, streaked with brown. Dark brown streak through eye; eye dark brown. Tail pale brown. Bill brown. Legs and feet dark grey

JUVENAL
Similar to adult female except feathers of upperparts edged with fawn

IDENTIFICATION
Medium-sized weaver-bird; generally "sparrow-like" in appearance. Predominantly streaky-brown. Male larger than female and has conspicuous white wing patch and bright yellow shoulder patch. During breeding season, male attains striking black plumage, still retaining conspicuous white and yellow of the non-breeding season

VARIATION
Four subspecies. Australian population probably E. a. albonotatus, identified on the basis of colour of shoulder patch (yellow) and length of tail

VOICE
Various brief twittering notes

HABITAT
In Africa, prefers dry grassland and low open woodland. In Australia, is found in swamps, waterways and cultivated areas where tall grasses occur

FOOD
Grain and seed. Diet of young birds supplemented with insects

HABITS
Conspicuous, diurnal, sedentary and gregarious. Polygamous; males have territories about 50 m square. Feeds on ground or in low vegetation. Usually seen sitting on grass heads, twittering, with long split tail spread. Reported to fly higher and faster than most other wydahs

BREEDING
Little known of breeding in Australia; probably early spring to late summer. Male constructs framework of nest, completed by female. Female incubates eggs and cares for young, male defends unattended nests

NEST: Spherical, with a large side entrance; made of woven grass and suspended between tall grass stems. Inner lining of thin grass heads, coiled round with a few projecting (as a sort of porch)

EGGS: 2 – 4; pale turquoise-blue, streaked and spotted with brown and grey; 19 mm x 14 mm

INCUBATION: 12 – 14 days; fledging period 11 – 14 days

DISTRIBUTION
Introduced to Australia in 1931. Native to south-eastern Africa: south-eastern Belgian Congo, Tanganyika, the eastern Transvaal and Natal. In Australia, isolated populations established in Windsor and Hawkesbury River district, near Sydney

STATUS
No confirmed sightings since 1975. Presumed extirpated

P. Rowland

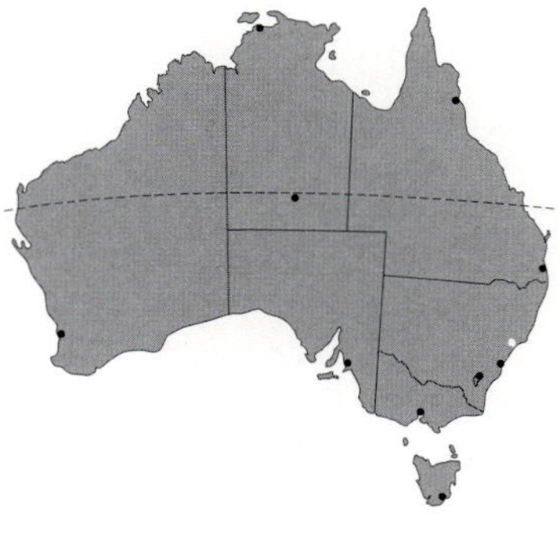

Red Bishop

Euplectes orix

A notorious pest of grain crops in its native southern Africa, the Red Bishop was first observed in Australia near Adelaide in 1926. By 1929 a breeding colony had firmly established itself at nearby Woods Point, in the Murray Bridge area. Following these early observations, reports flooded in from the surrounding areas and it was feared that this aggressive pest was going to become a major problem to the grain farmers of the region. Then, curiously, the spreading seemed to stop and the reports became fewer. By 1941 the birds had once again become confined to the Murray Bridge area, where they were apparently being assisted by certain local aviculturalists. The population thrived until 1956 when the Murray Bridge region was affected by a severe flood that destroyed most of the colony. No significant observations of Red Bishops have been reported since then and the species appears to have been extirpated.

The species was introduced to the Hawaiian island of Oahu in 1965, where it became firmly established for a number of years. Then, as in Australia, numbers dwindled fairly rapidly for no apparent reason and, during a bird survey in 1976, only one individual was observed. Attempts were also made to introduce the species to St Helena and Tahiti in 1929 and 1940 respectively. In both instances the species failed to become established.

As the name suggests the male Red Bishop is adorned with a striking scarlet plumage during the breeding season. For the remainder of the year it is a comparatively drab "sparrow-like" bird, virtually indistinguishable in plumage from the female. As in many weavers, the Red Bishop is polygamous, males having several mates in a single breeding season.

It has no song, its utterances having been likened to "a series of machine-like repetitive sounds". During the breeding season, however, the male makes a variety of calls as nest advertisement, in alarm, and to announce his presence to any prospective females.

Early in the breeding season, even before attaining full breeding plumage, the male begins to frequent a particular patch of reeds that he has chosen for a nesting site. Territories are not well marked and several males may share a single area, indicating a strong tendency toward colonial nesting. However, if a neighbouring bird approaches too close to an occupied nest, the resident male will chase the intruder off with an acrobatic aerial pursuit, seldom involving physical contact.

The male builds the framework for several nests in his selected area. The finishing touches, including lining, are added by the female. The male then sets about the task of attracting a mate. Interestingly, it has been noted that the male apparently has some difficulty in distinguishing between females and immature males which both share the same plumage, and he readily courts any "brown" Bishop that approaches. During courtship display, he performs a series of hopping movements in front of the other bird; his plumage is fluffed out, or ruffled, and his bill is depressed into his breast. He also often alights on a grass stem and swivels around on it. A common part of his courtship display is the "bumbleflight", in which he flies with plumage ruffled and head held forward. The wing beats are very slow and the action is comparable to the flight of a bumble bee. Throughout the courtship display, he utters a repetitive rattling noise. During the entire performance by the male, the female (or at least the bird being courted) remains very passive and does not respond to any of the male's antics, except for avoiding physical contact. Once mating has taken place, the female is left to incubate the eggs and to feed and care for the young birds. Meanwhile the male repeats his routine with another intended mate.

Outside the breeding season, in dull coloration, Red Bishops mix readily with other weavers to form large nomadic flocks that move around together in search of suitable feeding grounds and roosting sites.

P. Rowland

Male coming into breeding plumage; Taronga Zoo, N.S.W.; March 1977 (D. & M. Trounson)

Sponsors: H.G. & R.J. Mondel

RED BISHOP

Euplectes orix

Pair; Taronga Zoo, N.S.W.; March 1977 (D. & M. Trounson)

OTHER NAMES
Grenadier Weaver, African Grenadier Weaver, Red Bishopbird

MEASUREMENTS
LENGTH: *110 – 115 mm*
WING: *67 – 74 mm*
TAIL: *c 40 – 46 mm*
BILL: *c 14 – 16 mm*
TARSUS: *c 18 – 21 mm*

DESCRIPTION
ADULT
Male: (Breeding plumage) Forehead, cheeks, chin, breast and belly black; remainder of plumage scarlet. Wings and tail dark brown, with scarlet uppertail coverts. Eye dark brown. Bill black. Legs red-brown. (Non-breeding plumage similar to female)

Female: Most of plumage brown, streaked with darker brown and black; breast and flanks paler, also with dark streaking. Sides of head and around eyes pale buff, bisected by darker brown, forming indistinct pale eye-stripe. Bill horn-coloured. Eye dark brown

JUVENAL
Similar to adult female

IDENTIFICATION
Small "sparrow-like" bird with fairly drab brown plumage. During breeding season, male attains beautiful black and scarlet plumage

VARIATION
Four subspecies. Australian population is E. o. orix

VOICE
A variety of twittering notes, most commonly a "zik-zik-zik" or "chip-chip"

HABITAT
Savanna in Africa. In Australia, reed beds and tall grasses along watercourses

FOOD
Grain and grass seeds. Young birds also feed on insects

HABITS
In Africa, gregarious, forming large winter flocks with other small birds. Dispersive, wandering country in search of grass seeds. In Australia, more sedentary. Diurnal. Feeds on or near to ground. Nests colonially

BREEDING
Early spring to summer. Male builds framework of nest, completed by female. Males monogamous or polygamous; up to six mates, each rearing one or two broods per season. Australian breeding sites less colonial than in Africa. Incubation and rearing of young carried out solely by the females, with male in vicinity

NEST: Oval woven grass structure with a side entrance, suspended between upright stems of reeds and lined with soft grasses

EGGS: 3 – 4 (usually 3); turquoise; 18 x 13 mm

INCUBATION: 12 – 14 days; fledging period 12 – 16 days

DISTRIBUTION
Native to eastern and southern areas of central Africa. Small colony of escaped aviary birds established in portions of reed beds along Murray River, South Australia. Main colony at Woods Point, south of Adelaide

STATUS
Introduced. Rare in Australia, probably extirpated (possibly during floods of 1956)

P. Rowland

Sponsors: Sir John & Lady Proud

GRASS-FINCHES

The family Estrildidae comprises approximately 124 species of grass-finches (also known as weaverfinches or waxbills), occurring naturally in Australasia, Africa and southern parts of Asia. Deliberate introductions and escaped aviary birds have liberated numerous species throughout many other parts of the world but only three of these have been successful: the Zebra Finch, *Taeniopygia gutatta,* in Tahiti; the Chestnut-breasted Mannikin, *Lonchura castaneothorax*, in New Caledonia and the Society Islands; and the Red-browed Finch, *Neochmia temporalis*, in New Caledonia..

Australia has 18 native estrildids and two introduced species: the Black-headed Mannikin, *Lonchura malacca* from India and surrounding areas; and the Nutmeg Mannikin, *Lonchura punctulata* from south-east Asia. The Nutmeg Mannikin has been by far the more successful. Since its introduction in about 1930, it has rapidly increased in range and numbers, especially north of the Tropic of Capricorn, where conditions are very favourable to the species and it is able to breed almost all year round. It is unafraid of humans and is never found far from settled areas. Where its range overlaps with native grass-finches, it has become so competitive that it has forced many of them out. If it continues to spread at the current rapid rate, it could cause a major ecological disaster. In addition to its effect on native finches, it has become a major agricultural pest, feeding extensively on ripening grain crops. The Black-headed Mannikin has been the subject of scattered sightings and rare breeding records since the 1920s but it has not been reported since 1981.

The native species are remarkable for their plumage variation and their popularity as cagebirds has made the Estrildidae one of the best studied bird families in the world. However, in Australia, this study got off to a fairly slow start. John Gould is quoted by Gregory Mathews as enthusing over birds collected on the voyage of the *Beagle*:

> The specimens from which my description of this bird was taken are from the interesting collection placed in my hands by the late Mr Bynoe, whose great perseverance and assiduity have enabled me to add many species to the fauna of Australia. Indeed, many of the officers of the Beagle will have their names handed down to posterity in consequence of the attention they paid to this branch of science, independently of the legitimate objects of their various expeditions; among others I particularly allude to Charles Darwin, Captain Wickham, Captain Stokes and Mr Dring.

Adult female Red-browed Finch at nest; You Yangs, Vic; January 1975 (M. Seyfort)

Sponsors: Dr Mervyn & Mrs Susan Cobcroft

Mathews went on to comment:

> It is somewhat strange to note the extraordinary number of undescribed species of Australian birds Gould was able to find in English collections when he first turned his attention to the Australian avifauna. Undoubtedly it was this plenitude of novelties that determined his expedition to Australia to discover many more.

Since the turn of the century many native estrildids have shown a significant reduction in numbers for a variety of reasons, such as changes in habitat, illegal trapping and competition with introduced birds, such as the Nutmeg Mannikin. Lord (1956) noted the decline of the Diamond Firetail, *Stagonapleura guttatum*, "once plentiful all year ... only occasionally a pair or small flock is now seen"; and of the Plum-headed Finch, *Aidemosyne modesta* from "large flocks" to "small parties only occasionally being seen".

Another such species is the spectacular Gouldian Finch, *Erythrura gouldiae*, which has decreased dramatically in recent decades. The beautiful markings of this bird gained it much attention and, worldwide, it became a popular cagebird. Unfortunately, this has led to much illegal trapping. Moreover, the Gouldian Finch is very susceptible to the air-sac mite, *Sternostoma tracheacolum*, which gives the bird asthma-like symptoms, and dramatically reduces its fitness. Interestingly, neighbouring bird populations appear to be far less affected by the air-sac mite, even when sharing the same habitat.

Australian grass-finches feed mainly on grass seeds. Insects form a variable part of their diets. Some species become completely insectivorous in the breeding season; others merely supplement the usual diet with a few high protein insects.

The method of drinking varies between species of grass-finches. Some employ the usual method of dipping the bill into the water, then tilting the head back to swallow. In arid parts of australia most grass-finches drink by immersing the face and sucking water up through the open mouth.

The estrildid finches are mostly monogamous and many are thought to pair for life. The strength of the pair bond varies from species to species, but is especially strong during the breeding season, which usually coincides with rising temperature at the onset of summer. Courtship displays comprise three main stages: the introductory stage and two main phases of courtship "dance". Courtship behaviour varies between species and between particular habitats. In the more arid areas of Australia some species have become opportunistic breeders and may reproduce whenever suitable conditions occur.

The chicks of estrildid finches have distinct markings inside the mouth, characteristic of each species. These may be varied symmetrical patterns of spots, inverted U-shapes or a series of lines or blotches. The markings are usually black, but species of the genus *Erythrura* have iridescent turquoise markings that can be distinguished by the parents, even when there is little or no light. Estrildid finches undergo a complete post-juvenile moult. After they leave the nest they remain in the same plumage for up to six months, during which time they gradually moult their old feathers and replace them with the feathers of the adult type. Thus they bypass the immature stage of feather development.

P. Rowland

Red-browed Finch

Neochmia temporalis

Apart from the well-documented Zebra Finch, *Taeniopygia guttata*, the Red-browed Finch inhabits the widest variety of habitats of any Australian grass-finch. It thrives in the hot tropics as well as the colder southern areas where the temperatures may fall below freezing during the winter. The Red-browed Finch also inhabits open pastures with scattered groups of trees or bushes. In the many coastal districts of Queensland, it frequents the fringes of cane fields, as well as being a common bird around, and even inside, human settlements; it is a typical inhabitant of gardens and parks. In some settlements of northern Queensland, however, it has recently decreased quite markedly in numbers, being a victim of the highly competitive, introduced Nutmeg mannikin, *Lonchura punctulata*. Introduced to Western Australia, the Red-browed Finch was seen near pastures and orchards in the Darling Ranges, east of Perth for some years. It is now relatively scarce.

The bulk of its diet consists of ripe and half-ripe seeds. Like most other finches, it husks the seeds before swallowing them and it is pertinent to ask the value, if any, of this behaviour. It may save space in the crop or reflect the difficulty of digesting husks. Laboratory tests have shown that smaller seeds are husked more rapidly than larger seeds but higher-energy, intermediate sized seeds are consumed more frequently than larger or smaller ones. The varying risks of predation in different locations may predetermine which seeds are taken. Feeding sessions are interspersed with frequent visits to waterholes, where it drinks in the typical manner by scooping. If disturbed, a group flies with strong, slightly undulating flight to a nearby bush or tree.

Adult male at nest; Forestry Reserve, West Pennant Hills, N.S.W.; 1982 (J. Purnell)

Sponsored in memory of Charles & 'Bobbie' Benton

The Red-browed Finch is very social, forming large, noisy flocks of up to 200 or 300 birds outside the breeding season. Members of the flock keep in touch with constant high-pitched calling, both while feeding and in flight. Pair bonds seem to be very strong, bodily contact and preening being common between members of a pair, even when in a flock. During the breeding season, birds associate in pairs or small groups but they remain very sociable.

Several nests may be constructed in adjacent trees or bushes or, occasionally, in the same tree. Of all the Australian grass-finches the Red-browed Finch is the most trusting of humans, frequently visiting houses and gardens to feed on breadcrumbs and food scraps.

The courtship display of the Red-browed Finch has been described as follows by D. Morris:

> The male selects a long straw, holds it in his beak, advances to the side of the other bird, and begins to sing and dance with vertical movements. These are only slightly directed towards the other, the male being almost broadside-on. The body is now extremely erect, with the legs stretched and the head held back, the beak pointing to the sky. The vertical inverted-curtsy jumps are so vigorous that, as in the Gouldian Finch, the male's feet leave the perch each time. The tail is held more downward now and is strongly twisted towards the female. The display attracts her and she approaches the male. As she alights near him she throws back her head and, for a few moments, holds this position with her beak pointing upward like his.

The Red-browed Finch is an energetic, if somewhat erratic builder, starting to make many nests that are never finished. The typical nest is a bulky, flask-shaped construction of dried or green grass, usually placed in a spiky tree or bush, up to eight metres from the ground.

Both parents share incubation and both spend the night in the nest. Parents also share the feeding of the young. An interesting behaviour has been observed in young Red-browed Finches: while begging food, they quiver their wings which, while being quite common in other passerines, is very rare among grass-finches.

John Orell, an aviculturalist, reported a very interesting account of communal feeding of captive Red-browed Finches by wild Red-browed Finches. He had been hand-rearing six chicks from a very early age. One day he had to go out and so hung the cage in the porch of his house. He writes:

> On our return we were surprised to find that there was a constant stream of finches flying to and from the porch – all feeding the orphans through the bars of the cage. During the next week we placed the cage in a shady spot and allowed the local birds to rear the fledglings to flight stage by a communal effort.

This unusual case of "foster-parenting" appears to be an exaggerated incidence of "helpers" at the nest. Often, young birds and those that have not paired for the season will "help" parents of young birds, by fetching food and defending the nest, although this has not been well documented in grass-finches. Behaviour of this kind has also been observed in members of the tit family, *Paridae*, where an abandoned brood was raised by several other "foster parents" and also in the thrush family, *Turdidae*, where an "orphaned" Blackbird, *Turdus merula*, was raised by "foster parents" in a captive/wild situation.

P. Rowland

Red-browed Finch

Neochmia temporalis

Juvenile; Toowoomba, Qld; March 1988 (R.E. Viljoen)

OTHER NAMES
Temporal Finch, Red-head, Sydney Waxbill, Waxbill, Lesser Red-browed Finch, Redbill, Red-eyebrowed Finch, Redbrow, Red-browed Firetail

MEASUREMENTS
LENGTH: *100 – 122 mm*
WING: *47 – 55 mm*
TAIL: *39 – 51 mm*
BILL: *10.7 – 12.8 mm*
TARSUS: *13.3 – 17.4 mm*
WEIGHT: *8 – 13 g*

DESCRIPTION
ADULT
Male: Back and upper wing coverts dull olive-green; brighter olive-yellow at sides of hind-neck. Crown and nape grey. Broad crimson eyebrow extending to sides of nape. Ear coverts, cheeks, throat and underparts grey, washed pale buff. Rump and uppertail coverts crimson; tail feathers dark grey-brown. Bill red, with triangular black patch on culmen and under surface of lower mandible. Eye deep red-brown. Legs and feet flesh-coloured

Female: Similar to male, but eyebrow generally longer and more obviously tapered

JUVENAL
Head, cheeks and underparts ash-brown. Rump and uppertail coverts duller; tail feathers dusky-brown. Eyebrow shorter. Bill black. Eye brown. Legs and feet yellowish-brown

IDENTIFICATION
Medium-sized finch, slightly smaller than a sparrow. Upperparts olive-green, apart from red rump. Underparts grey. Conspicuous bright red eyebrow and red bill

VARIATION
Complex. Earlier thought to comprise six subspecies, now three: A. t. temporalis, in a broad coastal band from Younghusband Peninsula, South Australia, through Victoria and New South Wales, to Townsville in Queensland; A. t. minor from Rockhampton, north to Weipa on west coast of Cape York Peninsula, Queensland; A t. loftyi in localised area around Adelaide and Port Pirie in South Australia. In A. t. minor the grey of the head extends further down to the shoulders, the back is bright yellowish-olive and the throat and abdomen are white

VOICE
High-pitched, almost inaudible, piercing "seeee" or "ssitt", accompanying most movements; repeated incessantly in song

HABITAT
Varied: Normally near streams; frequenting grasslands, land cleared for agriculture, open forest, fringes of scrub and a regular visitor to parks and gardens in towns and cities

FOOD
Great variety of ripe and half-ripe grass seeds, normally on ground, loosely supplemented with small herbs, berries and insects. Young fed almost exclusively on insects and their larvae

HABITS
Diurnal, sedentary and nomadic. Gregarious, forming small family groups during breeding season and flocks of 10 or more during non-breeding season. Trusting and confiding in humans, easily observed at fairly close quarters. Young differ from most other grass-finches in shaking their wings when begging for food. Many dead birds are found during periods of heavy rain in colder parts of range, probably due to water-logging of feathers

BREEDING
In the northern part of range breeding takes place between January and April; in the south, from September to November (active nests have been found in every month except July and August). Often more than one brood per season. Both parents incubate eggs and care for young

NEST: Bulky, flask-shaped nest of dried or stiff green grass, lined with soft grasses and feathers (white feathers preferred). False spout, or small chamber often added to completed nest. Usually in spiky tree or bush, less than 2 m above ground

EGGS: 5 – 8, pure white; 16 x 12 mm; laid one day apart

INCUBATION: 12 – 15 days; fledging period 21 – 24 days

DISTRIBUTION
Eastern Australia from Cape York Peninsula south to Victoria and west to Adelaide and Kangaroo Island. A small population near Perth is descended from aviary escapees. Also introduced to New Caledonia

STATUS
Endemic. Common

P. Rowland

Star Finch

Neochmia ruficauda

The Star Finch is not really a ground-dweller, preferring tall grasses, reeds and the like, in close proximity to water. Here it feeds on ripe and half-ripe grass seeds. It drinks regularly by immersing its face and sucking up water. Because of its dependence on surface water, it has become a regular visitor to artificially irrigated areas and is regularly seen around the rice fields of northern and north-western Australia. Unlike some mannikins, it does not seem to do any damage to the rice crop, merely feeding on the seeds of grasses growing nearby. Although it occurs regularly in and around cultivated areas, it is a rather timid bird and is not often seen near homesteads or in and around towns.

The species has suffered a serious decline in numbers in recent decades, probably due to habitat modifications and trapping. Pastoralism has contributed greatly to its decline in the east, the rank grasses around permanent freshwater holes being trodden down by stock during times of drought. The effects of stock and feral animals such as pigs have contributed to its decline in the northern parts of its range and the exotic Rubber Vine has smothered much of the vegetation which it inhabits. Consequently, in contrast to the Double-barred Finch, *Taeniopygia bichenovii*, the Star Finch is now restricted to possibly less than half of its former range.

Flock; Fitzroy Crossing, W.A.; October 1969 (G. Chapman)

Sponsors: E.W. & M.J. Abrahams

Like mannikins, the Star Finch has two modes of flight. Individuals fly with slight undulations, whereas big flocks fly rapidly and directly, sometimes making sharp simultaneous turns. Additionally, the female Star Finch has an elaborate display flight. Holding a long piece of grass in her beak, she flutters around the perched male several times. Each time she passes him, the grass nearly touches his bill. The role of this display is not fully understood; it mostly occurs in unmated birds and may serve as a social bonding ceremony similar to the peering behaviour of the Plum-headed Finch, *Aidemosyne modesta*.

Courtship commences towards the end of the wet season. The male begins by grasping a long piece of green grass in his beak. Then, with his plumage strongly fluffed, he twists his tail towards the female and bobs up and down – so energetically that his feet occasionally leave the perch. During this bobbing, his beak is held horizontally, not depressed into the chest or raised in the air as in most other grass-finches. Each bob is followed by a deep bow by the male. Towards the end of the courtship display the bobbing becomes more active and only then is the beak pointed upwards. Throughout the performance the male utters his song phrase. The female signifies her willingness to copulate by quivering her tail, whereupon the male drops his piece of grass.

A curious behaviour occurs, mostly among young males, early in the breeding season. A "half-hearted" courtship dance is performed in the absence of a female partner. Tail-twisting and bowing are omitted, but all other movements are identical to the proper courtship dance.

The nest, which is placed in a tussock or low in a bush or tree, is a rounded mass of coarse green or dried grass stems. Both partners construct the nest but the male forages for the bulk of the material. Both birds also incubate the 3 – 6 eggs which hatch after a period of 12 – 14 days.

Star Finches are fairly gregarious, gathering outside the breeding season in flocks of 10 – 20 birds, occasionally larger. Flocking birds constantly utter high-pitched contact calls, and peering is also quite common in perched birds. Although the species is social, mutual preening and body contact are restricted to mated pairs and are uncommon outside the breeding season. Pair bonding is fairly strong, partners keeping together throughout the year

P. Rowland

Star Finch

Neochmia ruficauda

Adult female; Carawine Gorge, Pilbara, W.A.; June 1988 (M. Lochman/Lochman Transparencies)

OTHER NAMES
Red-tailed Finch, Red-faced Finch

MEASUREMENTS
LENGTH: *110 – 124 mm*
WING: *50 – 54 mm*
TAIL: *39 – 45 mm*
BILL: *11 – 12.5 mm*
TARSUS: *14 – 15 mm*
WEIGHT: *8.5 – 11 g*

DESCRIPTION
ADULT
Male: Upperparts generally olive-brown, slightly darker on wings; uppertail coverts dull crimson with large pinkish-white spot on each; two central tail feathers dull crimson, remainder dull crimson-brown. Front half of head crimson; ear coverts and upper throat crimson with white spots. Lower throat and flanks olive-grey with white spotting; abdomen plain yellow, undertail white. Eye light red. Bill scarlet. Legs and feet yellow

Female: Duller than male with less crimson on head

JUVENAL
Upperparts dull olive-brown, rump and tail darker; head and face grey; underparts brownish-buff; abdomen whitish. Bill blackish

IDENTIFICATION
Small to medium grass-finch. Generally yellow-olive in plumage, paler below, with conspicuous red face, slightly smaller in female

VARIATION
Two proposed subspecies: N. r. ruficauda in the south-east; N. r. clarescens in the north. Differ in size and colour intensity

VOICE
Call: Communication call a low "tlit" or "pslit"; loud identity call "sseet", uttered almost constantly, especially when in flight.

Song: soft, high-pitched twittering by male, audible only from a short distance

HABITAT
Brakes of rushes and tall, rank grasses around swamps and along rivers; never far from water

FOOD
Ripe and half-ripe seeds from heads of tall grasses, also fallen seed; insects taken in breeding season

HABITS
Sedentary. Diurnal, roosting communally in thickets (in breeding season female stays in nest overnight). Gregarious, forming flocks of up to 20 or more when feeding and drinking. Food taken on or near the ground, usually from tall grasses

BREEDING
December – August (mostly March and April); usually one clutch. Male forages for nesting material; both members of pair construct nest, incubate eggs and brood young

NEST: Rounded in shape and lacking entrance tunnel; constructed of green grass-stems or coarse dried grass, lined thickly with feathers; mostly placed in grass tussock or in low tree or bush; 3 – 6 m above ground

EGGS: 3 – 6; white; about 14 x 12 mm

INCUBATION: 12 – 14 days; fledging period 15 – 17 days

DISTRIBUTION
Northern Australia, from Pilbara to east coast of Cape York Peninsula and south to Rockhampton

STATUS
Uncommon. Declining in range due to loss of habitat and illegal trapping

P. Rowland

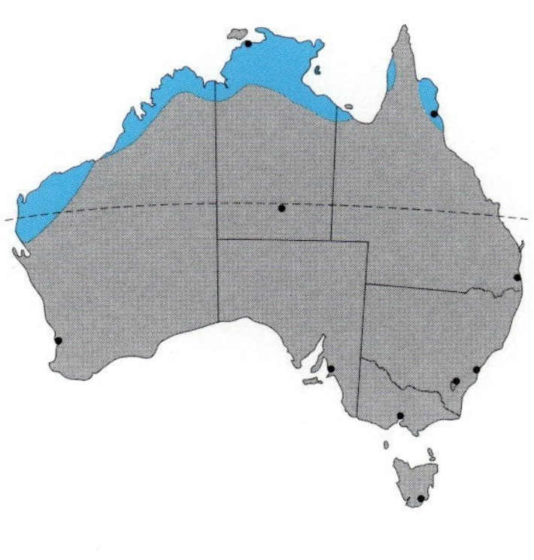

Sponsored in memoriam: Dr Judith Ross M.B., B.S., F.F.A.R.A.C.S.

Adult male; Lawn Hill, Qld; July 1990 (R. Brown)

Sponsor: Bessie Stevenson

Crimson Finch

Neochmia phaeton

This extraordinary finch was discovered at Raffles Bay, Northern Territory, by the French scientists, Jacques Hombron and Charles Jacquinot on board the *Astrolabe* and *Zélée*, and was described by them in Paris in 1841. Among the open-country grass-finches of Australia, it is unusual in spending very little time on the bare ground, only alighting occasionally at the edges of waterholes to drink. Food is generally gathered from the tops of seeding grasses and found by fluttering short distances from one plant to another. Nevertheless, instances of feeding on the ground have been reported. A general impression of its activity was written by Captain Samuel White to Gregory Mathews:

> Met with this glorious little bird on the Adelaide River, N.T., in 1922, June. There were numbers feeding on an open space near the river bank, and as they flew up and down from the ground to a low bushy tree they were like flashes of fire shooting out of the dry grass, their beautiful plumage shows so brilliantly in the strong sunlight.

The Crimson Finch has the most conspicuous tail movements of any Australian grass-finch, and when walking around it flicks its tail upwards to an almost vertical position, showing off its bright crimson rump and central tail feathers.

Of all the grass-finches of northern Australia, it is perhaps the most dependent upon surface water, generally restricted to tall rank grasses and pandanus trees near watercourses. This dependence on water, coupled with the increase of farms and homesteads in the area has led to it becoming increasingly common around homesteads and settled areas; it even places nests in roofs of buildings in certain parts of the range. It may indeed be becoming more dependent upon "imported" rather than natural surface water, for it is often seen drinking from wells and troughs at great distances from surface water and even from a dripping tap, rather than a nearby stream.

It is by far the most aggressive of all the Australian grass-finches. Wherever two birds which are not a pair are together, they squabble constantly, using their bills in an endeavour to drive one another from a perch. Even members of a pair do not perch in contact or preen each other. When a bird is fighting, the tail is fanned and one wing is raised vertically. Fighting is not confined to members of the same group; a Crimson Finch readily attacks much larger birds.

The Crimson Finch is the least social of all the Australian grass-finches, but the aggressiveness described above does not extend into territories. Several pairs may nest in close proximity but without social activity, apparently involving tolerance rather than acceptance. During the breeding season, birds mostly associate in pairs. Although flocks of 18 – 20 may form outside the breeding season, there is only a very loose social connection between the birds. The largest known flock, recorded by Bell in 1967 was "up to 100 individuals". He commented that "...the Balimo area was in the grip of the worst drought for 25 years and this may have caused the unusual concentration".

At the commencement of courtship, a male finds a piece of green grass and grasps it in his beak. He alights near a female and takes on a horizontal position with head and tail twisted towards her.

The female then twists her head and tail towards him. The male fluffs the feathers of his underparts, sides of head and back, and starts to bob up and down, with his feet actually leaving the perch. He then makes a series of bobs in an upright position, whereupon the female quivers her tail (not easily visible due to the length of the tail feathers) and copulation ensues. Like the Star Finch, *Neochmia ruficauda*, the male normally drops his piece of grass prior to copulation.

Selection of a suitable nesting site is an energetic and highly ritualised procedure, more complex than in any other Australian grass-finch. Klaus Immelmann wrote in 1982:

> As soon as a pair has found a prospective nest site both birds indulge in an elaborate bowing ceremony, with heads and tails twisted towards each other and the bodies almost touching. During this ceremony, which may be repeated many times, the birds frequently wipe their beaks. They utter high-pitched whining calls which are heard only in this situation. After several bouts of ceremonies, the male may fly off and quickly return with the first piece of grass for nest construction. With this in his bill he may again indulge in ceremonies; but with every piece of building material subsequently brought, the ceremony will become less intensive and there will be a gradual transition to the building activities.

The nest itself is a bulky structure of dried grass, bark and leaves, constructed by both birds. The male collects most of the material but both birds weave the nest, which is situated in a pandanus tree, or occasionally in a building, about 3 – 15 metres from the ground.

Habitat disturbance and illegal trapping has led to a slight decrease in Crimson Finch numbers over the past few decades – especially the white-bellied subspecies of northern Queensland. Populations do not seem to be under any great risk at the present time, but conservation measures may have to be adopted in the near future.

P. Rowland

Pair feeding on green algae; Riversleigh, Qld; August 1989 (G. Chapman)

Sponsor: Peddle Thorp & Walker

Crimson Finch

Neochmia phaeton

Juvenile; Fogg Dam, N.T.; July 1989 (R. Brown)

OTHER NAMES

Blood Finch, Cape York Crimson Finch, Pale Crimson Finch, White-bellied Crimson Finch

MEASUREMENTS

LENGTH: 123 – 136 mm
WING: 48 – 54 mm
TAIL: 57 – 70 mm
BILL: 12 – 13.9 mm
TARSUS: 14 – 17 mm
WEIGHT: 7 – 10 g

DESCRIPTION

ADULT

Male: Back and upperwing coverts grey-olive, washed with crimson; uppertail coverts and rump bright crimson. Central pair of tail feathers dull crimson, outer feathers dull brown, washed with crimson. Crown and neck dusky brown; face, ear coverts and under surface crimson, with fine white spots on flanks. Abdomen and undertail coverts black or white, depending on subspecies. Eye light yellowish-brown. Bill red, base of lower mandible white or blue. Feet yellow-brown

Female: Forehead and crown grey-olive; front of neck and breast ashy brown with white spotting on flanks; lower breast and abdomen lighter yellowish-brown. Crimson of head and tail duller

JUVENAL

Generally dull reddish-fawn above, with rump and tail dull crimson; breast pale brown; abdomen yellowish-brown. Bill dusky-black

IDENTIFICATION

Slightly larger grass-finch with characteristic long tapering tail and generally crimson plumage. Crown and nape grey-brown

VARIATION

Two distinct subspecies: N. p. phaeton in north-west and east Australia is distinguished by black abdomen; N. p. evangilinae in northern tip of Cape York Peninsula and New Guinea has a white abdomen, with occasional yellow suffusion and slight greenish hue to feathers of the head. Some past workers recognised a third subspecies, N. p. albiventer, of Cape York Peninsula, but this is now considered to be a synonym of N. p. evangelinae

VOICE

Call: Loud, characteristic, penetrating "tsee-tsee-tsee-tsee-tsee" or "che-che-che-che-che", two syllabled in contact, or loud and repeated rapidly in alarm or identification. Can be heard over several hundred metres

Song: Soft, rapid series of rasping sounds closing with three descending notes

HABITAT

Pockets of tall grassland and pandanus on streams and coastal plains; around settled areas with wells or troughs

FOOD

Half-ripe and ripe seeds of a wide variety of grasses and herbs and, during breeding season, many insects

HABITS

Sedentary. Diurnal, sleeping individually on perches in thickets. Not especially gregarious, usually alone, in pairs or family groups, but may mix with other finch flocks when feeding or drinking. Territories not aggressively defended and several pairs may nest close to one another. Little body contact within pairs; individuals frequently squabble over perches. Feeds mostly on tall grasses well above ground, constantly moves in search of food

BREEDING

September – May in Queensland; January – April in Western Australia and Northern Territory (towards latter half of rainy season). Male collects material for nest, but both members of pair construct it. Both incubate, brood and feed young

NEST: A bulky structure without entrance tunnel, varying in shape according to position, made of dried grass, bark and leaves; thickly lined with feathers; built in pandanus palm, or in and around buildings when available. Usually 3 – 15 m above ground

EGGS: 5 – 8; white; 14 x 12 mm

INCUBATION: 12 – 14 days; fledging period 20 – 21 days

DISTRIBUTION

North-western and north-eastern Australia, from about Broome, Western Australia, to Mackay, Queensland; and west coast of Cape York Peninsula, Queensland. Also in southern New Guinea.

STATUS

Common

P. Rowland

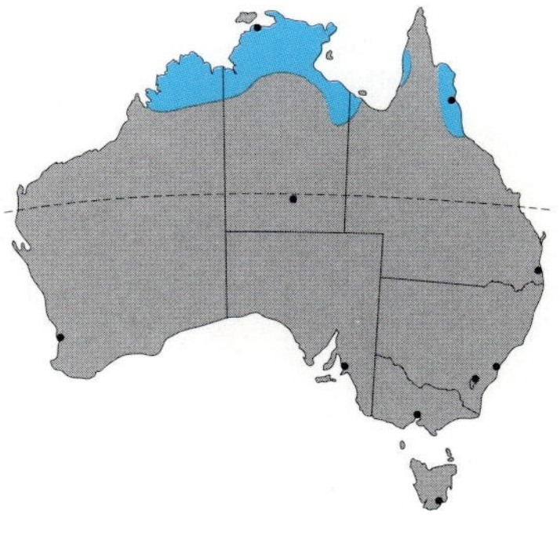

Beautiful Firetail

Stagonopleura bella

The Beautiful Firetail is found in the south-eastern part of mainland Australia and is the only native Australian grass-finch to reside in Tasmania, where it still occurs in reasonable numbers. Nevertheless, it is one of Australia's rarest grass-finches, being the victim of considerable predation by the cat and having lost much of its habitat to human settlement. The Tasmanian naturalist Frank Littler wrote to Gregory Mathews early in the twentieth century:

> This is the only Finch found in Tasmania, where it frequents dense scrubby country and the undergrowth near springs, the food being seeds and small snails. When flushed it rises with a loud whirring noise like a small Quail, the flight being rapid, though not long sustained ... Its note is a low, plaintive whistle, which sounds very mournful when all else is still.

Littler's observations are pertinent but, in fact, the Beautiful Firetail is a fairly strong flier. When crossing open plains within its habitat, it flies straight and rapidly, with scarcely any of the characteristic undulations exhibited by many other grass-finches.

It is less solitary than the closely related Red-eared Firetail, forming flocks of up to 12 individuals, (small in comparison with most grass-finches). In respect of his experiences in 1838, John Gould wrote:

> I generally observed it in small communities varying from six to a dozen in number, searching on the ground for the seeds of grasses and other small plants which grow on the plains and open parts of the forest. It also frequents the gardens and pleasure grounds of the settlers, with whom it is a favourite, few birds being more tame or more beautifully coloured than this little finch.

Pairs of Beautiful Firetails have very large foraging areas – up to several square kilometres – which contain a breeding area, usually towards the centre. The pair works this area in a regular daily routine. Soon after dawn they leave the roosting site (normally a loosely constructed ball-shaped roost nest built in a tussock or low shrub) and set out for a particular feeding site to forage for food, mainly seeds and a few insects. They remain in this spot for a few hours, resting during the heat of the day and returning late in the afternoon to resume feeding. Food is taken mostly on the ground, the birds staying near to or under cover, hopping rapidly over large distances. Such behaviour leads them to often be mistaken for small mice scuttling through the undergrowth. A single feeding site may be used for several days before the birds move off to another. In this manner they gradually move through the entire foraging area. Feeding is interrupted by frequent bouts of bathing and drinking, water being obtained by scooping, both from dew and pools. Other Beautiful Firetails that have strayed into the feeding area are tolerated but if one wanders too close to the nest area, particularly during the breeding season, the resident male drives it away. He does this with a threat display, consisting of tail-fanning to display the bright crimson rump as, with open wings, he hops at the intruding bird. Aerial chases may occasionally ensue. One bird normally acts as a lookout and, if danger approaches, it utters the alarm call, "tup-tup-tup". If birds are disturbed, they usually try to hide between grasses, where they remain until the danger has passed.

With the onset of the breeding season at the beginning of summer, the male finds a long piece of grass with which to begin his courtship. With the grass in his bill he attracts his mate to an area low in the ground-cover by nodding his head, usually in sequences of three. When his mate approaches, he assumes a crouched posture, leans forward, fans his tail and lowers his wings to show off his bright crimson rump. This is followed by ritualised bouncing up and down, again in bouts of three, while he emits a cricket-like "caw-caw-caw". The female responds by moving closer, with her tail facing towards him. Unlike other grass-finches she does not quiver her tail.

P. Rowland

Adult; Tianjara, N.S.W.; April 1987 (G. Chapman)

Sponsors: John & Jean Broinowski

Beautiful Firetail

Stagonopleura bella

Adult at nest; Medlow Bath, N.S.W.; 1979 (J. Purnell)

OTHER NAMES
Firetail Finch, Firetailed Finch

MEASUREMENTS
LENGTH: 107 – 128 mm
WING: 56 – 63 mm
TAIL: 44 – 50 mm
BILL: 11 – 13 mm
TARSUS: 16 – 17.3 mm
WEIGHT: 9 – 15 g

DESCRIPTION
ADULT
Male: General colour above olive-brown, finely barred with blackish-brown, broader and more conspicuous on lower back; rump and uppertail coverts bright crimson; tail feathers dark brown, with blackish cross-bars, central ones more blackish with crimson on basal portions. Black mask extending through lores and around each eye. Most of underparts light grey, finely barred with blackish-brown. Centre of abdomen and undertail coverts dull black. Eye hazel with pale opal-blue eye ring. Bill red. Legs and feet pink-cream. In breeding birds, plumage darkens and wing tips whiten

Female: Similar to non-breeding male but centre of belly has barring like rest of underparts

JUVENAL
Duller than adults, with indistinct barring of feathers; eye-patch smaller. Bill blackish.

IDENTIFICATION
Generally brown-grey, with fine black and white barring and contrasting crimson rump and bill. Conspicuous black face-mask and pale blue eye-ring

VARIATION
No significant geographical variation but much individual variation

VOICE
Call: Identity call a mournful, floating whistle "weee", with no obvious beginning or end. Alarm call, a loud "tup-tup-tup". Contact call, a low-pitched "cherrr-it"

Song: Cricket-like "caw-caw-caw", accompanied by bobbing in male

HABITAT
Thick coastal belts of open swamps and waste heathlands, lightly forested country, tea-tree scrub and casuarinas; never far from water

FOOD
Mainly grass-seeds, also seeds of trees and shrubs. During breeding season, psyllids and other insects eaten

HABITS
Sedentary. Associates in pairs or small family groups. Diurnal, roosting at night in specialised roost nest. Feeds on or near ground, in heath and fringing grassland, hopping rapidly along ground in high jumps. After bathing, birds often seen in small clusters, preening themselves and partners

BREEDING
September – January. Usually one clutch per season. Both birds construct nest, incubate eggs and care for young. Non-incubating adult roosts at night in cup-shaped vestibule of entrance tunnel

NEST: A bulky dome, with spherical egg chamber, shaped like a bottle on its side with a long side entrance tunnel, and cup depression adjacent to chamber; constructed of thin green grass stems and a few leaves, lined with plant-wool and (usually) white feathers; placed in thick foliage of bush or tree, generally within 2 – 6 m of ground

EGGS: 4 – 8 (usually 4 – 5); smooth, pure white; about 18 x 12 mm

INCUBATION: 20 – 22 days; fledging period about 23 days

DISTRIBUTION
South-eastern mainland from near Newcastle and west to Kangaroo Island; Tasmania and offshore islands

STATUS
Endemic. Uncommon on mainland and Kangaroo Island; more plentiful in Tasmania and offshore islands

P. Rowland

Diamond Firetail

Stagonapleura guttatum

Next to the Gouldian Finch, *Erythrura gouldiae*, the most strikingly coloured of the Australian grass-finches must be the Diamond Firetail of eastern and south-eastern Australia. It inhabits open forest, mallee and savanna grasslands, preferring the vicinity of watercourses, but is able to withdraw farther from water than the Red-eared Firetail, *Stagonopleura oculatum*, and the Beautiful Firetail, *S. bellum*. Early in the twentieth century, Frank Howe remarked that the sweet crescendo "Kweet" of this finch is heard throughout the district, but always in the open and grassy country that has a good covering of saplings.

Ripe and half-ripe grass seeds form the bulk of its diet; insects and their larvae are also taken. Food is gathered on the ground and the Diamond Firetail often searches among horse-droppings on country roads. Feeding is interrupted by frequent bouts of drinking, when flocks gather around a waterhole to quench their thirst. Water is taken by immersing the face and sucking in a pigeon-like manner. If alarmed, birds fly in a small tight group to the nearest tree, their bright crimson rumps looking like flaming sparks shooting up from the ground. Flight is rather strong, with characteristic grass-finch undulations. As the species does not make a flight call, the only sound usually heard is the whirring of wings.

The Diamond Firetail is a moderately social species but, due to the low productivity of its habitat, flocks in the non-breeding season comprise only about 30 birds. During the breeding season, loose colonies are formed, any nests being constructed in adjacent trees or even in the same tree. As well as this relatively "safe" nesting policy, the Diamond Firetail often constructs its own nest in the base of a hawk's nest. John Gould wrote:

> The nest is frequently built among the large sticks forming the under-surface of the nest of the smaller species of Eagles, and that too during the time that the Eagle is incubating, both species hatching and rearing their progeny in harmony.

The Diamond Firetail has a very elaborate courtship display. At the beginning, the male selects the longest piece of green grass available (occasionally more than a metre long) and grasps it in the end of his bill. He then flies to a dead branch in a tall tree where he sits near the female. The plumage of his underparts is fluffed, especially the heavily spotted flanks, and the neck, body and legs are stretched vertically, with the bill held just off the chest. In this position, he begins to bob up and down by alternately bending and stretching his legs, his feet remaining in contact with the perch. Throughout the routine, the male utters his long, raspy song phrase. The female approaches the male, whereupon he twists his neck round and opens his bill, in a posture similar to that of the begging young. The female accepts the male by quivering her tail, and mating then takes place. As in many grass-finches, the courtship routine is subject to much individual variation, especially in the nature of the grass held in the male's bill and in the use of begging postures.

The nest is a bulky sphere with a long entrance tunnel of green grass and leaves. It is constructed by both birds but only the female does the weaving. The five or six eggs are incubated by both parents, who change over at regular intervals. The relieving bird calls from outside the nest, and waits for a reply before entering. The eggs hatch in about 12 to 15 days and the young are ready to leave the nest after a further 22 to 25 days. Occasionally two or more broods are raised in quick succession.

Many attempts have been made to introduce the Diamond Firetail to New Zealand, Tahiti, Fiji and the Hawaiian Islands. All have been unsuccessful although flocks were noted, in most places, for a number of years after the release of the birds.

P. Rowland

Adult; Near Stanthorpe, Qld; November 1982 (R.E. Viljoen)

Sponsor: Anne Byrne

Diamond Firetail

Stagonopleura guttatum

Juvenile acquiring adult bill and eye colour; Wee Jasper, N.S.W.; March 1986 (G. Chapman)

OTHER NAMES
Diamond Finch, Diamond Sparrow, Spotted-sided Finch

MEASUREMENTS
LENGTH: 120 – 130 mm
WING: 63 – 68.5 mm
TAIL: 37 – 43 mm
BILL: 9.3 – 11 mm
TARSUS: 17.5 – 19.3 mm
WEIGHT: 15 – 20 g

DESCRIPTION
ADULT
Male: General colour above ash-brown; crown, forehead and hind-neck grey; rump and uppertail coverts bright crimson; tail black. Lores black; throat, lower breast, abdomen and undertail coverts white, with occasional red mottling; black band across fore-neck, extending back along flanks where each feather has a sub-terminal white spot. Eye and bare eye-ring red. Bill red-pink, base tinged with lilac. Legs and feet dark grey

Female: Similar to male. Usually smaller; lores browner; black band of fore-neck narrower

JUVENAL
Similar to adults. Back and wings olive-brown, head greyer; rump and uppertail coverts crimson; tail feathers brownish-black; underparts white with dull wash, occasional feathers of breast-band having blackish bases; flanks barred dull white and blackish-brown. Bill grey-black

IDENTIFICATION
Head grey; remainder of upperparts generally brown, except conspicuous red rump; underparts white, with contrasting black on chest and flanks; flanks heavily spotted white

VOICE
Identity call a long drawn-out, plaintive, whistled "twooo-heee", with first syllable ascending and second descending in pitch, higher pitched and longer in the male. Also low, snoring calls between birds at nest relief

Song: Long series of very low, raspy notes by male in display

HABITAT
Open grassy woodland, mallee and forest, in vicinity of watercourses

FOOD
Ripe and part-ripe seeds and occasionally, insects and their larvae

HABITS
Sedentary. Diurnal, roosting at night in nests built specifically for this purpose, or in dense shrubbery. Feeds exclusively on ground in a hopping manner, frequently jumping from the ground to grasp overhead grass seed heads. When disturbed rises to the trees. Gregarious, forming winter flocks of up to 150, and breeding in small colonies

BREEDING
August – January, sometimes later; normally only one clutch per season, but in wetter areas may produce two or three. Nest construction by both partners but only female does the weaving. Both incubate eggs and care for the young; both spend the night in nest

NEST: Bulky structure with spherical nesting chamber and long entrance tunnel (very variable in length and almost absent in some nests). Constructed of long green grass blades and stems and occasionally twigs; lined with fine grasses and white feathers; placed in bush or tree in thick foliage, or in mistletoe clump, occasionally in base of a hawk's nest. Nests have been recorded 1.7 m to more than 27 m (usually 2 – 10m) above ground

EGGS: 4 – 9 (usually 5 or 6); pure white; about 18 x 13 mm

INCUBATION: 12 – 15 days; fledging period 22 – 25 days

DISTRIBUTION
South-eastern Australia, from Carnarvon and Expedition Ranges, Queensland, to Eyre Peninsula and Kangaroo Island, South Australia. Range once extended north to Atherton Tableland, Queensland

STATUS
Uncommon throughout most of its range; vulnerable

P. Rowland

Adult; Pallinup River, W.A.; January 1971 (G. Chapman)

Sponsors: John & Jo Allison

Red-eared Firetail

Stagonopleura oculatum

The Red-eared Firetail, named and described by the french naturalists Hombron and Jacquinot, is found in the dense scrubs and heavy jarrah forests of the south-western corner of Western Australia, and in paperbark swamps along the coast, between Perth and Esperance. It is officially listed on the state's fauna list as "rare" due to reduction of its habitat, large areas of land having been cleared for agriculture or destroyed by construction for water storage or bauxite mining. However, studies carried out in 1980 – 1981 by O.G. Nichols and his colleagues suggested that:

> There appears to be little conflict between bauxite mining and the conservation of the Red-eared Firetail. Firetails were never observed to inhabit upland or mid-slope areas of forest where almost all mining is carried out. Juvenile birds were observed within 150 metres of active mine pits.

John Gilbert who collected birds in Western Australia in the early 1940s reported to John Gould that:

> It is a solitary species and is generally found in the most retired spots in the thickets, where its mournful, slow drawn-out note only serves to add to the loneliness of the place ... The natives of the mountain districts of Western Australia have a tradition that the first birds of this species speared a dog and drank its blood, and thus attained its red bill.

The Red-eared Firetail is the most solitary of the Australian finches. Pairs usually remain in their fairly large territories (100 – 200 square metres) all year round. As with the closely related Beautiful Firetail, *S. bella*, only the nesting area is strongly defended.

The Red-eared Firetail feeds on the seeds of grasses and sedges found in the low ground cover of its territory, manoeuvring in and around the bushes and trees in a low, weaving manner; it is possibly the most efficient flyer of all the Australian grass-finches. Flight is silent, steady and effortless, with hardly any undulations. When moving along a horizontal branch, the bird pivots from side to side in an arc-like manner, with wing and tail movements kept to a minimum, occasionally gleaning a few insects and spiders from the foliage as it goes. The members of a pair may forage up to 100 metres apart, keeping in contact with a long mournful "oowee". They spend the night in a small, unlined roost nest, which is frequently reconstructed.

Courtship, which takes place high up in a tree, is not often observed. The courting male begins by uttering at the nest site the call "oooweeee-a-a-a-a-a" while holding a long piece of grass at the tip of his bill. The grass used is usually green and about 20 – 45 centimetres long. When the female approaches close by, the male begins to jump to and fro between two branches while continuing to utter more nest site or identity calls. Eventually he moves nearer to the watching female and begins his courtship dance. Standing upright, with body stretched and bill pointed upward, he bobs repeatedly up and down on the same spot. Plumage, particularly on the abdomen, is generally fluffed out. The female shows her willingness to copulate by quivering her tail, and the male drops his piece of grass.

The nest, constructed of green grass stems, which soon turn brown, is bottle-shaped, with a long, narrow entrance tunnel. It is usually placed 5 – 15 m above the ground. Parents share incubation of the 4 – 6 eggs, changing over at intervals of 90 – 120 minutes. The male generally brings a feather with him as he relieves the female, and the pair remain together in the nest for up to 30 minutes at each changeover. Both feed the young on spiders and insects. As with the Beautiful Firetail, the male tends to dominate in feeding the young after they have left the nest. Juvenal moult is generally completed within six months of hatching and, if seasons are favourable, two broods may be reared in quick succession.

P. Rowland

Juvenile acquiring adult plumage (A.Y. Pepper)

Sponsor: R.M. Gibson

Red-eared Firetail

Stagonopleura oculatum

Adult (A.Y. Pepper)

OTHER NAMES
Red-eared Finch, Red-eared Firetail Finch

MEASUREMENTS
LENGTH: *115 – 130 mm*
WING: *56 – 59 mm*
TAIL: *47 – 50 mm*
BILL: *11 – 13 mm*
TARSUS: *15 – 18 mm*
WEIGHT: *c 17.5 g*

DESCRIPTION
ADULT
Male: Head, back and wings olive-green, finely barred with blackish-brown; bars become closer toward head; rump and uppertail coverts crimson; central tail brown-grey barred, faintly dusky-black and margined on outer webs with dull crimson; remainder of tail lighter olive and more widely barred. Lores and around eyes black; crimson patch behind eye, becoming darker in breeding. Chin, throat and upper breast buff-brown, finely barred with black; lower breast, belly and undertail black, spotted white. Eye red; naked blue eye-ring. Bill scarlet. Feet dark brown

Female: Similar to non-breeding male. In breeding, ear patch is orange-scarlet

JUVENAL
Upperparts dull brown-black, underparts grey-brown; lower breast and abdomen have alternate light and dark brown striping; no crimson ear-patch. Eye-ring has inconspicuous blue-tinge. Bill black

IDENTIFICATION
Upperparts generally brown-grey with fine black barring; underparts generally black with heavy white spotting. Conspicuous red patch behind eye and on upper tail

VARIATION
No geographical variation

VOICE
A wide variety of calls: identity call a mournful whistled "oowee" with seemingly no beginning or end, often singly or repeated; intimate nest call "twit-twit" on approach to nest at changeover, answered with "tweet-tit-tit-tit" from bird within nest; communication call a quiet, quickly repeated "quirk" or "quark"; harsh clucking in alarm. No flight call

HABITAT
Dense scrub and uncleared gullies; heavy forest and moist paperbark swamps along coast and nearby ranges

FOOD
Mainly seeds of grasses and sedges, plus occasional casuarina "nuts" and a few insects and spiders gleaned from foliage

HABITS
Sedentary. Diurnal, roosting at night in specially built roost nest. Seldom seen on the ground, even when feeding. Hops and flies among low branches and fallen twigs. Not gregarious. Forms permanent territories, but only immediate vicinity of nest is strongly defended. Much time spent in pairs or small family groups; occasional temporary flocks

BREEDING
August – January, usually in later months. In good season, two broods may be reared in succession. Both birds incubate eggs, at 90 – 120 minute intervals; both remain in nest at night. Both brood and care for young

NEST: Horizontal, bottle-shaped structure with a two part spherical nesting chamber connected to a long, narrow entrance tunnel. Outer walls strongly built of green grass stems lined with plant down and feathers; placed 5 – 15 m from ground in outer branches of trees and shrubs, favouring lower levels

EGGS: 4 – 6; white; about 18 x 12 mm

INCUBATION: 12 – 14 days; fledging period 21 – 24 days

DISTRIBUTION
South-western mainland in coastal belt from Darling Range to Esperance, Western Australia

STATUS
Vulnerable

P. Rowland

Painted Firetail

Emblema pictum

The harsh arid areas of northern and central Australia are home to the Painted Firetail. It is normally found, year-round, near permanent deep waterholes formed in the stony, spinifex-clad hills. Although it is also found in open spinifex plains it is never far from water which it drinks by scooping and tilting its head back in the usual avian manner, having failed to develop any of the specialised drinking styles of other grass-finches found in similar arid environments. Permanent residency around waterholes is associated with small flocks, normally less than 30 birds. Its mostly sedentary behaviour has also created an interesting pattern of distribution and, although the range of the species is quite extensive, groups may be separated by gaps of hundreds of kilometres, even within the middle of their geographic range.

The Painted Firetail has a characteristic long bill which it uses to search through the ground litter and in and around the rocky nooks and crannies of its habitat for the fallen grass seeds that constitute its diet. In flight, it exhibits the usual grass-finch undulations and, although it tends to fly very close to the ground, it manoeuvres among rocky outcrops and spinifex tussocks with very little effort.

Not surprisingly, courtship takes place on the ground. At the commencement, both the male and female pick up various items such as small stones, twigs and any other available nesting material, dropping these almost immediately. The behaviour is very simple in comparison to that of most other grass-finches, the only resemblance to a dance being a few occasional hopping movements as the male approaches the female prior to copulation. During courtship, the male adopts an upright stance with the feathers of the head and belly fluffed so as to highlight his scarlet face mask and breast patch. He pivots his head rapidly form side to side while he sings to the female. The amount of beak-wiping is subject to much individual variation and is commonly employed by both male and female. Tail-twisting is used to a much greater extent than in other grass-finches and each partner twists its tail towards the other during the entire courtship routine. The female signifies her willingness to copulate by quivering her tail in the typical grass-finch manner.

Pair bonding is not very strong between mated Painted Firetails and there is no mutual preening or close social contact, even during the breeding season. Breeding takes place in small, loose colonies. Nests tend to be scattered throughout the territory, especially if suitable nesting sites are hard to find. The nest is a more or less globular construction of any accessible materials such as bark, rootlets, grass stems and spinifex stalks, and is placed on a platform of earth, stones and bark, situated near the ground in a spinifex tussock. The nest material is gathered by both sexes, but the female carries out most of the construction.

The three to five plain white eggs, which are more rounded than in other grassfinches, are incubated by both parents who change over at regular intervals of between 30 – 60 minutes. Only the female remains in the nest overnight. The young hatch after 19 – 25 days and are cared for by both adults.

P. Rowland

Adult male at nest; Chambers Creek, Flinders Ranges, S.A.; October 1992 (R. Brown)

Sponsors: Peter & Jill Sayers

Painted Firetail

Emblema pictum

Pair; Chambers Creek, Flinders Ranges, S.A.; October 1992 (R. Brown)

OTHER NAMES
Emblema, Mountain Finch, Painted Finch

MEASUREMENTS
LENGTH: *110 – 120 mm*
WING: *57 – 60 mm*
TAIL: *36 – 39 mm*
BILL: *13.2 – 14.1 mm*
TARSUS: *13.1 – 14.9 mm*
WEIGHT: *c 14 g*

DESCRIPTION
ADULT
Male: Crown dusky; back and wings pale brown; rump and uppertail coverts scarlet; tail dusky brown. Face, lores, forehead, chin and upper throat scarlet. Fore-neck, breast, abdomen and flanks black with white spots; centre of belly scarlet; undertail coverts black. Eye buff-white. Upper mandible black with red tip; lower mandible red with pale blue base. Feet flesh-coloured

Female: Scarlet of head confined to area around bill; under surface duller black with smaller scarlet patch on belly; more heavily spotted on breast and flanks

JUVENAL
Similar to adult female but much duller; no scarlet on head and breast; white spots on under surface smutty; bill black above, grey below with whitish base

IDENTIFICATION
Characterised by long, pointed bill. Upperparts chestnut-brown; underparts generally black with white spotting. Conspicuous red patch in centre of belly

VARIATION
No known geographical variation

VOICE
Call: Among loudest of Australian finches. Identification call a harsh chirped "chek" or "chek-did-did-dit", especially in flight. Female utters loud rattling "terrait" when alarmed

Song: Courtship song a wheezy chattering followed by long whistles, "che che che-che-che-che-werreeeee-oooeeeee", repeated two or three times by male in each song bout, while part-opened bill is pointed upwards.

HABITAT
Spinifex country, including rocky gorges and stony hills, also in open areas; never far from water

FOOD
A wide variety of fallen grass seeds

HABITS
Sedentary. Diurnal, roosting at night near each other under low bushes or tussocks; during the breeding season, only female roosts in the nest. Gregarious, generally forming small groups of 5 – 30. Pair bonds not strong and most body contact generally avoided. Water taken by scooping and tilting head back (not by sucking, as in most other finches)

BREEDING
Season variable, according to rainfall. Both members of pair supply material for nest but female begins construction. Both feed and care for the young

NEST: Globular, with short entrance tunnel or open side entrance; built of twigs, bark, rootlets and spinifex stems (almost any available material), on platform of earth, twigs, small stones and bark; inner parts of softer grass stems, lined with hairy seeds, plant down and feathers. Located near the ground in small cavity in a spinifex tussock

EGGS: 3 – 4 (occasionally 5); plain white; more rounded than other species; about 14 x 11 mm

INCUBATION: 15 – 19 days; fledging period 19 – 25 days

DISTRIBUTION
Northern interior, from south Kimberley and Pilbara, through southern Northern Territory to far western Queensland, south to Everard and Birksgate Ranges, South Australia; erratically reaches Flinders and Gawler Ranges, South Australia

STATUS
Locally common in northern and western parts of range, more erratic in the south

P. Rowland

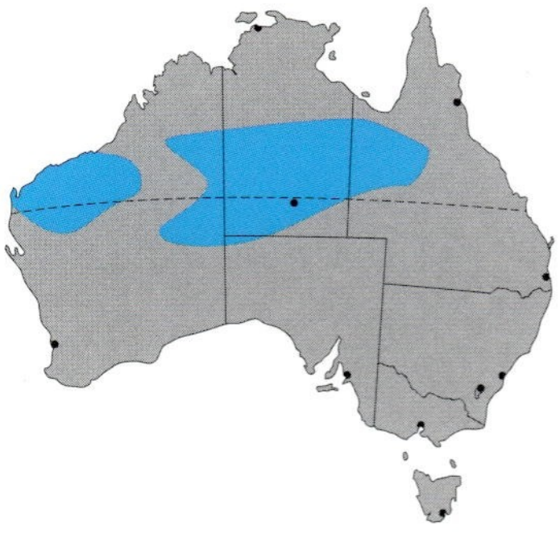

Zebra Finch

Taeniopygia guttata

Within its vast extent, the Zebra Finch is found in habitats ranging from open savanna woodland to mulga and spinifex grassland. Early in the twentieth century, John Mellor commented upon its abundance:

> The bird, called the Zebra Finch, is the commonest and most widely spread of the Australian Finches, being found in every imaginable locality from the verdant country with rich grass and flowing streams with good, fertile soil, all along the sandy coastline and shores of the sea, right into the arid interior of the continent.

The Zebra Finch needs to drink regularly, as often as every hour, and its distribution has become concentrated around natural and artificial watering places throughout its range. Tom Carter wrote:

> In hot weather they are seldom far from water, as they seem to have an insatiable thirst, and keep drinking throughout the day. At shallow wells and the troughing attached to other wells, they are quite a nuisance, as numbers fall in the water daily and pollute it with their decaying bodies. It was usual at shallow wells on many stations to provide a sort of ladder in the corners to enable the birds that had fallen in to climb out again.

In view of this behaviour it was rather surprising to discover that, in experimental situations, the Zebra Finch can be gradually habituated to a very small daily ration of water and that some individuals have survived for more than 250 days without any water other than that contained in their diet of seeds (about 10% of the weight of the seeds).

Adult female at nest; Mildura, Vic; 1974 (J. Purnell)

The range of the species seems to have become extended following the introduction of artificial sources of water. It is now found in areas where it was formerly unable to penetrate, becoming quite tame and trusting in humans, though not to the same degree as the Red-browed Finch, *Neochmia temporalis*. In northern Queensland, another factor has affected the distribution of the Zebra Finch. The introduced Nutmeg Mannikin, *Lonchura puntulata*, has forced the Zebra Finch, as well as many other grass-finches, out of the areas where their ranges overlap. The Zebra Finch is also found in Timor and on the chain of lesser Sunda islands in Indonesia. The Australian subspecies is slightly larger and has more obvious black barring on the breast and throat of the male. The species was introduced to New Zealand and Tahiti but did not become established. Isolated sightings are made from time to time but are thought to be escapees from avicultural stocks.

The Zebra Finch is very social; large sedentary flocks of up to a hundred birds regularly form outside the breeding season. However, in drier parts of the range, populations become partially nomadic as they continually search for water. When water is scarce, several flocks may meet around the few remaining waterholes, forming massive, noisy aggregations of several hundred, or even a

Adult male; Birdsville, Qld; August 1991 (R. Brown)

Sponsors: R.C. & K.P. Butler

thousand or so birds. Pair bonding is very strong and birds probably mate for life. A pair keep together throughout the year, even when in the large flocks of the non-breeding season. Close body contact and mutual preening are very common between partners. The large flocks of the non-breeding season do not completely disintegrate in the breeding season. Small colonies of 20 to 30 pairs nest in close proximity, even in the same tree when suitable nesting places are scarce. The nest is dome-shaped, with a side entrance tunnel. Poorly constructed of grass stems, it has a large spherical nesting chamber, lined with feathers or rabbit fur.

Courtship usually takes place among dead branches of a bush or tree, but has also been seen to occur on the ground. Unlike many other grass-finches, the male Zebra Finch does not hold a piece of grass in its bill. Courtship begins with the male and female hopping between two branches with their tails twisted towards each other, frequently stopping to wipe their beaks. After a minute or two, the female stops and the male moves towards her in a series of pivoting hops, singing as he goes. The feathers of the belly, back and sides of the head on the male are fluffed to accentuate the black and white areas of its body. As the male approaches the female, she begins to quiver her tail and mating then takes place.

Only the male sings. His loudest note is the distance call, which has been likened to the sound of a knife striking a tautly strained wire fence. This call is given in a variety of situations including flight, alarm and courtship.

It has been strongly suggested that individuals in a breeding colony are able to recognise each other by their calls. Neighbouring pairs have frequently been observed to "visit" each other at their nests, while strangers are driven off by the occupants. During the breeding season, Zebra Finches usually defend the area around their nest quite vigorously, but couples looking for a suitable nest have been observed to enter another pair's nest while the owners were away. An instance of intraspecific brood parasitism has been observed where a visiting female entered an unattended nest and deposited an egg. Upon their return, the owners were apparently oblivious of this addition and continued to incubate the clutch as before. This habit of laying eggs in other birds' nests has led to the Zebra Finch being credited with clutch sizes varying from three to seven eggs. Eggs are incubated by both parents, and both remain in the nest overnight.

The breeding season is related to rainfall. In places where rainfall is sparse and infrequent, the Zebra Finch's response is so rapid that courtship and nest building can commence within hours of rain. Sexual maturation is reached earlier than in other Australian grass-finches; birds no more than 11 or 12 weeks old can successfully rear a clutch.

P. Rowland

Zebra Finch

Taeniopygia guttata

Adults; Canberra, A.C.T.; January 1989 (C. Andrew Henley/Larus)

OTHER NAMES
Chestnut-eared Finch, Pimp

MEASUREMENTS
LENGTH: 100 – 115 mm
WING: 51 – 57 mm
TAIL: 32 – 40 mm
BILL: 10 – 12.4 mm
TARSUS: 12.4 – 15 mm
WEIGHT: 9 – 13 g

DESCRIPTION
ADULT
Male: Head silvery-grey; back and wings pale grey-brown; rump white, black at sides; uppertail coverts barred black and white; tail dusky-black. Face white, with black "tear marks"; sides of face and ear coverts rusty-yellow. Throat and upper breast pale grey, with fine black barring, forming broad black band on upper chest; flanks chestnut with obvious white spotting; abdomen and undertail coverts white. Eye red. Bill red. Legs and feet orange-yellow

Female: Like male but cheeks and ear coverts grey-white; throat, breast and flanks grey; abdomen and undertail coverts may have pale yellow suffusion. Bill paler

JUVENAL
Similar to female; lacks clear black and white markings on head. Eye grey-brown. Bill black

IDENTIFICATION
Small grass-finch. Unmistakable. Plumage generally grey, more so in female, with obvious chestnut cheeks, male only. Characteristic black "tear-drop" and "zebra-like" black and white barring on rump and uppertail coverts.

VARIATION
Two subspecies: Australian population is T. g. castanotis, distinguished by black barring on breast and depth of colour on back

VOICE
Calls: Considerable variety. Loud identity call, often in flight, a nasal "tiah"; soft "tet-tet" in close contact; hissing "wsst" in chases

Song: Harsh, ventriloquial trilling, repeated several times in series of one to two second phrases

HABITAT
Wide variety, mainly dry wooded grassland, bordering watercourses; also common around cultivated areas

FOOD
Fallen and ripening grass seeds and insects (particularly when feeding young)

HABITS
Sedentary. Very gregarious, living in social flocks of 10 – 100 or more throughout year. Diurnal, pairs roosting at night in roost nests, which may be specially built or unused old nests. Pair bonding permanent. Feeds on ground; never pulls seed heads down with feet

BREEDING
Usually October-April, but may breed whenever rainfall triggers appropriate conditions. Nest site selected by female from several sites chosen by male. Male supplies nesting material and female constructs nest. Both incubate eggs and care for young

NEST: Dome-shaped with spherical nesting chamber and side entrance tunnel, poorly constructed of grass stems and lined with feathers or rabbit fur; placed in bush or low tree, about 1.5 – 3 m from ground; occasionally in hollows or on ground

EGGS: 3 – 7 (usually 4 – 5); white; about 16 x 11 mm

INCUBATION: About 12 – 16 days; fledging period about 3 weeks

DISTRIBUTION
All over mainland Australia except Cape York Peninsula, Arnhem Land, south-west and south-east corners. Also Timor and Lesser Sunda Islands

STATUS
Common

P. Rowland

Double-barred Finch

Taeniopygia bichenovii

The Double-barred Finch of northern and eastern Australia can be found in a wide variety of habitats from pandanus and long grasses adjacent to streams and lagoons, to dry plains intersected with scrubby trees and bushes. It is seldom encountered more than a few kilometres from water. Like the Zebra Finch, *Taeniopygia guttata*, the Double-barred Finch drinks briefly at fairly regular intervals, immersing its face in the water and sucking – a technique used by many Australian grass-finches.

In the course of two centuries, the Double-barred Finch appears to have changed its choice of habitat and now prefers the company of humans. Consequently, it is one of the few native finches that has actually increased its range. In northern Queensland it has become a frequent visitor to parks and gardens and the margins of cornfields.

However, the Double-barred Finch is diminishing in numbers within its range, due to competition from the introduced Nutmeg Mannikin, *Lonchura punctulata*. This may partially account for the apparent movement of the Double-barred Finch to the coastal regions of northern Queensland, especially after the late 1970s, but many other factors should be considered. An influx of birds to coastal areas has been noted in southern Queensland and the species is also reported on some offshore

Adult at nest; Eastern race; Brisbane, Qld; November 1982 (R. Slater)

Sponsors: Tony & Carole Britton

islands where it was previously unknown. One such apparent self-colonisation occurred on North Stradbroke Island where it was first recorded between 1973 and 1980 and is now reasonably well established.

The Double-barred Finch is an efficient climber but spends a great deal of time on the ground. Food is usually gathered by hopping energetically along the ground and picking up the fallen seeds of various grasses and herbs, but an individual may occasionally flit up and pluck ripening seeds from an overhead seed head. Unlike a mannikin, it does not clamp a stalk down with its feet in order to pluck out the ripening seeds. Insects, which make up a substantial part of the Double-barred Finch's diet, are picked up from the ground litter, rather than captured in flight. Although rather trusting birds, Double-barred Finches dart into the nearest thicket if disturbed, registering disquiet by constantly flicking their tails in a semicircle and uttering their "tiaat" call.

Pair bonding of Double-barred Finches is extremely strong. A pair remains together for the whole year, members of the pair frequently preening each other and sharing the same roost nest. Even within members of a group, which usually comprises 6 – 24 birds, social preening is a regular activity. Such small groups keep together throughout the year; larger flocks may be formed if seasons are exceptionally dry. The birds remain highly social, even during the breeding season, when several pairs may nest in the same bush or tree.

In contrast to most other grass-finches, the Double-barred Finch has rather subdued courtship behaviour. No nesting symbol is used and the characteristic bobbing dance is absent. D. Morris has described the courtship procedure:

> The male assumes a courting posture in which the plumage is fully fluffed so that the body appears spheroid, with no special differential erection of the feathers of the type found in most other species. The bird crouches over the perch, parallel to the female, with his head twisted towards her. In this position he sings and beak-wipes. In order to perform the latter he does not have to move as far as he would if his courting posture was similar to his more general stance. It seems as if the posture is itself part of the beak-wiping; the bird performs the action so many times that he never, so to speak, quite straightens up again, until after the whole courting phase is over.

This account was based on observations of aviary birds, the actual "wild" courtship is virtually identical, the only major difference being that the male hops towards the female during his bouts of singing, and rotates his body 180 degrees with each hop.

On the whole, the Double-barred Finch is a very poor nest builder. The nest is an untidy sphere, constructed of dry grass stems, with a short entrance tunnel at one side. Quite often, a pair will take over and renovate a nest of another bird – usually a babbler – rather than construct a nest of their own. Both birds contribute to the construction, the male bringing most of the material, while the female does the weaving.

The four or five eggs are incubated by both parents, changing over in intervals of about two hours, this period being longer than most other grass-finches. Nestlings beg with both wings raised in the air and are fed by both adults.

P. Rowland

DOUBLED-BARRED FINCH

Taeniopygia bichenovii

Adults; Western race; Kimberley, W.A.; circa 1980 (M. & I. Morcombe)

OTHER NAMES
Bicheno's Finch, Black-ringed Finch, Black-rumped Double-bar, Double-bearded Finch, Owl-faced Finch, Ringed Finch, Banded Finch, Double-bar

MEASUREMENTS
LENGTH: 97 – 118 mm
WING: 46 – 56 mm
TAIL: 40 – 49 mm
BILL: 9.5 – 11.6 mm
TARSUS: 13.3 – 16.5 mm
WEIGHT: 7.5 – 11 g

DESCRIPTION
ADULT
Male: Upperparts from crown to lower back pale brown-grey with fine dark-brown barring, more conspicuous on lower back; rump and uppertail coverts white in eastern T. b. bichenovii, black in north-western T. b. annulosa; tail black. Lower portion of wings black, finely spotted white. Face white-cream, bordered by a thin black line, thicker on forehead; throat and chest white, tinged with buff, with another thin black bar across lower breast. Lower breast and belly cream; undertail black. Eye dark brown. Bill bluish-grey. Legs and feet grey

Female: As male, but black barring around face and breast often narrower

JUVENAL
Similar to adults but duller. Upperparts brown, tinged olive-grey with buffish-white markings on wings. Breast bands less distinct

IDENTIFICATION
Small grass-finch. Characteristic white face and cheeks, bordered with a broad black line; upper breast also white, bordered with a broad black line (confined to bottom and lower sides); remainder of underparts buff-white with contrasting black undertail. Upperparts generally mid-brown with conspicuous fine white spotting

VARIATION
Two subspecies, both in Australia, separated on basis of rump colouration: black-rumped T. b. annulosa in north-west; white-rumped T. b. bichenovii in east; some hybridisation where ranges overlap at Gulf of Carpentaria

VOICE
Call: High, plaintive "tiaat" in identification, alarm and flight, audible over a great distance; communication call a soft "tat"; louder high-pitched squeaking at nest

Song: Almost absent; more common in breeding season; soft repetition of identification notes

HABITAT
Pandanus and pockets of grass near waterways, in north and north-west; in east, found on fringes of cane fields and gardens and parks

FOOD
Half-ripe and fallen seeds of grasses and herbs; some insects may be taken, but not in flight

HABITS
Mostly sedentary, but locally nomadic flocks arise in times of drought. Very social and gregarious, keeping in flocks of up to 30 or 40 throughout year. Pair formation permanent. Diurnal, roosting at night in a nest specially built for this purpose (up to six birds are often crammed into one nest). Feeds mostly on ground; birds frequently jump up to pull down heads of grasses

BREEDING
August – November and January – May in south-east; June – November in Queensland and January – March in north-west (towards the end of wet season when seeding grasses plentiful). Male brings most of nesting material, female constructs nest. Disused nests of other birds (especially babblers) sometimes taken over and renovated. Both birds incubate eggs and care for young

NEST: Poorly constructed; almost spherical, with short side entrance tunnel; constructed of dry grass stems, coarser outside; lined copiously with feathers (in the east) or soft grasses (in the north-west) placed in twiggy branchlets of small shrub or tree about 1 – 4 m above ground, usually near wasp nest. Occasionally in stump, hollow or under eaves of building

EGGS: 4 – 8 (usually 4 – 5); white; about 16 x 11 mm

INCUBATION: About 12 – 14 days; fledging period about 21 days

DISTRIBUTION
Thick coastal and near inland band through northern and eastern Australia, from about Broome, Western Australia to Murray River, New South Wales

STATUS
Common, expanding in settled south-east

P. Rowland

Adult male; Kimberley, W.A.; circa 1980 (M. & I. Morcombe)

Sponsor: David R. Butler

Masked Finch

Peophila personata

Although the Masked Finch was first collected at Port Essington, the *Atlas of Australian Birds*, which maps the distribution of birds from 1977 to 1981, has no record of the species from the Cobourg Peninsula. It ranges across northern Australia from the Kimberley Range to Cape York Peninsula; there is a narrow disjunction across the Gulf of Carpentaria separating the nominate western subspecies from the distinctive, white-cheeked, eastern subspecies, *P. p. leucotis*. Overall, this distribution corresponds closely to the combined range of its two close relatives: the Long-tailed Finch, *P. acuticauda*, west of the Gulf; and the Black-throated Finch, *P. cincta*, to the east. The three species are similar in overall appearance.

Klaus Immelmann, whose patient field studies are models for all students of bird behaviour, confessed that he was frequently brought very close to desperation by the bird's "delaying tactics":

> Even at the peak of building activities they never bring more than about a dozen pieces per day. The actual selection of nest-building material takes a lot of time. A bird will pick up a piece of grass from the ground, investigate it by nibbling movements of the bill and then drop it. This is repeated dozens of times. Finally the bird seems to have got the right one and takes off for the nest. After its arrival it does not even attempt to work the grass in but sits quietly at the nest-entrance as if to recover from the strenuous flight. For as long as 15 minutes the bird may perch there, completely motionless, while a desperate human observer sits beneath an adjacent tree, trying to get rid of hundreds of flies and waiting for the bird to do at least something. Finally, it drops the piece of grass and flies off again!

Such behaviour contrasts with the steady, busy activity of most grass-finches and, indeed, most nesting birds. Consequently, the construction of a breeding nest usually takes several weeks. Both of the partners bring material to the nest and each builds it into the structure independently. As

Adult female at nest; Kimberley, W.A.; circa 1980 (M. & I. Morcombe)

Immelmann remarks, "only occasionally does the female stay in the nest to dedicate herself to the refinement of the inner construction". The nest is a relatively large structure with a wide opening, hardly any entrance tunnel, and a spacious egg chamber.

The Masked Finch may make a nest in a tree, often on a broad horizontal branch of a low eucalypt and commonly where there is an outgrowth of extra twigs. Many make a nest on the ground, usually under a grass tussock and often on a thick layer of dried grass. Nests may also be built in the cavities of terrestrial termite mounds. Tree nests may be quite conspicuous, whereas those in tussocks can be hard to find.

Breeding is usually from March to June, after the wet season. Nevertheless, early clutches may be lost due to falls of rain, which may occur almost up to the winter solstice. On the other hand, once the rains have ceased the ground vegetation soon dries off or is burned over large areas and the half-ripe grass seeds that are the preferred food for the young disappear. Apparently there is usually time between these two hazards for a pair to raise, on average, a brood of five eggs each year. Immelmann remarks that, despite this constraint:

> The astonishing thing is that the Masked Finch is one of the most common finches in northern Australia. It is, in fact, as abundant as the Gouldian Finch, which raises two or three broods with up to eight young, and thus raises three to four times as many young per year as the Masked Finch.

Even more surprisingly, populations of the Gouldian Finch have decreased considerably in recent decades while the Masked Finch has not.

The Masked Finch is extremely sociable. It breeds in groups of about a dozen pairs, building nests up to 45 metres apart, but assembling together several times a day – away from the nesting area – to bathe together and preen each other. Eggs and nestlings are neglected for a while during these assemblies. Socialising may also involve meetings of one pair with another, outside the breeding area. To quote Immelmann again: "They greet each other with a long sequence of chattering calls, which is heard only on such occasions. Meanwhile, all birds queue their tails in a vertical direction".
As is abundantly clear from the accounts of other grass-finches in this volume that tail-quivering is characteristic of estrildid finches, almost unique to them, and usually the signal given by a female that she is ready to accept copulation. Males rarely make the signal but young of both sexes employ it as part of their begging technique. In the Masked Finch, it has become a greeting. The calls made by two socialising pairs often attract others. As each couple arrives, it is greeted. At first, each pair stays side-by-side, self-preening or mate-preening, and at a small distance from the others. After several minutes, one bird will suddenly leave its mate, hop to a nearby pair, nestle against one of the birds and start to preen it. Immediately, the whole group starts to re-form, almost all participants swapping their preening partners, sometimes forming dense rows of six or eight birds in which the central birds may be preened by two neighbours. This mixing may continue for 5 to 30 minutes, after which the pairs match up again and return to their nests. The significance of this behaviour, which appears to be an extension of pair bonding, is not clear. It could perhaps serve to synchronise breeding.

W. Filewood

Masked Finch

Poephila personata

Adult male nest-building; Cape York Peninsula, Qld (C. & D. Frith/Frithfoto)

OTHER NAMES
Masked Grass-finch, White-eared Finch, White-eared Grass-finch (last two names apply only to Cape York race)

MEASUREMENTS
LENGTH: 110 – 135 mm
WING: 56 – 59 mm
TAIL: 46 – 59 mm
BILL: 10 – 13.4 mm
TARSUS: 11.3 – 16.2 mm
WEIGHT: 9.5 – 14 g

DESCRIPTION
ADULT
Male: Head, back and wings light cinnamon-brown, slightly darker on front of crown; upper rump black; lower rump and uppertail coverts white; tail black. Front of forehead, face, chin and upper throat and flanks black; cheeks and majority of underparts sandy-buff, whiter on lower abdomen and undertail (more white on cheeks and flanks in Cape York Peninsula race). Bill yellow. Eye brown. Legs and feet red-orange

Female: Similar to male. Generally smaller; less black on face and throat

JUVENAL
Similar to adult but generally duller; face mask blackish-brown and slightly smaller

IDENTIFICATION
Slightly larger grass-finch. Upperparts generally reddish-brown; black band on upper rump; lower rump and uppertail white. Underparts pale buff, whiter on cheeks; contrasting black on flanks. Conspicuous black face mask

VARIATION
Two subspecies, both in Australia: western P. p. personata has more buff on cheeks and flanks; P. p. leucotis of Cape York Peninsula has white on cheeks and flanks

VOICE
Call: Identification call "tiat" or "twet-twet" in close contact and a hissing "wsst" in chases

Song: Series of monotone loud, rapid notes

HABITAT
Lightly wooded grassland with dense ground cover, in the vicinity of water

FOOD
Ripe and half-ripe grass seeds. Insects frequently taken, mainly in flight, particularly in the breeding season

HABITS
Sedentary. Diurnal, roosting at night in specially constructed roost nests. Gregarious, forming small to medium-sized flocks of about 30 in non-breeding season. Feeds mainly on or near to ground. Fairly confiding in humans, often seen in vicinity of homesteads and feeding on side of road

BREEDING
March – June, also August – December; usually only one brood per season. Both members of pair collect materials, build nest, incubate and brood young

NEST: A bulky flask shape without entrance tunnel, only side entrance with platform. Constructed of grass blades and stems, lined with finer grass, feathers and fur, occasionally pieces of charcoal; placed in a low bush, grasses or tree about 2 – 8 m above ground (occasionally on ground)

EGGS: 4 – 6 (usually 5); white; about 17 x 12 mm

INCUBATION: 13 – 16 days; fledging period about 22 days

DISTRIBUTION
Northern Australia from Derby, Western Australia, to Cape York Peninsula, Queensland

STATUS
Endemic; common

P. Rowland

Long-tailed Finch

Poephila acuticauda

The Long-tailed Finch is a typical savanna bird, being mostly restricted to areas of tall eucalypts along watercourses through tropical north-western Australia. Unlike the Masked Finch, *Poephila personata*, the Long-tailed Finch tends to avoid the close proximity of human settlements and only approaches homesteads in times of severe drought.

Like the Black-headed Mannikin, *Lonchura malacca*, the Long-tailed Finch has the habit of bobbing its head up and down quickly in various situations. Although this head movement is used to a large extent during social activities, it seems to have originated as a mechanism for helping to maintain balance after landing. In newly fledged young, the head bobbing movement is slower and repeated only once or twice.

The Long-tailed Finch is very social. Outside the breeding season, flocks of about 2 – 30 birds occur; in very dry years much larger flocks are formed as the birds search for waterholes. These partially nomadic flocks occasionally number thousands during extended droughts, generally no more than a metre apart on the ground and almost as close in flight. On alighting in a tree, one bird invites its mate to preen with a series of the characteristic head-bobs and by pointing its bill. Each partner takes turns to preen its mate, the ritual undoubtedly reinforcing bonding.

Birds drink in the morning and evenings, usually by immersing the face in water and sucking. However, they are often seen catching raindrops. Their diet consists mainly of ripe and half-ripe grass seeds picked up from the ground, but birds often jump up and pull down seed heads of growing grasses. Insects are occasionally taken on the wing, especially during the breeding season. John Rodgers reported to Gregory Mathews:

> On 15 December these birds were feeding on flying ants; they were hopping about on the ground and when an ant passed overhead a bird would shoot straight up at it, and they always took their prey from below, seizing it with the bill, then returned to the ground ... The ants flew about twenty feet from the ground and were in swarms.

Courtship is elaborate, occasionally preceded by a special dance performed by the male. Holding his body vertical and grasping a long grass stem in his bill, he bobs up and down on his perch with his tail twisted towards the female. Some observers suggest that the piece of grass is not normally used in this phase, being only occasionally employed by the younger males. In the introductory phase of courtship, both birds hop to and fro between two branches, bobbing their heads each time they land. This may be accompanied by beak-wiping. As soon as the female settles on a branch, the male starts his courtship dance: he bounces towards the female, bowing frequently and constantly bobbing his head up and down. He fluffs his body feathers, more so on the head, highlighting the black on his throat, and makes his striking black bib more conspicuous to the female by raising the feathers towards the bottom of his throat more than at the top.

While the male is dancing, the female indulges in frequent bowing and head-bobbing, and invites copulation by quivering her tail. After copulation, the male sits next to the female, turning his head to one side and presenting his black throat from below. Apart from regular cackles, courtship may be silent or accompanied by song; this is a matter of individual variation.

Like the Black-throated Finch, the Long-tailed Finch breeds in loose colonies, usually with only one nest per tree. As the preferred nesting site is a tall eucalypt, distances between nests are sometimes quite large. Making various nest-building motions, the male proposes a number of sites to the female. She will disagree with most of these but eventually acknowledges her approval by sitting alongside the male, joining him in mandibulations and various whining calls.

Both parents incubate the eggs, changing over at regular intervals of about one and a half hours. The bird in the nest does not wait for the other to enter the nest, but leaves as soon as it lands in front of the entrance. Both brood and care for the young. The young birds share a peculiar characteristic with the Black-throated Finch in that the black feathers of the throat take longer to grow than the other feathers, sometimes as long as a fortnight. The bottle-shaped nest is constructed of dried grass stems and lined with feathers and, occasionally, bits of charcoal. As with the Black-throated Finch, the male gathers most of the material while the female does the weaving.

P. Rowland

Adult; Yellow-billed form; Broome, W.A.; May 1980 (G. Chapman)

Sponsor: Clive M. Hall

LONG-TAILED FINCH

Poephila acuticauda

Adult; Red-billed form; Kimberley, W.A.; circa 1980 (M. & I. Morcombe)

OTHER NAMES
Black-heart Finch, Heck's Finch, Long-tailed Grass Finch, Orange-billed Finch, Orange-billed Grass Finch

MEASUREMENTS
LENGTH: 151 – 160 mm
WING: 57 – 61 mm
TAIL: 59 – 78 mm
BILL: 11 – 12 mm
TARSUS: 14.5 – 15.9 mm
WEIGHT: 13.5 – 14.5 g

DESCRIPTION
ADULT
Male: Crown and nape blue-grey; mantle fawn; lower back and wings brown; uppertail coverts white; black bar on rump; tail black and pointed. Lores, throat and upper chest black; rest of underparts cinnamon-fawn; undertail coverts white. Eye brown. Bill yellow in western race, orange-red in east. Legs and feet bright red

Female: Very similar to male; black "heart" throat patch tends to be smaller

JUVENAL
Similar to adults, but duller and more buff; throat patch blackish-brown. Bill black

IDENTIFICATION
Head blue-grey with black patch in front of eye and black throat; upperparts fawn-brown with white uppertail and long black tail; underparts lighter fawn. Distinguished from Black-throated Finch by long thread-like tail

VARIATION
Two supposed subspecies, both in Australia: eastern P. a. acuticauda has orange-red bill; western P. a. hecki has yellow bill

VOICE
Call: Two very distinct main calls: identity call a loud whistling "thwirrr"; conversational call a short "tet"

Song: Series of soft flute-like whistles in two-second phrases, used more frequently in breeding season

HABITAT
Dry savanna country; also common in and around eucalypts along creeks and rivers

FOOD
Ripe and half-ripe seeds taken from ground; occasionally by jumping up and pulling spikes of seeds. Insects, such as termites and ants, taken by hawking, especially during breeding season

HABITS
Sedentary. Diurnal, gregarious, forming loose flocks of 10 – 15 pairs. When landing, head is bobbed several times. Strong pair bonding keeps mates close together. Feeds on or near ground. If disturbed, flies direct to nearby trees. Water obtained by sucking, which can be performed while hanging upside down; much water obtained by sucking drops of dew from blades of grass and even by catching raindrops

BREEDING
January – May, sometimes earlier; up to three broods in a season. Female selects nest site from many offered by male. Male supplies most of nesting material while the female weaves nest. Both share incubation and brooding

NEST: Bottle-shaped, with long side entrance tunnel. Constructed of dried grass stems, lined with down and feathers, and occasional pieces of charcoal; built in twigs at tops of eucalypts, often in clumps of mistletoe and sometimes pandanus, 3 – 20 m above ground

EGGS: 4 – 7; white; about 16 x 12 mm.

INCUBATION: 13 days; fledging period 22 days

DISTRIBUTION
Northern Western Australia to north-western Queensland, from Kimberley Range to Gulf of Carpentaria

STATUS
Endemic; common

P. Rowland

Black-throated Finch

Poephila cincta

The Black-throated Finch is a very close relative of the Long-tailed Finch, *Poephila acuticauda*, which it replaces geographically at the eastern end of its range. Their habits and way of life are almost identical. The Black-throated Finch is very social, normally living in small flocks of 12 – 30 birds, although vast nomadic flocks of several hundred individuals may congregate around waterholes during droughts. Most social activities take place in and around the nesting area and most members of the group gather in the uppermost branches of a dead tree several times each day. As a bird arrives, it gives a series of deep jerks of the head and is greeted by bowing, mandibulation, and short songs from each bird nearby. A bird moves around the group, displaying to several other members before leaving.

The characteristic habitat is timbered country with few, but usually fairly high, eucalypts and a somewhat dense ground cover of grass and scrub, normally in the vicinity of a watercourse. The southern subspecies *P. c. cincta*, which has a white rump, appears to prefer slightly drier country than the black-rumped forms. It has come under increasing threat from the effects of farming on its habitat. At all times, but particularly during drought, farm stock congregate around waterholes and trample down the surrounding vegetation. Rabbits have also destroyed areas inhabited by the Black-throated Finch through their excessive grazing habits. These factors, coupled with illegal trapping, have reduced the status of the Black-throated Finch to "vulnerable".

The diet comprises ripe and near-ripe grass seeds, gathered on or near the ground. It mostly eats fallen seeds but a bird may pull down the seed head of a grass and hold this with one foot while plucking out the seeds. Insects form a regular part of the diet and are either caught in mid-air or picked up off the ground. The Black-throated Finch drinks by sucking, but often only twice a day, in the morning and evening. When moving from one feeding area to the next all the birds in a flock tend to take off at about the same time and in the same direction. When in flight, they form separate groups, about one or two metres apart.

Around August, the permanently mated pairs search together for a suitable nest site. In a highly ritualised routine, the male finds a site and "proposes" it to the female by making various courtship and nest-building gestures. If she accepts the site, she joins the male and sits in contact with him, both birds making mutual mandibulations and whining calls. Constructed of dried grasses, the nest is bottle-shaped, with a long entrance tunnel at the side. Both male and female contribute to nest building, the male finding the bulk of the material while the female sits in the nest, rearranging the grasses. Nests are placed in the outer branches of trees, about 12 metres from the ground, and are constructed from the bottom upwards, stems being added from the inside.

Courtship is short and uncomplicated, consisting of a preliminary dance by both birds. The male fluffs his throat feathers and hops towards the female, bowing as he goes. He may or may not sing. Shortly afterward the female invites copulation by quivering her tail. Between four and nine eggs are laid, one per day. Incubation begins after the fourth egg is laid. Both parents incubate the eggs, changing over at intervals of 30 to 40 minutes. The relieving bird carries a soft feather which it pushes into the wall of the nest on entering. Both birds brood and feed the young.

Nestlings beg for food in the three postures characteristic of grass-finches. The main position is the "head-down-neck-twist" in which the neck is twisted so that the head is held to one side, pointing the open beak upwards. Others are the "neck-down" position where the bird lies with its neck on the floor and the head turned upwards; and the "neck-up" position where the neck is angled towards the entrance tunnel, exposing the gape. Young birds fledge after about three weeks but remain dependent on the adults for a further two weeks, and remain in contact with their parents, even as they raise a second brood.

P. Rowland

Pair; Georgetown, Qld; May 1990 (G. Chapman)

Sponsor: J.H. Dutton

Black-throated Finch

Poephila cincta

Adult; Mt. Isa, Qld; September 1970 (K. Ireland)

OTHER NAMES
Banded Grass Finch, Black-rumped Finch, Black-rumped Grass Finch, Black-tailed Finch, Black-throat, Diggles Finch, Parson Finch

MEASUREMENTS
LENGTH: 103 – 129 mm
WING: 60 – 63 mm
TAIL: 36 – 49 mm
BILL: 12 – 14.3 mm
TARSUS: 14.4 – 15.4 mm
WEIGHT: 11.5 – 17 g

DESCRIPTION
ADULT
Male: Crown and nape blue-grey; front of head, ear coverts and cheeks bluish-white; hind-neck and upper back pinkish-brown; lower back and wings fawn-brown; uppertail coverts black in northern forms, white in southern; black bar on rump, visible only in white-rumped subspecies; tail black. Lores, throat and upper breast black; lower breast, belly and flanks cinnamon-fawn; abdomen and undertail coverts white. Eye dark brown. Bill black. Legs and feet orange-red

Female: Similar to male; no characteristic differences. Head may be duller grey; black throat patch slightly smaller in some cases

JUVENAL
Similar to adults but duller and more buff. Throat patch brownish-black

IDENTIFICATION
Head grey with conspicuous black patch in front of eye; throat black; upperparts pinkish-buff with black tail; underparts pinkish-fawn. Distinguishable from Long-tailed Finch by shorter tail

VARIATION
Three subspecies, all present in Australia: P. c. cincta, white-rumped, from about Inverell, New South Wales, to Innisfail, Queensland; P. c. atropygialis is black-rumped, from Cairns to Coen and across to Normanton, Queensland; P. c. nigrotecta, also black-rumped but with more brown on breast than the other and without any red suffusion, far north of Cape York Peninsula, Queensland. Some hybrids occur between P. c. cincta and P. c. atropygialis where their ranges overlap

VOICE
Call: Identification call a loud, hoarse whistling "teeweet"; a hissing "wsst" in chases; a soft "tet" in close contact

Song: Very musical variety of soft squeezing notes, generally ending in long mournful whistle

HABITAT
Open timbered grasslands with dense grassy ground-cover, usually in vicinity of water

FOOD
Ripe and part-ripe grass seeds, mostly taken from ground, but often by jumping up and pulling down a spike. Insects form small proportion of diet

HABITS
Sedentary. Diurnal, roosting at night in specially built roost nest. Gregarious, forming loose flocks of 10 – 30. Very social, often drinking, bathing and resting together. Pair bonds are permanent; mates greet each other with head-bobbing and cackling. Drinks (by sucking) at irregular intervals, often only once in morning and once in afternoon

BREEDING
September – May in the north (during wet season); August – January in the south. Nest site chosen by female, from several sites pre-selected by male. Male supplies most of material while female weaves it. Both incubate eggs and brood young

NEST: Bottle-shaped with long side entrance tunnel; constructed of dried grass; lined with feathers and occasionally charcoal; normally placed in outer eucalypt twigs up to 12 m from the ground, but also in hollows or termite mounds and often in the base of a hawk's nest

EGGS: 4 – 9 (usually 4 – 5); white; about 17 x 12 mm

INCUBATION: 12 – 14 days; fledging period about 21 days

DISTRIBUTION
North-eastern Australia from Cape York and eastern side of Gulf of Carpentaria to Inverell, New South Wales

STATUS
Endemic. Black-rumped race abundant; white-rumped race declining in range

P. Rowland

Plum-headed Finch

Aidemosyne modesta

In its natural habitat, the Plum-headed Finch prefers rank grasses and reeds fringing rivers and tree-studded savanna with an undergrowth of various bushes. It is a gregarious species, usually seen in pairs or small flocks, although flocks of several hundred birds may congregate outside the breeding season and mix with flocks of other grass-finches, especially the Double-barred Finch, *Taeniopygia bichenovii*. Within a flock, bodily contact is usually restricted to mated males and females (which pair for life) but even this is infrequent outside the breeding season. Large flocks are rather noisy congregations, the individuals uttering a constant murmur of soft "tlips" while feeding. This sound is replaced by a chorus of louder, drawn out, "pyiiits" when the birds take off.

Feeding takes place on the ground and in low vegetation. When feeding on ripe or half-ripe seeds on a grass stem, the finch bends the top of the stem over and clamps it to the ground with one foot, while plucking the seeds out with its bill. In addition to seeds of grasses and other plants, insects make up a small portion of the diet, the birds generally taking them from the foliage or off the ground, rather than in the air. The Plum-headed Finch is moderately social and "peering" is a frequent occurrence within a flock. If a bird is singing on its own, another member of the flock may approach it until their bills are nearly touching and peer into its face, giving the impression of listening most intently. This behaviour occurs in a few other grass-finches, especially the Nutmeg Mannikin. Peering is thought to be a form of social bonding but it is often performed with slight aggression and may possibly be a form of threat display.

The Plum-headed Finch is fairly tolerant of humans and allows approach to within about 10 metres if sudden movements are avoided. When disturbed, the birds rise in compact groups with a strong, slightly undulating flight, to the top open branches of nearby trees.

There is some individual variation in courtship but stem-display is generally constant. The male grasps the tip of a long grass stem in his bill and flies up to a perch. Sitting with his bill pointing steeply upwards, he begins to display with well spaced and deliberate upward jerks. This is accompanied by a rapid hoarse piping and, although this part of the courtship is generally regarded as merely a pair bonding ritual, it is occasionally followed by copulation. Courtship proper is a repetition of the above routine, followed by the male dropping the piece of grass and fluffing his belly feathers. Singing continues while he carries out ritualised beak-wiping and body-shaking. The female accepts the male by quivering her tail.

The nest, which is a small compressed sphere, is constructed by both sexes, but the male is chiefly employed in fetching nest material while the female does the weaving. This is one of the smallest nests of any grass-finch and is well concealed in tall grass stems. Parents share the incubation, changing over at regular intervals of about 45 minutes. Plum-headed Finches are prone to desert a nest if disturbed to any great degree. Young leave the nest for the first time after about 21 days but return to roost there for the next few nights.

P. Rowland

Adult male; Near Dubbo, N.S.W.; September 1981 (T. & P. Gardner)

Sponsor: Michael F. White

Plum-headed Finch

Aidemosyne modesta

Pair; Gara River, N.S.W.; April 1993 (B. Shepherd)

OTHER NAMES

Cherry Finch, Diadem Finch, Modest Grass Finch, Plum-capped Finch, Plumhead, Plain-coloured Finch

MEASUREMENTS

LENGTH: 107 – 120 mm
WING: 52 – 59 mm
TAIL: 43 – 51 mm
BILL: 8.2 – 12.1 mm
TARSUS: 14.5 – 16.9 mm
WEIGHT: 9 – 14 g

DESCRIPTION

ADULT

Male: Upperparts, including wings, generally deep olive-brown; wing coverts and inner secondaries tipped white; rump and uppertail coverts barred white. Tail black, with white spot on tip of outer feathers. Top of head, forehead and chin dark claret-red; lores black; ear coverts white, streaked brown; cheeks white. Underparts white with pale olive-brown bars from throat to sides of abdomen; centre of breast, belly and undertail plain white. Eyes brownish-black. Bill black. Legs and feet pink-brown

Female: Similar to male; red of face duller than in male and absent on chin; white line above eye; less barring on under surface

JUVENAL

Similar to female; no claret-red head markings; lighter brown above, with white markings on wings; underparts greyish-white with almost no barring

IDENTIFICATION

Small to medium grass-finch with dark claret-red forehead and crown. Mostly brown above with some white spotting. Underparts buff-white with darker brown barring

VARIATION

No obvious geographic variation

VOICE

Call: Identification call a loud tinkling "pyiiit", often in flight; softer "tlip" or "tleep" in close contact

Song: Very high-pitched and almost inaudible gargling trills which become louder and more flute-like towards end

HABITAT

Restricted to immediate vicinity of waterways in tree-dotted savannas, and reeds and bushes along margins of creeks and rivers

FOOD

Ripe and part-ripe seeds of grasses and herbs

HABITS

Seasonally nomadic, moving about in relation to the availability of water. Gregarious throughout the year, but flocks become smaller in breeding season. Diurnal, roosting at night on twigs and grass stems in thickets. Pairing is permanent; pairs avoid close contact with other members of flock. Feeds on or near ground. If disturbed, birds rise as a compact group to higher open branches of nearby tree. Tame and confiding in humans, so long as sudden movements are avoided

BREEDING

September – January in the south; August to March in the north. Both members of pair supply material for the nest but female responsible for most of construction; both incubate eggs but only female spends night in nest. Both brood young

NEST: One of the smallest nests of the Australian grass-finches, consisting of small, round, laterally compressed chamber of mostly green grass, with side entrance but without tunnel; lined with feathers (if at all); concealed in tall grass stems with some blades built into nest walls

EGGS: 4 – 7 (usually 5 or 6); white;about 17 x 12 mm

INCUBATION: 12 – 14 days; fledging period about 21 days

DISTRIBUTION

Eastern Australia on western (less often eastern) slopes of Great Dividing Range, north to Rockhampton, Queensland, south to Lachlan Valley, New South Wales, and Canberra. Occasional sightings as far north as Cairns, Queensland

STATUS

Endemic. Generally uncommon and patchy throughout range

P. Rowland

Pictorella Mannikin

Heteromunia pectoralis

The mobility of this species was remarked upon by Frederick Berney to Gregory Mathews early in the twentieth century:

> This is an uncertain visitor. One year or another I have seen them during all four seasons, but they generally pick the good times ... They invariably frequent the peabush (*Sesbania aculeata*) flats and other low-lying spots, where the grass and herbage grows rather rank.

The Pictorella Mannikin inhabits the grassy plains, which are dotted sparsely with trees, of northern Australia. It is a true open country and savanna bird and has probably adapted better to life in the arid areas than any other mannikin. Being somewhat nomadic, it suddenly arrives in places where it may not have been seen for several years, and disappears just as quickly. In the wet season, small groups can be seen scattered all over the range. During long dry spells, flocks of up to several hundred individuals

Adult female; Karunjie, W.A.; May 1982 (G. Chapman)

Sponsor: Kathleen A. Blofeld

move towards the coastal areas where they may mix freely with flocks of Chestnut-breasted and Yellow-rumped Mannikins. Although the species appears not to be in any present danger, overgrazing and an increased incidence of fire throughout its range may have detrimental effects on populations in the years to come.

The Pictorella Mannikin tends to drink only once or twice a day and is often found far from water. It usually imbibes water by sipping in rapid succession, much faster than any other Australian grass-finch, thus minimising the time spent exposed at a waterhole. Birds have been seen to suck up water in the manner of some other grass-finches. It is less gregarious than the Chestnut-breasted or Yellow-rumped Mannikin; the large flocks formed outside the breeding season are loose aggregations and lack any strong social structure. During the breeding season, it is usually found in pairs or in small flocks of up to 12 members.

Breeding occurs between January and May. Just prior to this, with the coming of the wet season, birds move inland in small groups. Courtship behaviour is basically similar to that of most mannikins but is characterised by adaptations to the arid environment and its ground-dwelling habits. As in the Painted Finch, *Emblema pictum*, courtship takes place on the ground. In the lead-up to the courtship display, both male and female make picking gestures on the ground, behaviour that has probably replaced the beak-wiping seen in other grass-finches. Occasionally, small sticks or stones are picked up briefly then dropped. The courtship dance consists of two parts, both accompanied by tail-fanning and wing-grinding while singing a constant nasal "g'ee".

First, the male holds a piece of grass in his beak, though this is subject to individual variation and is sometimes omitted. He fluffs all his body feathers and fans his tail. Then, with bill pointing skyward, he hops to and fro in front of the female in an arcing pattern. The length of the arc decreases and the intensity of the hopping increases until he hops up and down on a spot in front of the female. If a piece of grass has been held, it is dropped at the commencement of the second part of the routine, when the male hops almost completely around the female in a circular pattern, stopping when he is alongside her. In this position, with the tail still fanned out, and slightly turned towards the female, he makes a deep bow. This is then reversed and the male circles around behind the female until he is on her other side and, again, makes a deep bow. This part of the courtship may be repeated several times until the female signifies her acceptance by quivering her tail.

The nest is an untidy, bottle-shaped construction with no entrance tunnel. Made from dry grass stems and twigs, it is usually placed in a low bush or spinifex tussock. Both partners share in the incubation, changing over regularly. The female remains in the nest overnight while the male roosts in a nearby tussock. The young leave the nest after about three weeks but return there to roost for the first few nights. They are fed by the parents for a further four to six days and become fully independent after another 10 or 12 days.

P. Rowland

Pictorella Mannikin

Heteromunia pectoralis

Adult male (M. Carter)

OTHER NAMES
Pictorella, Pictorella Finch, Pectoral Finch, White-breasted Finch

MEASUREMENTS
LENGTH: 109 – 120 mm
WING: 58 – 62 mm
TAIL: 35 – 41 mm
BILL: 13 – 14.7 mm
TARSUS: 14.4 – 16.2 mm
WEIGHT: 13.5 – 15 g

DESCRIPTION
ADULT
Male: Upperparts, including forehead and crown, warm greyish-brown; rump and tail duskier; wing coverts very finely spotted white at tips. Face and throat glossy purplish-black edged around ear coverts with pale cinnamon-buff. Breast white, feathers have black barring; remainder of underparts pale purple-buff, undertail coverts slightly duskier. Eye dark brown. Bill pale blue-grey. Legs and feet flesh-coloured

Female: Similar to male; face and throat brown-black; broader black barring on breast

JUVENAL
Similar to adults; generally dull grey-brown above, darker on the wings; underparts pale-pinkish-buff; face and throat dusky-grey. Bill brownish-black

IDENTIFICATION
Medium-sized mannikin. Generally grey-brown above, with fine white spotting on wings; face and throat black; breast white, with black flecking; rest of underparts purple-buff

VARIATION
No evident geographical variation but much individual variation

VOICE
Call: Communication call a very low sparrow-like "chip"

Identity call a piercing "tlit", also used as a lost call, but more drawn out "tleet"

Song: Greatly reduced; occasionally soft, disyllabic "g'ee", repeated at two-second intervals

HABITAT
Open, arid, acacia-dotted savannas and spinifex

FOOD
Fallen and half-ripe grass seeds, also insects

HABITS
Nomadic, following food; usually moving south in the wet and north in the dry season. Gregarious, mixing readily with other finches, forming flocks of several hundreds. Diurnal, roosting at night in tussocks or thickets. Food taken on or near the ground

BREEDING
January – April; incubation and brooding carried out by both members of pair but only female remains in the nest overnight

NEST: Untidy, bulky, bottle-shaped structure without side entrance tunnel; constructed of dry grass stems, twigs and rootlets; lining (if any) consists of only a few feathers or fine grasses; placed in low bushes or tussocks of spinifex, generally within half a metre of ground

EGGS: 4 – 6; white; about 16 x 11 mm

INCUBATION: Unknown; fledging period about 20 – 24 days

DISTRIBUTION
Range fluctuates due to nomadic tendencies; generally, northern Australia, from Kimberley Range to Normanton and Cloncurry, Queensland

STATUS
Endemic. Moderately common

P. Rowland

Chestnut-breasted Mannikin

Lonchura Castaneothorax

Rank grasses and reeds bordering streams, swamps and mangroves of northern and eastern Australia provide the habitat of this grass-finch. With the introduction of irrigation in many areas, the Chestnut-breasted Mannikin has thrived, increasing in range and numbers. Although a frequent visitor to the cane fields of the north, it also visits rice fields to feed on the half-ripe seeds. Together with the closely related Yellow-rumped Mannikin, *L. flaviprymna*, it has become a considerable pest, not only eating the seeds, but also bending and breaking the rice stems. In New South Wales the Chestnut-breasted Mannikin is known locally as the Barley-bird because it eats the ripening seeds of barley plants.

In recent years the Chestnut-breasted Mannikin has met with serious competition in northern Queensland from the introduced Nutmeg Mannikin, *L. punctulata*. This species has found the conditions very favourable and, if it continues to expand at the current rapid rate, it may well replace its native relative in this area.

The Chestnut-breasted Mannikin has itself been the subject of introductions. Over the years, it has been released in New Caledonia where it is now commonly found in gardens and cultivated areas, feeding on seeds on the ground. It is also established on Tahiti, Moorea, Raiatea, Tahaa, Bora Bora and Huahine, where it inhabits the coastal borders to the interior valleys, in groups of 50 – 100, often accompanying the Red-browed Finch, *Neochmia temporalis*. In these areas it is in danger of reaching pest numbers and control measures have been adopted in New Caledonia to reduce numbers and protect crops. It has been introduced into New Zealand on several occasions but apparently has failed to become established.

Within Australia, the Chestnut-breasted Mannikin is very social. Outside the breeding season, large, slightly nomadic flocks of Chestnut-breasted and Yellow-rumped Mannikins gather where their ranges overlap, mainly in the Kimberley Range, and Northern Territory. These flocks bathe, feed, drink and breed together, the Yellow-rumped Mannikins being outnumbered by about two to one, sometimes as much as three to one. When the creeks and waterholes dry up, they move in the direction of the coast, returning inland at the onset of the wet season (drought has to be quite severe to force the birds to actually reach the coast).

Pair bonds are not very strong and partners are not easily recognisable within these flocks. Even during the breeding season, the species is very social, establishing small breeding colonies with nests as little as one metre apart. It is interesting that, when moving in a large flock, the flight pattern changes from the usual undulating mode of grass-finches, to a direct and very rapid progress with no undulations at all. Birds in a flock are capable of very sharp simultaneous turns.

The degree of relatedness of the Yellow-rumped and Chestnut-breasted Mannikins is still uncertain. Many authors have argued that the Yellow-rumped Mannikin is no more than the desert form of the Chestnut-breasted Mannikin, but it has been shown that they both enjoy the same habitat. The fact that they so readily interbreed with each other adds fuel to the discussion.

A courting Chestnut-breasted Mannikin never holds a feather or piece of grass in its bill. The courtship sequence is a song and dance routine divided into two main parts, subject to much individual variation, especially in the amount of beak-wiping and body-shaking. The male starts his song phrase with the bill wide open, head bent down, and feathers of the head, breast, belly, and rump fully fluffed out. While in this position, he hops up and down on a twig, then makes a series of bobbing movements by alternately stretching and bending the legs. This may be accompanied with bursts of song. Gradually, the male approaches the female until they are touching. The second part of the courtship begins once both birds are in close contact (very rare among grass-finches). Both birds bow their bodies a number of times, stretching from horizontal to vertical (though the female may simply turn her tail towards the male instead of copying his movements). The performance is repeated several times, usually in silence, while the tails are directed towards each other, and the feathers of the belly are fluffed out to a greater extent than in the first part of the behaviour. Copulation usually follows. Occasionally one of the parts of the routine may be omitted altogether, the courtship consisting of only one performance.

The nest is a bulky sphere of green or dried grass, without an entrance tunnel. It is constructed by both birds and usually placed in a clump of sugar cane or similar tall grassy plants, within two metres of the ground. Both birds share the incubation of the eggs, the female remaining in the nest overnight. The four to six eggs hatch after about 12 or 13 days and both birds feed the young. In a favourable season, two broods may be reared in quick succession.

P. Rowland

Adult at nest; Stanwell, Qld; March 1975 (K. Ireland)

Sponsor: Peddle Thorp & Walker

Chestnut-breasted Mannikin

Lonchura castaneothorax

Flock; Goomboorian, Qld; March 1989 (C. Webster)

OTHER NAMES
Chestnut-breasted Finch, Chestnut Finch, Barley-bird, Barley Sparrow, Dun-coloured Finch

MEASUREMENTS
LENGTH: 111 – 114 mm
WING: 51 – 56 mm
TAIL: 31 – 42 mm
BILL: 11 – 13.9 mm
TARSUS: 13.6 – 16.7 mm
WEIGHT: 11.5 – 16 g

DESCRIPTION
ADULT
Sexes similar, female may be slightly paler than male. Crown and back of neck grey-brown with dark streaks; back and wings cinnamon-brown; rump, uppertail coverts and central tail feathers straw-brown; outer tail feathers brown with yellow margins. Face, chin, throat and undertail coverts black; breast pale chestnut with black band on lower chest (band sometimes narrower in female); abdomen white; flanks marked with cinnamon and black. Eye brown. Bill blue-grey. Legs and feet leaden-grey

JUVENAL
Uniform olive-brown above; flight feathers and tail ashy-brown; head and throat dusky, with indistinct dark barring on throat; underparts pale buff with brown-buff breast

IDENTIFICATION
Thickset mannikin with broad chestnut breast-band, bordered below with black; conspicuous black face and throat; remainder of underparts white, with contrasting black undertail; upperparts rich-chestnut. Crown grey

VARIATION
Five subspecies, Australian population is L. c. castaneothorax

VOICE
Call: Seemingly only one main call, "teet" or "tit", which may be either bell-like or long and drawn-out, used in a variety of situations

Song: Long, high-pitched song with a duration of up to 12 seconds "weeeee-eeeeeee-tuee-tuee-tuee-tuee-tuee-tuee-tuee-tuee-tuee-tuee-cheeouk-cheeouk-cheeouk-cheeouk-cheeouk-ching-ching-ching-ching-ching"

HABITAT
Rank grasses; reeds fringing rivers, swamps and mangroves

FOOD
Mainly half-ripe grass seeds, rarely taken from the ground; also termites, taken on the wing at the onset of the breeding season

HABITS
Locally nomadic, more so in the north. Gregarious, forming flocks of several hundreds. Very social – eating, drinking, roosting and breeding together, but not preening one another; fighting rare and ritualised. Diurnal, roosting at night among reeds. Feeds in low grasses. An occasional pest of crops

BREEDING
At any time. In the north, during the second half of the wet season, mainly December to April (sometimes earlier); in the south, breeds in both spring and autumn. Breeding is colonial, with nests only a few metres apart. Nest-building shared by both partners. Both also incubate and rear the young, but only the female remains in the nest overnight

NEST: Bulky sphere, compressed at sides, without entrance tunnel although entrance may have a hood; constructed of either green or dried blades of grass; lined with fine grass stems; placed in clump of grass, sugar cane or reeds, within 2 m of ground

EGGS: 4 – 6 (occasionally 7 – 8); white; about 17 x 12 mm

INCUBATION: 12 – 13 days; fledging period 21 – 22 days

DISTRIBUTION:
Around northern and eastern coast, from Kimberley Range to Sydney. Also New Guinea

STATUS
Resident. Common

P. Rowland

Yellow-rumped Mannikin

Lonchura flaviprymna

John Gould, who described half of the currently recognised species of Australian grass-finches, was quite familiar with many of these but he wrote in 1865 of the Yellow-rumped Mannikin:

> A single specimen, and the only one I have ever seen of this pretty Finch, was presented to me by the late Mr Binoe, who procured it on the banks of the Victoria River during the late surveying voyage of H.M.S. Ship Beagle. It is very nearly allied to the *Donacola castaneothorax* [Chestnut-breasted Mannikin], but is specifically distinct from that as well as from every other known species of this now numerous tribe of birds. I regret to add that nothing whatever is known of its habits or mode of life.

Adult; Taronga Zoo, N.S.W.; February 1972 (D. & M. Trounson)

Sponsor: Robert L. Morrow

Somewhat more is now known of its biology. It is confined to the more arid areas of northern Australia and where its range overlaps with the Chestnut-breasted Mannikin, *L. castaneothorax*, vast flocks of several hundred birds occur. It is reported to move towards coastal areas during periods of excessive drought and to return inland with the onset of the wet season. Klaus Immelmann writes:

> I was able to observe these birds during my stay at the Kimberley Research Station (about 96 kilometres south-east of Wyndham) – thus including the peak and end of the dry season as well as the whole of the wet season – and in all months I observed Chestnut-breasted and Yellow-rumped Finches together, usually in mixed flocks.

The flocks consist predominantly of Chestnut-breasted Mannikins, outnumbering the Yellow-rumped Mannikins about two or three to one. Members of both species feed, bathe and roost together, and the close association between the two species continues during the breeding season. Since they usually nest in mixed colonies, one might expect half to two-thirds of the nests to be those of the Chestnut-breasted Mannikin but the situation is complicated by the formation of mixed pairs. Approximately 10% of a flock has been reported as having the markings of both the Chestnut-breasted and Yellow-rumped Mannikins. Interbreeding is a regular occurrence, the usual pairing being a Yellow-rumped Mannikin male with a Chestnut-breasted Mannikin female.

This hybridization, coupled with competition from the Chestnut-breasted Mannikin, has affected the population stability of the Yellow-rumped Mannikin where the ranges of these two species overlap. Understandably, the specific status of the Yellow-rumped Mannikin is in question. Many authors have argued that it is merely a desert form of the Chestnut-breasted Mannikin, and the incidence of hybridization lends weight to this view. In the absence of conclusive evidence, and taking a conservative attitude, the Yellow-rumped Mannikin is treated here as a distinct species.

Given the success of interbreeding, it is not surprising to find that there are only small differences in the courtship behaviour of the two species. The entire courtship performance is more intense than in the Chestnut-breasted Mannikin, especially in beak-wiping and feather-fluffing. The first part of the courtship display is generally combined with the second (the male performing his bobbing and bowing movements at the same time) or omitted altogether. Beak-wiping is performed between each bob and the male usually sings throughout the performance.

The nest is a bulky sphere without an entrance tunnel, constructed by both partners from green or dried grass. The eggs are incubated by both birds, but only the female resides in the nest overnight. The young are cared for by both adults, being fed on a mixture of insects and ripening seeds. Young birds resemble those of the Chestnut-breasted Mannikin and are virtually indistinguishable at this age. If seasons are favourable, two broods maybe reared in quick succession.

P. Rowland

YELLOW-RUMPED MANNIKIN

Lonchura flaviprymna

OTHER NAMES
Yellow-rumped Finch, Yellow-tailed Finch, White-headed Finch, Yellow-rumped Munia

MEASUREMENTS
LENGTH: *100 – 115 mm*
WING: *48 – 58 mm*
TAIL: *32 – 37 mm*
BILL: *12.5 – 13.5 mm*
TARSUS: *15.5 – 17.5 mm*
WEIGHT: *c 14 g*

DESCRIPTION
ADULT
Sexes similar, underparts of female occasionally paler than male (insufficient for positive sexing). Crown pale grey; back and wing coverts cinnamon-brown; flight feathers greyish-brown, edged with chestnut; rump, uppertail coverts and central tail feathers golden-straw; edges of outer tail feathers blackish-brown edged with pale brown. Face plain cream-buff, becoming fawn-buff over breast and abdomen; undertail coverts black. Eye dark brown. Bill, legs and feet pale blue-grey

JUVENAL
Back generally buff-olive; wings, head and throat darker grey-brown, with indistinct dark barring on throat; underparts buff-grey, pale grey on breast; rump dirty yellow

IDENTIFICATION
Similar in size and habits to Chestnut-breasted Mannikin. Conspicuous golden-straw coloured rump. Head and hind neck cream; back and wings dark chestnut; underparts fawn-buff, contrasting black undertail

VARIATION
Monotypic. Hybridizes with Chestnut-breasted Mannikin

VOICE
One main call type, very similar to that of Chestnut-breasted Mannikin; bell-like "tit" or "teet", with variation in length and intensity

Song: Very similar to that of Chestnut-breasted Mannikin. Introduced by rhythmic clicks of the bill, followed by harsh toneless phrases, each preceded by silent mandibulation

HABITAT
Rank reeds and grass bordering streams, also grassy plains

FOOD
Ripe and part-ripe seeds of a variety of grasses. Also flying insects such as termites

HABITS
Both sedentary and nomadic tendencies. Very social, often constituting about one-third of mixed flock with Chestnut-breasted Mannikins; feeding, sleeping, bathing and occasionally interbreeding with these. Diurnal, sleeping at night in reeds

BREEDING
January – March or April (in latter half of wet season, when food abounds). Occasionally two broods in quick succession. Both parents incubate and feed young, only the female remains in nest at night

NEST: Bulky sphere, compressed at its side entrance, without a tunnel; constructed of dry blades of grass; lined with fine grass stems and feathers; situated in tall grasses and reeds

EGGS: 4 – 7 (usually 4 – 5); white; about 17 x 12 mm

INCUBATION: About 12 – 13 days; fledging period about 20 – 22 days

DISTRIBUTION
Northern Australia, from Derby, Western Australia, to western Arnhem Land and Northern Territory; moves southward during wet season to breed

STATUS
Endemic. Uncommon

P. Rowland

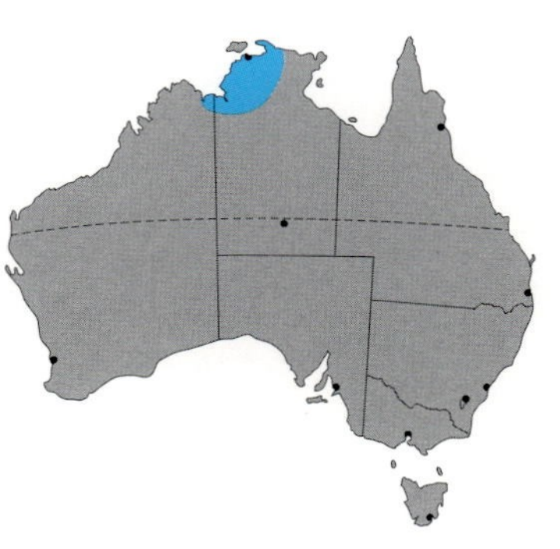

Black-headed Mannikin

Lonchura malacca

On August 1912, Mr Robert Kemp captured a single specimen of a "new" finch. Medium-sized, with a heavy bill, and predominantly chestnut-brown with a black head, it was named *Munia atricapilla novana.* This was a mistake, for the bird was in fact a specimen of the well-known Black-headed Mannikin, *Lonchura malacca*, later determined to be the subspecies *L. m. atricapilla*.

The Black-headed Mannikin established itself in Australia from escaped aviary birds, early in the twentieth century. By 1929 it was found breeding in Centennial Park, Sydney, and numerous other sightings in subsequent years indicated that it had become well established. However, the last recorded sighting was in Victoria in 1981. It may have been extirpated but it is too early to be certain of this.

The Black-headed Mannikin is very social, forming small to large flocks outside the breeding season, but only pairs or small groups during the breeding season. It usually travels in small groups, meeting with other parties to feed and roost. Roost nests are often constructed and 10 birds have been recorded in one of these. The normal food consists of seeds of grasses and weeds and cultivated grain crops. In its native range of Nepal and northern India, through Bangladesh to Burma and Yunnan, it is a major pest of rice crops.

The song is a peculiar arrangement of notes of such high frequency as to be almost inaudible to humans. It is uttered in multiples of three notes, with bill clicking following each song phrase. The most common vocalisation is a "peekt" or "pee-eet", apparently a contact call.

There appears to be no regularity in the breeding times of the Black-headed Mannikin within Australia. Breeding pairs have been recorded in May to November in India, June to September in Burma, and December to October in Malaya. As a prelude to courtship, a male and female bird fly around excitedly, carrying pieces of nesting material in their bills. The male usually utters a song phrase as he flies. When this stage has finished, the male lands near the female and courship proper commences. With his body stiff and slightly angled, he fluffs the feathers of his body and bends his head down, twisting it slightly toward the female. If the male has not already dropped his nesting material, he does so now and begins to sing, while jumping up and down vigorously, his feet leaving the perch. As is usual in grass-finches, the female responds with a quivering tail. Following copulation, the pair engages in mutual preening.

The typical nest of the Black-headed Mannikin is a large loose ball of coarse grass or leaves, constructed by both birds and placed in a clump of tall grasses or sugar cane. Small, loose colonies are occasionally formed when several pairs of mannikins build nests in relative proximity. These colonies are not restricted to members of the same species and often incorporate the nests of weavers and other finches. The five or six eggs are incubated by both birds and young hatch after 12 to 13 days. The young birds are fed by both parents and fledge at about 22 to 28 days, returning to the nest each night for a further one to three weeks.

P. Rowland

Adult; Taronga Zoo, N.S.W.; January 1976 (D. & M. Trounson)

Sponsors: Sir Vincent & Lady Fairfax

BLACK-HEADED MANNIKIN

Lonchura malacca

Adult; Taronga Zoo, N.S.W.; January 1976 (D. & M. Trounson)

OTHER NAMES
Black-headed Munia, Eastern Black-headed Munia, Tri-coloured Munia, Chestnut-breasted Munia, Chestnut Munia

MEASUREMENTS
LENGTH: 98 – 120 mm
WING: 53 – 59 mm
TAIL: 34 mm
BILL: 9.5 – 14.5 mm
TARSUS: 11 – 15.5 mm
WEIGHT: 9 – 15 g

DESCRIPTION
ADULT
Sexes similar. Head, throat, nape, centre of belly and undertail black; remainder of underparts rich chestnut; upperparts slightly lighter chestnut; wings more brown; tail brownish chestnut; central two tail feathers more reddish. Eye dark brown. Bill pale grey-blue. Legs and feet slate-grey. Female may be slightly paler; bill smaller

JUVENAL
Upperparts generally pale reddish-brown, darker and greyer on head; underparts light buff. Bill and legs duller in first few weeks

IDENTIFICATION
Medium-sized Mannikin with short tail. Most of upperparts rufous-chestnut; rump more chestnut; head, neck and upper breast black; remainder of underparts brown, with white patch through centre of abdomen. Bill slate-grey

VARIATION
Australian subspecies is L. m. atricapilla

VOICE
Variety of almost inaudible calls. Contact call: "peekt" or "pee-eet"; flock communication call, "pwi"

HABITAT
Grassland and cultivated areas. In Australia, tends to prefer swamps and areas around human settlements

FOOD
Seeds of grasses and weeds; cultivated grain crops, mostly rice

HABITS
Sedentary. Gregarious, sometimes forming flocks of up to 100; mixes with other flocks. Diurnal, large communal roosts formed at night. Breeds in small loose colonies. Feeds on or near ground

BREEDING
In natural range, May – November, mostly June – September. In Australia, little information but several records of breeding at scattered times throughout year. Both parents share incubation and care for young

NEST: Large, loose, spherical; constructed of coarse grass or leaves; lined with finer grass. Usually placed in reeds in and around watercourses, or sugar cane, within 2 m of ground. Nest built by both parents

EGGS: 5 – 6; white; 16.3 x 11.5 mm

INCUBATION: 12 – 13 days; fledging period 22 – 28 days

DISTRIBUTION
Occurs naturally in Nepal, northern India through Bangladesh to Burma and Yunnan. Introduced to Australia, Moluccas, Hawaiian Islands and Micronesia. In Australia, scattered sightings reported in and around Sydney and Nowra, New South Wales and in Lower Plenty, Victoria. Single specimen captured at Utingu, Cape York, Queensland

STATUS
Introduced. Rare, not recorded in Sydney since 1960; single sighting in Nowra, New South Wales, in 1977 and Lower Plenty, Victoria in 1981. No records since. Common in natural range

P. Rowland

Nutmeg Mannikin

Lonchura punctulata

The Nutmeg Mannikin, formerly known as the Spice Finch, was first sighted in Brisbane in about 1937. Since the birds had escaped from aviaries, they were not given much attention and it was thought that they would not become established. However, by the mid- to late-1950s the Nutmeg Mannikin had become established throughout northern coastal Queensland and some birds had been seen around Sydney. The climate of the whole area was suitable and, having a high tolerance of human surroundings, the species flourished. Preferring settled areas, the Nutmeg Mannikin is seldom seen very far from towns and homesteads.

Its rapid expansion is related to its ability to eat the seeds of almost any grass and herb and to breed continuously, particularly north of the Tropic of Capricorn, where the climate permits nesting

Adult; Townsville, Qld; December 1994 (D.R. Rowland)

Sponsor: J.O. Baldock

throughout almost all of the year. If expansion of its range and numbers continues at the past rate, the Nutmeg Mannikin may ultimately replace many native grass-finches.

Rather gregarious, the Nutmeg Mannikin readily associates with the Chestnut-breasted Mannikin, *Lonchura castaneothorax*, and Zebra Finch, *Taeniopygia gutatta*. Instances of hybridization with the Chestnut-breasted Mannikin have been reported. Population pressure from the Nutmeg Mannikin has led to a steady decrease in numbers of the native finches and, in Townsville, a local bird-watcher, Miss Nancy Hopkins, reported in the mid-1950s:

> The Zebra Finch has been crowded out; I can be fairly definite about this. Since the advent of the Spice Finch [Nutmeg Mannikin] the Zebra Finch has disappeared from the localities which are constantly under my eye, that is the neighbourhood of my home and my place of work. The Spice Finch is now breeding in both areas.

As well as posing a threat to native birds, the Nutmeg Mannikin is a potential pest of Australian crops. In the Hawaiian Islands, India and the Philippines, it is a pest, principally of rice. However, rice crops are no longer extensively grown in the Hawaiian Islands and the bird is regarded as an interesting wayside species. In addition to its regular diet of half-ripe seeds, the Nutmeg Mannikin often picks at the remains of road-killed animals. It may be that it eats insects that are attracted to the dead bodies, but insects are not considered to be a part of the usual diet.

The Nutmeg Mannikin is very social, regularly forming close-knit flocks of several dozen, or even hundreds, of birds which persist year-round. At night, up to 20 birds may share a single large roost nest. Even during the breeding season, birds do not give up their social life. Breeding colonies are established with nests so close that occasionally they are in contact. When in flocks, Nutmeg Mannikins fly in a straight and very rapid manner, often making sharp and simultaneous turns. Individuals fly in the typical undulating grass-finch fashion. When moving along the ground or a branch, and before take-off, they exhibit very marked tail movements, accompanied by rapid wing-flicking.

Breeding can occur at any time of the year, especially in the northern parts of the range. At the beginning of courtship, both birds pick up dried grass and hop to and fro between two branches or fly around excitedly for several minutes; the grass is dropped before the courtship dance gets under way. The male perches near the female and leans towards her. Bouts of singing ensue while the tail is drooped and the body feathers are fluffed out. During his performance, the male constantly moves his head and body from side to side and bobs up and down on the perch. Beak-wiping is frequently part of courtship but it is irregular in pattern. Finally, the female signifies her acceptance by assuming a hunched posture and quivering her tail.

The nest is a large spherical construction of coarse green grass and pieces of bark, generally placed in the centre of a bush or tree about two to nine metres from the ground. Nests have also been found in the eaves of buildings. Both birds share in nest building duties, incubating the eggs, and caring for the young – feeding them by regurgitation, the typical grass-finch fashion. The young fledge after 12 to 14 days but continue to roost in the nest with the parents for some time. Occasionally, more than one female lays her eggs in a single nest, and clutches of up to 10 eggs are common.

P. Rowland

Nutmeg Mannikin

Lonchura punctulata

Adult; Adelaide Zoo, S.A. (F. Lewitzka)

OTHER NAMES
Nutmeg Finch, Spice Finch, Spotted Munia, Scaly-breasted Munia, Common Munia, Ricebird

MEASUREMENTS
LENGTH: 109 – 121 mm
WING: 50 – 55 mm
TAIL: 33 – 40 mm
BILL: 12 – 13 mm
TARSUS: 14 – 16 mm
WEIGHT: 11 – 16 g

DESCRIPTION
ADULT
Sexes similar. Crown, back, wings and rump chestnut-brown with pale whitish streaking. Tail and uppertail coverts grey-yellow. Head and throat reddish-brown; feathers of underparts and undertail coverts white, heavily scalloped dark brown. Bill blue-grey; culmen blackish. Eye brown-red. Legs and feet blue-grey

JUVENAL
Generally pale brown above, rump slightly darker; underparts dull yellow-brown and lacking scalloping

IDENTIFICATION
Upperparts and throat generally chestnut-brown; underparts white with fine dark-brown scalloping. Bill heavy. Possibly confused in the field with immature Chestnut-breasted Mannikin

VARIATION
Twelve subspecies; Australian population is L. p. topela

VOICE
Call: Wide variety of calls. Identification call a loud, penetrating "kit-teeee", the first part being almost inaudible; contact call a soft "chip-chip" or "chippy-chippy"; alarm call a sharp, creaking "tret-tret" or "ki-ki-ki-ki-ki-teeee"

Song: A jingling single "tee" or "tit", repeated several times, accompanied with bill rattles, ending with long descending whistle

HABITAT
Settled areas, seldom far from habitation. Long grasses and rank herbage along roads and river banks, canefields, crops around farms. Similar role to the House Sparrow

FOOD
Staple diet is half-ripe seeds of grasses and herbs but some birds have been seen scavenging around farms and garbage dumps, others picking flesh off road-killed animals. Few insects are taken

HABITS
Gregarious, in groups of 20 – 50, sometimes more. Does not readily mix with other species but occasionally does so. Diurnal, roosting communally at night in special roost nests in bushes and trees (as many as 15 – 20 birds in a nest). Very social, birds often flying straight up to, and peering into, the faces of solitary members of the flock. Confiding in humans

BREEDING
Mainly October – April but at any time, particularly in north-eastern Queensland, where many broods reared in a year. Breeds colonially. Both parents share nest building, incubation, and care of young

NEST: Globular with large chamber and short side entrance tunnel; usually constructed of coarse green grass and leaves (occasionally bark); lined with finer grass stems and seed heads; placed near centre of shrub or tree, 2 – 9 m above ground

EGGS: 4 – 8 (usually 6); white; about 17 x 12 mm

INCUBATION: About 12 – 13 days; fledging period 20 – 22 days

DISTRIBUTION
Coastal and near-coastal areas of eastern Australia, between Cooktown, Queensland, and Sydney, New South Wales. Native to south-east Asia

STATUS
Introduced and established. Common

P. Rowland

Blue-faced Finch

Erythrura trichroa

The range of this beautiful little finch extends from Burma, through Indonesia and New Guinea, to islands of the western Pacific. It also occurs as a rare bird in northern Queensland. Throughout its extensive range, various subspecies inhabit tropical forests from sea level to an altitude of about 3000 metres. It is also found in dense vegetation adjacent to rainforests or mangroves. Until 1944, when Jock Marshall found a breeding pair in a patch of rainforest in the foothills of Mount Fisher, on the Atherton Tableland, the Blue-faced Finch was known in Australia from only three specimens, two from the Atherton region and one from Lloyd Island (off Iron Range, Cape York). Shot in 1914, the Lloyd Island specimen was "feeding in grass and took refuge in mangroves". There has been no subsequent Australian record from mangroves or from Cape York.

The rarity and isolation of the Australian population stimulated debate whether (as suggested by Keith Hindwood), the Blue-faced Finch is a recent colonist or (in Marshall's view) that the Australian population is the remnant of a stock that once extended continuously from the rainforests of northern Australia to those of New Guinea. Marshall's palaeoclimatic interpretation, published in 1948, was ahead of its time but would now be regarded as orthodox.

Today, Cape York Peninsula is harsh dry land, dominated by eucalypts, savanna and flat tea tree scrublands. It was once covered by a vast "super wet" rainforest where abundant vegetation provided an ideal home for a bird like the Blue-faced Finch. When the land bridge to New Guinea subsided in the last glacial epoch, the area changed to the harsh arid land that we know now and the rainforests withdrew to a few scattered pockets. Thus, it is quite possible that the Blue-faced Finch, which is not a strong flier, has merely withdrawn from former, larger habitats. Clearing of rainforest has reduced its potential range in Australia but, despite its rarity, there is no indication that the Australian population is endangered.

Courtship is fairly ritualised and complex. The male begins with his body stiff and upright, and tail twisted toward the female. He then engages in a series of hops and the female responds by twisting her tail towards him. She then flies off and the male gives chase, uttering loud calls. He repeats the performance each time she lands and each time she flies off again. This behaviour may be repeated as many as 20 times before she eventually allows him to approach her. She then signifies her acceptance by quivering her tail.

The nest is a bulky, pear-shaped construction, with an angled entrance tunnel at the side. Both adults share in nest building, the male finding the bulk of the material while the female does the weaving. Made predominantly from grass and vines, the nest is placed in the fork of a tree up to six metres from the ground. The female carries out most of the incubation, but both birds spend the night in the nest. The eggs hatch in about 12 to 14 days and are cared for by both birds. Heumann made the interesting observation that the young are not fed until they are 24 hours old, probably to await the development of the iridescent turquoise beads (glands), two on each side of the base of the bill. This enables the parents to feed the babies in rotation, even when they fill the narrow entrance hole and their bodies shut out most possible light.

The Blue-faced Finch usually moves about in pairs but groups of three to six are not uncommon and flocks of up to 20 to 30 birds sometimes form outside the breeding season.

P. Rowland

Adult; Taronga Zoo, N.S.W.; April 1972 (D. & M. Trounson)

Sponsor: Hazel B. Keating

Blue-faced Finch

Erythrura trichroa

Adult; Julatten, Qld (H. & J. Beste)

OTHER NAMES
Blue-faced Parrot-Finch, Tricoloured Parrot-finch

MEASUREMENTS
LENGTH: 109 – 130 mm
WING: 60 – 63 mm
TAIL: 41 – 50 mm
BILL: 10 – 13.7 mm
TARSUS: 14.6 – 20 mm
WEIGHT: 10 – 13.5 g

DESCRIPTION
ADULT
Male: Upperparts generally rich grass-green; rump and uppertail coverts dull carmine; tail feathers olive-brown edged with dull carmine. Face, throat and front of head dark cobalt-blue washed with mauve; underparts light grass-green tinged with golden-olive on thighs. Eye brown. Bill black. Legs and feet brown
Female: Generally duller, blue face patch duller and smaller than in male

JUVENAL
Similar to female but duller; underparts greenish-buff; blue face-patch normally absent. Bill pale grey

IDENTIFICATION
Medium-sized finch with generally green plumage and dull scarlet rump and tail; striking cobalt-blue face

VARIATION
Nine subspecies. Only four specimens taken in Australia, tentatively assigned to E. t. sigillifera

VOICE
Song: High-pitched trilling, followed by a single whistle
Call: High-pitched "tseet-tseet" and various low trilling notes

HABITAT
In Australia, recorded from sea level to high altitudes and from rainforest to open grassy areas and mangroves

FOOD
Mainly seeds found in low vegetation or on ground, supplemented with insects and their larvae

HABITS
Still disagreement whether Australian records are of "vestigial survivors" or "recent colonists". Diurnal. Gregarious outside breeding season, forming flocks of 20 – 30. Largely arboreal, usually near to ground

BREEDING
Very few records of breeding in Australia; nests have been found in November and March. Both members of pair construct nest; male brings material, female fits it into the structure. Female incubates eggs but both parents roost in nest during night. Both feed and care for the young

NEST: Bulky, pear-shaped, with angled entrance tunnel protruding from one side. Constructed of grass, vines and mosses, lined with coarse dead grass. Usually placed in fork of a tree up to 6 m from the ground

EGGS: 3 – 6; pure white; swollen oval; 15 x 10 mm, slightly blunter at egg-sac end

INCUBATION: 12 – 14 days; fledging period 21 days

DISTRIBUTION
From Celebes to New Guinea, Micronesia, The Bismark Archipelago, The Solomon Islands, New Hebrides and The Loyalty Islands. Australian records to date confined to the east coast of Cape York Peninsula, from Cape York to Atherton Tablelands, as far south as Ravenshoe, Queensland

STATUS
Uncommon

P. Rowland

Gouldian Finch

Erythrura gouldiae

In a rare acknowledgment of the labours of his wife Elizabeth, John Gould introduced his description of this species as follows:

> It was with feelings of the purest affection that I ventured to dedicate this lovely bird to the memory of my late wife, who for many years laboriously assisted me with her pencil, accompanied me to Australia, and cheerfully interested herself in all of my pursuits. The dedication of this bird to Mrs Gould's memory will surely then receive the sanction of every scientific ornithologist.

It so happens that only scientific ornithologists (and not all of these) would recognise that the specific name *gouldiae* is feminine; most people regard the species as John Gould's finch. This aside, Gould could not have picked a more beautiful finch with which to commemorate his wife. Unfortunately, its brilliant colours have not only contributed to the bird's fame, but to a rapid decrease in its numbers through illegal trapping. This, together with the effects of grazing, fire, and infestation from the air-sac mite, *Sternastoma tracheacolum*, has forced the bird to withdraw from nearly half of its range, a decline that has placed it on the endangered list. Another factor contributing to a certain extent is mining, where trees have been knocked over in explosions and nesting sites have been flooded by storage dams. Surprisingly, predation has not markedly contributed to the decline of the Gouldian Finch although the Brown Tree-snake, *Boiga irregularis*, and a small goanna, *Varanus timorensis*, have been found in hollows were Gouldian Finches have nested. Sonia Tidemann and her colleagues observed in 1992:

Adult males; Red-headed form; Canberra, A.C.T.; January 1989 (C. Andrew Henley/Larus)

Sponsor: Alan Atwill

Adult male; Black-headed form; Keep River, N.T. (B. & B. Wells)

Sponsor: Renison Goldfields Consolidated Limited

Observations of predation are rare in relation to its occurrence. There is no evidence, however, to suggest that predation is causing the numbers of Gouldian Finches to remain low.

The Gouldian Finch is seldom far from water, preferring open savanna plains and the edges of mangroves. It was first found by John Gilbert on Greenhill Island at the head of Van Diemens Gulf where it inhabited the edges of mangroves and thickets. When disturbed it invariably flew to the topmost branches of the loftiest gums, a habit I have not before observed in any other member of the genus. Nevertheless an early sighting of Gouldian Finches away from permanent water was reported to Gregory Mathews by George Barnard from the McCarthur River in the Northern Territory.

This handsome Finch was often seen on dry ridges at long distances from water. It resorts to small hollows in trees for nesting purposes and several pairs nest in the same hollow. An instance of this came under my notice while collecting on the McCarthur. In a swamp gum growing on the bank of the river, three nesting hollows were found at heights varying from 20 to 40 feet from the ground. One hollow contained five eggs, or two distinct types; a second contained no fewer than fourteen perfectly fresh eggs, of three distinct types; the third hollow was beyond reach.

The Gouldian Finch is the only Australian estrildid to nest almost exclusively in the hollows of trees or termite mounds. If a pair is forced to build a nest elsewhere, it is poorly constructed of dry grass and placed in a bush or tree. Both parents share in its construction and two or three broods may be reared in a season. Both parents incubate the four to eight eggs, changing over at regular intervals, with only the female remaining in the nest overnight. When the male relieves the female, he brings a feather, or other piece of nesting material, which the female places in the nest.

Courtship can be divided into two fairly distinct routines, but there may be a great deal of individual variation, especially in the amount of bobbing movements. No nesting symbol is used. The male initiates the routine by sitting in front of the female with his tail lifted and his body almost horizontal. He then begins to move his head up and down rapidly and performs various degrees of beak-wiping. The initial head movements are distinct from the beak-wiping stage as the beak does not touch the branch. The second routine opens with the male jerking his body up and down, while commencing his song phrase. When singing, his body is generally held vertical, with the beak pointed downwards and the head moving from side to side. The entire performance may be repeated several times. During each performance, the feathers of the male are fluffed, displaying his brilliant colours to the fullest extent. Occasionally the female joins the male in the beak-wiping section of the ceremony. After courtship, the birds usually fly to the nest and mating always takes place inside it.

The Gouldian Finch is very sociable. Outside the breeding season, flocks become smaller as the total population decreases. Even during the breeding season, social bonds are maintained and many pairs breed in close proximity to each other, sometimes nesting in the same hollow. Pair bonding is not strong and mated pairs are not identifiable in the flocks that gather outside the breeding season. Like the Long-tailed and Black-throated Finches, Gouldian Finches have daily "social meetings" during the breeding season, normally late in the afternoon. At this time, virtually all members of a colony gather to feed, bathe, or simply sit together in a tree, uttering loud calls.

P. Rowland

GOULDIAN FINCH

Erythrura gouldiae

Adult female; Yellow-headed form; Elanora Heights, N.S.W.; August 1993 (A. Foster)

OTHER NAMES

Painted Finch, Gouldian Grass-finch, Rainbow Finch, Purple-breasted Finch

MEASUREMENTS

LENGTH: 113 – 125 mm (excluding "streamer" of tail)
WING: 60 – 67 mm
TAIL: 32 – 47 mm (males with "streamers" up to 60 mm)
BILL: 12.7 – 13.7 mm
TARSUS: 13.4 – 15 mm
WEIGHT: 12 – 15 g

DESCRIPTION

ADULT

Male: Most of upperparts, including wing coverts, grass-green; flight feathers brown; edged with dull green; rump and uppertail coverts bright cobalt-blue; tail black and pointed. Crown and face-mask black (about 75% of individuals) or deep crimson (about 25%). Rare individuals have ochre-yellow face-mask (due to lack of red pigment). Red and yellow-faced forms have mask bordered by narrow line of black, followed by a band of bright cobalt-blue. Chin and throat black; upper breast lilac, with narrow yellow-orange band below; abdomen rich yellow; centre of abdomen and undertail white. Eye dark brown. Bill ivory with red tip. Feet pale yellow

Female: Duller than male

JUVENAL

Dull ashy-grey over head and hind-neck, shading to olive over back and tail; brown-white below, paler on chin, and belly has faint yellow tinge; upper mandible blackish, lower mandible pinkish-white, tipped with red; pale nodules at gape

IDENTIFICATION

Most colourful of Australian grass-finches. Some (75%) have full black face-mask; others have red face-mask, bordered black; rare mutants have yellow face-mask. Upperparts grass-green, cobalt-blue rump; breast purple; belly and flanks lemon

VARIATION

Dimorphic or trimorphic (see above) but no apparent geographical variation

VOICE

Call: Mostly silent. Main call is a reedy "ssitt" in one to four syllables, also "ssriih" and variations

Song: Extremely high-pitched hisses and whines, inaudible beyond a few metres

HABITAT

Edges of mangroves, thickets and savannas dotted with trees; never far from water

FOOD

Wide variety of ripe and half-ripe grass seeds plus flying insects. Almost completely insectivorous during breeding season

HABITS

Partly migratory, moving towards coast as inland grasses die off, returning with the rains to breed. Diurnal. Gregarious, in flocks of tens or hundreds. Very social, drinking and feeding together, and indulging in frequent bouts of sunbathing. Rarely seen on bare ground, favouring tall grasses. If disturbed, flock rises as group and sits at top of nearby tree. Feeding takes place in tall grass

BREEDING

January to April, occasionally earlier. Nesting is social in loose colonies, occasionally several pairs sharing the same hollow. Two or three broods may be reared in a season. Parents share incubation and brooding

NEST: Nest rarely constructed. If built, is shabby, of harsh, dry grass in a bush or tree. Preferred nesting site is in hollow in a branch or termite mound

EGGS: 4 – 8; white; about 17 x 13 mm

INCUBATION: About 12 – 13 days; fledging 21 days

DISTRIBUTION

Patchy through tropical northern subcoastal areas from Derby, Western Australia, to Gulf of Carpentaria and thinly to central Cape York Peninsula, Queensland

STATUS

Endemic. Locally common in patches of north to north-western parts of range. Withdrawn from nearly half of range

P. Rowland

Sponsors: Sir John & Lady Proud

Starlings and Mynas

There are just over 100 recognised species in the family Sturnidae, which extends over most of Africa and Eurasia and well into the Pacific Islands. The Common Starling was deliberately introduced to eastern North America, from which it has spread over most of the continent. Native American birds known as starlings are not closely related to the Sturnidae.

Some past authorities have regarded the group as comprising two families, but the Eulabetidae, or "true" mynas, typified by the well-known Hill Myna, or Grackle, are not now even accorded subfamily status. There seems no doubt that the peculiar African oxpeckers are closely related, and these are nowadays usually placed in a subfamily of the Sturnidae. Controversy still surrounds the affinities of the peculiar African sugarbirds; these are often included within the Honeyeaters (family Meliphagidae), but no Australasian ornithologist seriously entertains this idea and, although the sugarbirds may be aberrant starlings, it seems prudent to leave them in their own small family for the time being.

As with other vernacular names, "starling", "minah" and "grackle" are not very definite in application; in many cases two of them are used for members of the same genus or even the same species. "Starling" refers particularly to the common English species, and its name is a diminutive form, implying affection. "Myna" appears to be a Hindi term of affection, applied particularly to the Hill Myna, *Gracula* or *Eulabes religiosa*, a very familiar town bird and "talking" pet of southern Asia. The name "grackle" is also given to this species and to other larger arboreal starlings (or mynas), particularly those with naked areas of colourful skin or wattles about the face. Wattles are not restricted to arboreal species; the Common Myna, now widely spread throughout Australasia, is perhaps even more terrestrial than the Common Starling and has distinctive spectacles of naked flesh.

Americans prefer to reserve the name "grackle" for certain quite unrelated members of the family Icteridae, but Old World usage is against this. Adding to the confusion of names, the only genus of "grackle" in New Guinea, is called *Mino*; and various members of the genus *Sturnus* are officially known as starlings, mynas and – by way of variation – pastors. Pastor is a rather good name for many species which haunt pastures and associate with cattle or other open country herbivores. They make a good living from the abundant and diverse insects that thrive on well-grazed and well-manured grasslands. Since extensive grasslands and the faunas they support are relatively recent in earth history,

Adult Metallic Starling; Cairns, Qld; 1984 (M. & I. Morcombe)

Sponsor: C.H. Warman

it seems likely that tree-starlings are comparatively conservative members of the family whereas the terrestrial species tend to be the more adventurous in an evolutionary sense.

It is no coincidence that the Common Starling has been successful in the cooler lands of the Northern Hemisphere at the same time as grass seed-eating sparrows and of course cereal-eating, cattle-grazing, horse-taming humanity. The Common Myna fares equally well in the densely populated towns and intensive agriculture of southern Asia and beyond. Both thrive wherever they have been transplanted to lands which have been more or less drastically transformed to meet human needs. However it is not only these transplants that benefit from human activities. The New Guinea counterpart is the Coconut Starling, *Aplonis cantoroides*, much more terrestrial and opportunistic than its close relative, *A. metallicus*, the only species that actually reaches Australia naturally. In many plantations, the Coconut Starling is the most obvious bird, while in many towns in New Guinea it nests in buildings, electricity supply poles and other artifacts. Birds of other families may also use human-made structures but Starlings at least rank alongside sparrows and pigeons in being considered noxious because of this behaviour.

The family has more species in Africa than anywhere else, but the diversity is greater in intertropical Asia, and includes large and small fruit eaters which have spread through Wallacea to the Solomons (larger "grackles"), or opportunistically throughout many Pacific Islands (*Aplonis*). One widespread tree-starling has successfully colonised the north-eastern area of Queensland, where it has increased in range and numbers in historical times. Another, *A. fuscus*, has reached Norfolk Island and is presumably a representative of the widespread *A. tabuensis* group.

W. Filewood

Metallic Starling

Aplonis metallica

The Metallic (or Shining) Starling is Australia's only native starling, other than the now extinct Tasman Starling of Norfolk Island and Lord Howe Island. Although some remain in Australia throughout the year, most are summer migrants from New Guinea and nearby islands. The Metallic Starling travels in large, noisy flocks to the north-eastern coast of Australia, arriving from August to September to breed.

Like most other starlings, it is gregarious by nature. It is a colonial nester which roosts communally both in and out of the breeding season. After breeding, several thousand birds may gather at dusk to roost in large trees on rainforest fringes, in town parks or on overhead wires. Nesting colonies comprise several hundred individuals; one, recorded by the RAOU Nest Record Scheme contained 188 nests. A.J. North published some notes taken by E.J. Cairn and Robert Grant while collecting for the Trustees of the Australian Museum:

> Calornis [Aplonis metallica] used to nest in a large tree right in the centre of the track leading to the Pyramid Sugar Plantation, every branch, from the lowest to the highest, containing clusters of nests. I have seen branches of this tree lying on the ground eight inches or more in thickness, broken off, I presume, by the weight of the nests built on them, as the break was fresh and free from flaws. I am certain that some of these broken branches contained over a hundred weight of nests. The chattering of the birds was deafening, and throughout the day, from sunrise to sunset, flocks could be seen coming from or going to the tree. On one occasion I fired a shot into one of these flocks, when just returning to the tree, and picked up fifteen dead birds.

A tree containing such a colony is a phenomenal sight. The canopy throbs with life. Groups arrive and depart to and from feeding sites while others jostle for position on a sunny perch. A cacophony of sound surrounds the tree. Adults maintain contact with a constant wheezy chatter, others engage in trilling, and young at all stages of development beg for food. As many as three broods may be raised by each pair in a season. Underneath the tree is another dependent world. Fallen branches, nests and droppings accumulate, mites abound, and snakes and other reptiles frequent these sites, searching for freshly fallen eggs and nestlings.

On a recent trip to north Queensland I observed a breeding colony situated opposite a bus depot in a town park. There was a pungent aroma but the ground below the nesting tree was surprisingly free of debris – the local council gardeners had the thankless task of cleaning up after these untidy birds.

Unlike the two introduced starlings, the Metallic Starling is not an urban scavenger of human refuse. It feeds largely on native rainforest fruits but often raids soft fruit orchards. It is a clumsy feeder; large groups converge on a tree and clamber through the branches, randomly picking at fruit as they go, knocking to the ground as much as they consume.

Once breeding is finished, the Metallic Starling gathers in flocks of thousands to depart for wintering grounds in New Guinea, leaving behind an eerie tranquillity.

T. Ivison

METALLIC STARLING

Aplonis metallica

Adult at nest; Daintree, Qld; 1966 (J. Purnell)

OTHER NAMES
Glossy Starling, Shining Starling, Whirlwindbird

MEASUREMENTS
LENGTH: 210 – 240 mm
WING: c 109 mm
TAIL: c 101 mm
BILL: c 18 mm
TARSUS: c 24 mm
WEIGHT: 58 – 66 g

DESCRIPTION
ADULT
Sexes similar. Head and neck glossy black with purple iridescence; rest of body glossy black with green iridescence. Neck feathers long and pointed; tail longish and sharply tapered. Eye bulging and bright red. Bill and legs black

IMMATURE
Upperparts as adult; underparts creamy white with bold brown streaking. Eye greenish orange

JUVENAL
Dull and brownish above; underparts creamy white with bold dark brown streaks and black breast band; blanks mottled grey-brown. Eye olive-yellow

IDENTIFICATION
Medium-sized bird with dark metallic plumage and long, pointed tail. Conspicuous; usually seen in large noisy flocks

VARIATION
5 subspecies; one in Australia

VOICE
Wheezy chattering; song is a short canary-like trilling

HABITAT
Outskirts of lowland rainforest and mangroves; also urban areas; visits orchards, parks and gardens to feed

FOOD
Mainly native and cultivated fruits and berries; also insects and their larvae

HABITS
Migratory; arrives Australia July – September to breed and returns to New Guinea February – April. Gregarious. Communal nester

BREEDING
Communal; August – January; one or two, sometimes three clutches per year

NEST: In colonies; bulky and globular, made from vine tendrils, shredded leaves and other hair-like plant fibres; lined with fine grasses; suspended from branch with entrance at side placed in close proximity to one another in tall, isolated tree; 2 – 20 m or more above ground

EGGS: 2 – 3; pale blue to pale green-blue with red-brown spots and flecks, often concentrated at larger end; tapered oval; 30 x 21 mm

INCUBATION: Not recorded

DISTRIBUTION
In Australia, north-east coast from Cape York to Repulse Bay, inland to Atherton Tablelands, Queensland. Also Solomon Islands, Moluccas and New Guinea

STATUS
Regular breeding migrant; common to 750 m

T. Ivison

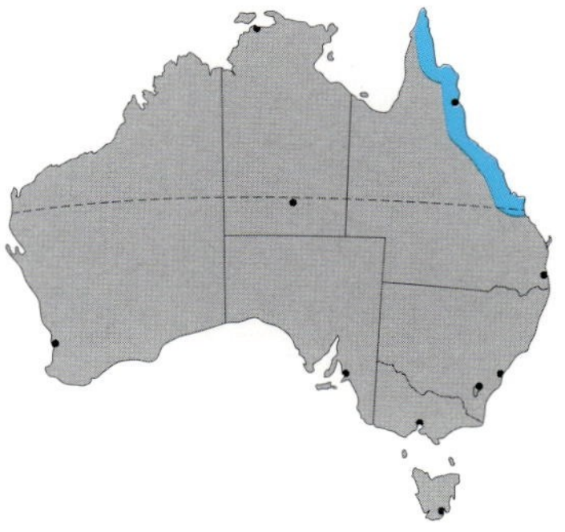

Tasman Starling

Aplonis fusca

The Tasman Starling, now extinct, once inhabited Norfolk and Lord Howe Island in large numbers and was considered to be an agricultural pest by local fruit growers. With the introduction of rats to Lord Howe Island in 1918 numbers quickly dwindled and within a few short years the species had disappeared. On neighbouring Norfolk Island, the last sighting was in 1923.

The reason for the extinction of the Norfolk Island subspecies is largely a mystery, since rats had not yet been introduced when the starlings disappeared. This suggests that rats may not have been the sole cause for the demise of the species on Lord Howe Island. A disease or other environmental factor may have contributed, but all we know is that a common species became extinct almost overnight, a dramatic and sudden extinction that can be compared with that of the Passenger Pigeon, *Ectopistes migratorius*, in America.

Rats are thought to have invaded Lord Howe Island when a ship ran aground on Ned's Beach in the winter of 1918. It was nine days before the ship was repaired, leaving plenty of time for the unwelcome stowaways to come ashore. In 1919, Allan McCulloch wrote:

> Within two years this paradise of birds has become a wilderness, and the quietness of death reigns where all was melody. One cannot see how the happy conditions are to be restored. The very few birds remaining are unable to breed, being either destroyed upon their nests or driven from them by rats, and their eggs eaten. One can scarcely imagine a greater calamity in the bird world than this tragedy which has overtaken the avifauna of Lord Howe Island.

The nest of the Tasman Starling was particularly vulnerable, a flimsy structure made from sticks and lined with dry leaves, placed fairly low to the ground in a tree hollow, which was easily accessible to rats.

Locals called the bird "Cudgimeruk", an interpretation of its loud and distinctive call. In this and most other respects the Tasman Starling was a typical metallic starling belonging to the same genus as the Metallic Starling, which, in contrast, still prospers on the mainland. Although the Tasman Starling never became the urban pest that the introduced Common Starling or Common Myna have, flocks of them did wreak havoc on orchards. Edgar Waite visited Lord Howe Island in 1902 and wrote of the bird:

> The commonest bird is the Cudgimeruk a kind of Starling which possesses many of the habits of the true starlings, it is a great pest in the garden, destroying enormous quantities of peaches and other fruits. At the same time it is the cheekiest of birds allowing itself to be caught and feeding out of the hand directly afterwards.

The Tasman Starling was about the same size as the Common Starling but grey-brown with a metallic sheen on its upper body and a bright orange eye. Little is known of its biology.

T. Ivison

Tasman Starling

Aplonis fusca

OTHER NAMES
Red Eyes, Cudgimeruk, Sooty Starling

MEASUREMENTS
LENGTH: 196 – 204 mm
WING: 100 – 106 mm
TAIL: 63 – 71 mm
BILL: 13 – 15 mm
TARSUS: 25 – 27 mm
WEIGHT: Not recorded

DESCRIPTION
ADULT
Male: Head, neck and nape dark grey, glossed green; upperparts, uppertail coverts and wing coverts dark grey with faint green gloss; primaries dark ash brown with black shafts; underparts sooty grey

Female: Head and neck same as male but upperparts duller and paler below. Eye orange-red. Bill and legs black

IMMATURE
Head blackish grey; neck and mantle dark grey with whiteish striations; rest of upperparts dark grey; wings, tail, throat and chest blue-grey with indistinct striations; underparts light buff. Eye greenish-yellow. Bill black. Legs grey-black

JUVENAL
Similar to immature but fluffier

IDENTIFICATION:
Sooty coloured bird, about the same size as the Common Starling, Sturnus vulgaris

VARIATION:
Two subspecies: A. f. fusca *on Norfolk Island and* A. f. hullianus *on Lord Howe Island*

VOICE:
Sharp distinctive cry "cudgimeruk"; probably other harsh starling-like wheezes and whistles

HABITAT:
Presumably most forested areas, also urban gardens and orchards

FOOD:
Mainly fruits, both native and cultivated; probably also insects and their larvae

HABITS:
Typical starling, bold and inquisitive; gregarious, usually in flocks; orchard pest

BREEDING
NEST: An open cup made from grasses and small twigs placed in hollow tree branch or tree-fern trunk; often close to ground

EGGS: 2 – 5; pale blue blotched with red-brown; tapered oval; 26 x 19 mm

INCUBATION: Not recorded.

DISTRIBUTION:
Formally Norfolk and Lord Howe Island

STATUS:
Extinct

T. Ivison

Adult; Non-breeding plumage; Boat Harbour, Tas.; May 1972 (T.A. Waite)

Sponsors: Lloyd & Rene Hartigan

Common Starling

Sturnus vulgaris

This bird was originally confined to the Northern Hemisphere, ranging through Europe to Russia, with a migratory path to the Mediterranean for winter. It has undergone some marked changes in its behaviour and, now, only birds in northern and eastern Europe are migratory. Some European populations consist of both migrant and resident birds; the rest of the European and worldwide populations are resident or sedentary. Resident birds are frequently able to raise more than one brood per season while constraints of energy and time make this impossible for migratory Starlings. As remarked by an early observer:

> The Starling's physical fitness for the struggle for supremacy is seen at once ... It is exceedingly tough and wiry, and the bill, its principal method of offence and defence ... is nearly straight, long and heavy, tapering, and nearly as keen as a meat axe ... Mentally the starling is superior to the sparrow, and while brave and active in the face of any foe that it can master, it shows the acme of caution and intelligence in its relations with man or any other creature too powerful to master.

Due to introductions by humans and its own amazing adaptability, the Common Starling now has the widest distribution of any of the world's birds. In second place is the House Sparrow, *Passer domesticus*.

The first introduction to Australia was in Melbourne in the late 1850s. This was followed by Brisbane in 1869 and South Australia in 1881. It had spread through New South Wales by 1926. The species is now widely distributed in towns and rural areas throughout the eastern states, including Tasmania.

In Queensland it has been unable to move further north than Brisbane, being unsuited to tropical climates. In Western Australia it has been declared a noxious species and its expansion across the Nullarbor Plain has so far been prevented by the notorious "Starling Patrol", a handful of shooters based on the border between Western and South Australia who are employed by the Agriculture Protection Board to kill any Starlings seen trying to cross the border via the railway line or road. Entry from the north is barred by vast expanses of desert.

Despite its status as a pest, the Common Starling is an attractive bird. The juvenile is uniformly fawn brown, with a dark bill, quite different in appearance from the adult. The adult plumage consists of characteristic sleek black feathers, each of which has a large white tip; the wing coverts have dark buffy edges. A fully plumaged adult is a striking black bird, prominently spotted with white. When the breeding season arrives, birds have become completely black, not because of replacement of feathers but due to their abrasion. As the plumage wears, the white spots and buff edges, which are structurally weaker than the black parts of the feathers, are lost. When the annual moult occurs, the fresh feathers have their adornment of spots and the process of abrasion begins again.

Liberal in its feeding habits, the Common Starling is an opportunistic omnivore. Working in groups, it struts boldly across open ground in search of seeds, insects, worms, beetles and other invertebrates. Although consuming many insects, it is regarded as a serious pest to fruit growers. Large flocks converge on orchards and fruits that are too large to swallow are punctured with the sharp, pointed bill, thus devastating crops. In some parts of North America, the Starling has found a seemingly endless source of food in cattle feedlots. This grain is a welcome addition to its diet, particularly in winter, at which time the Starling and several species of native New World blackbirds (Icteridae) converge on feedlots in huge numbers. The consequent loss of grain is of significant economic cost. Roosting flocks of Starlings and blackbirds may comprise thousands or even millions of individuals.

Although millions of dollars are spent annually on control measures, these have had limited success. Large-scale programs employing chemical repellent and poisons have been used, but Starlings have developed resistance to some of these substances. Winter roosts have been sprayed with soapy water, which removes the waterproofing of the feathers and soaks the plumage so that birds succumb to exposure to the winter cold. Perhaps the most effective and simple method has been the construction of large plastic covers over the feedlots.

In the evening, Starlings gather at communal roosting sites. The eaves of old buildings are a favoured site, as are cocos palms and large, bushy trees. There is a cacophony of raucous squeaks and rattles as latecomers jostle for position. While there is still light, small feeding groups fly to a staging point and merge into a large flock that flies together back to the roost. The droppings from such enormous aggregations can present a serious human health hazard. Respiratory ailments, the most prominent of which is histoplasmosis, can be transmitted in this way.

Flocks disperse in the breeding season and loose breeding colonies are formed, often with smaller temporary roosts nearby. Starlings nest in hollows and utilise almost any crack or crevice for this purpose. Ceilings, abandoned warehouses, cracks in buildings and rocks, are all potential nesting sites. When nesting in the roof of a house, Starlings often introduce red mites; although these cannot live long away from their avian host, they can cause considerable irritation to humans. But it is on the outskirts of towns and forest edges where the Starling causes most damage. Here it competes with less aggressive native species for precious tree hollows, sometimes even forcibly evicting them. The nest is an untidy structure, loosely constructed from any available materials, including grasses, bark, leaves, wool and even urban refuse. Four to six eggs are laid and one or two (sometimes three) broods are raised a season, perpetuating the spread of these "winged rats".

T. Ivison

COMMON STARLING

Sturnus vulgaris

Immature (T. Howard)

OTHER NAMES
English or European Starling, Starling

MEASUREMENTS
LENGTH: 210 – 230 mm
WING: 122 – 132 mm
TAIL: 64 – 68 mm
BILL: 25 – 27 mm
TARSUS: 28 – 30 mm
WEIGHT: 55 – 96 g

DESCRIPTION
ADULT
Sexes similar. Glossy black with green and purple iridescence; new plumage (autumn) spotted; feathers of upperparts tipped buff, underparts tipped white; tail and wing feathers edged buff; as feathers wear, spots and edging decrease and are absent by spring when birds are mostly glossy black with duller flight feathers. Eye brown. Bill and legs yellow in breeding season, dull brown for rest of year

IMMATURE
A mixture of light brown and glossy black plumage

JUVENAL
Light grey-brown, slightly paler below; throat whitish;

Bill and legs black

IDENTIFICATION:
A familiar bird throughout eastern Australia; longish pointed bill and short, blunt tail feathers. Usually in large flocks

VARIATION:
11 subspecies; one introduced to Australia

VOICE:
Wheezes, clicks, soft rattles and characteristic downward whistle are combined to make a song. Alarm notes harsh "tcheer" or short sharp "dik", "dik-ik-ik". Also excellent mimic

HABITAT:
Diverse, mainly open and disturbed areas; cities, urban parks and gardens, rural towns, pastures, orchards, and fringing bushland

FOOD:
Opportunistic omnivore; insects, beetles and their larvae, fruit both native and introduced, human refuse

HABITS:
Highly gregarious, usually seen in large flocks; when travelling, flocks rise and fall; walks jerkily and stabs ground with bill in search of food; communal roosts, often situated on ledges or in abandoned buildings; noisy and conspicuous; in Australia sedentary but populations dispersive

BREEDING
August – January; 1 – 3 clutches per season; both sexes contribute to incubation and care of young

NEST: Untidy heap of grasses, leaves, feathers and rubbish; placed in any available hollow, including roofs, crevices in old buildings and disused machinery, fence posts and natural tree hollows; often evicting native birds

EGGS: 4 – 8 (usually 4 – 5); glossy pale blue; tapered oval; 30 x 21 mm

INCUBATION: 10 – 12 days (slightly longer in Europe); young fledge about 21 days after hatching

DISTRIBUTION
Introduced to Australia, now widespread throughout south-eastern Australia from central Queensland to Tasmania and west to Eyre Peninsula; still spreading. Natural range through Europe to Russia with a migratory path to the Mediterranean

STATUS
Introduced and abundant; agricultural pest

T. Ivison

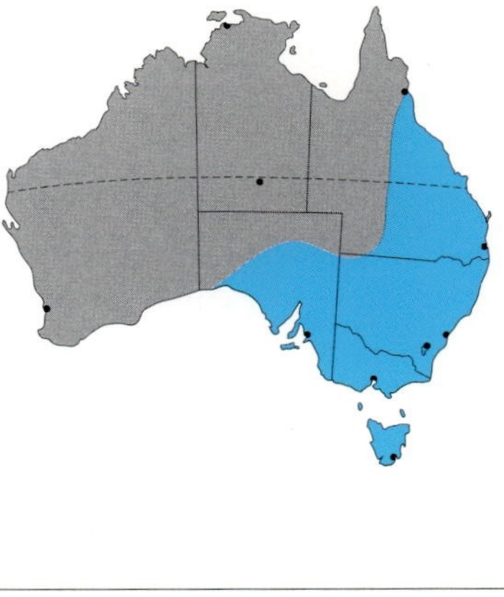

Common Myna

Acridotheres tristis

The aggressiveness and bold nature of the Common Myna, *Actridotheres tristis*, coupled with its omnivorous appetite, has made it a highly successful colonist. Together with the Common Starling, *Sturnus vulgaris*, and the House Sparrow, *Passer domesticus*, it is one of the most successful commensals of humans.

Introduced to Melbourne from South-east Asia between 1862 and 1872, it quickly became established. In 1883, birds were liberated in the cane fields of north-eastern Queensland to combat insect pests, particularly plague locusts and cane beetles. The Common Myna is now common in Townsville, throughout the cane growing areas and Cairns.

The attempted introduction of Common Mynas at Launceston in 1900 and in 1955 failed, although some sightings have occurred since, probably of birds that have reached there from the mainland. Releases occurred throughout eastern Australia, and by the 1940s and 1950s the species was established in many metropolitan areas. In New South Wales it was recorded at Lane Cove and North Ryde in 1950, and by 1964 it was firmly established, and has since become increasingly common along the New South Wales coast. Today it ranges from Sussex Inlet in the south, north to Tweed Heads and west to Marulan.

The Common Myna is an accomplished scavenger, feeding on almost anything from insects, scraps, fruits and vegetables to fledgling sparrows. In southern Asia it is not generally considered to be a pest, due to its penchant for insects and their larvae; flocks follow the plough to feast on the insects and grubs turned up with the soil. However, due to their fruit-eating habits, the species is regarded as being a pest of fruit, especially figs, and is often responsible for picking off seedlings in market gardens.

Common Mynas mate for life. During the breeding season – normally from October to March – there is usually considerable competition for nesting sites, which may be in the walls and ceilings of buildings (thus becoming a nuisance to humans) or in hollows used by our own native birds. During the breeding season violent battles often erupt between occupants of nesting sites and the couple that wishes to evict them. Each partner grapples with its opposite number in a vicious fight, the contestants dropping to the ground secured in each other's claws, jabbing ruthlessly with their bills to the accompaniment of loud calls. Finally, the defeated couple leaves to search for another site.

Outside the breeding season, Common Mynas keep in family groups of five or six individuals, occasionally forming flocks of many hundreds of birds. At night, large communal roosts form within the roofs of abandoned houses, in large trees, in cane fields or under bridges, occasionally numbering several thousand birds. The noise emitted from these roosts is deafening until the birds settle down to sleep; even then, small sections of the huge congregation erupt into noisy chattering during the night.

T. Ivison

Common Myna

Acridotheres tristis

Adult; Sydney, N.S.W. (K. Griffiths)

OTHER NAMES
Calcutta Myna, Indian Myna or Myna

MEASUREMENTS
LENGTH: *230 – 250 mm*
WING: *138 – 153 mm*
TAIL: *81 – 95 mm*
BILL: *25 – 30 mm*
TARSUS: *34 – 42 mm*
WEIGHT: *100 – 140 g*

DESCRIPTION
ADULT
Sexes similar; head and neck glossy black; upper parts chocolate brown; wings brown-black with large white patch evident in flight; tail brown-black, broadly tipped white; underparts chocolate brown, becoming paler towards lower chest; abdomen and undertail coverts creamy white. Eye amber. Bill and legs bright yellow

IMMATURE
As adult

JUVENAL
As adult but duller

IDENTIFICATION
Robust, ground-feeding bird; hops energetically; chocolate brown with bright yellow bill, colour extending to bare skin patch behind eye; large white wing patches evident in flight

VARIATION
Two subspecies, one introduced to Australia

VOICE
Raucous rattles and growls often strung together as song; liquid note on taking flight; harsh "skaair" in alarm

HABITAT
Urban areas and fringes; cities, towns, parks and gardens

FOOD
Scavenger; eats a wide variety of insects, their larvae, fruits, berries and human refuse

HABITS
Gregarious and garrulous; roosts communally in large noisy flocks; pairs disperse in breeding season; hollow nesters but any crack or crevice may be utilised; aggressive, often evicts native species from natural nest hollows; sedentary

BREEDING
October – March; two, occasionally three, clutches per season

NEST: Untidy gathering of twigs, leaves , grasses, feathers and human refuse; placed in a crack or crevice in a building or other man-made structure; tree hollows or thick vegetation such as palm trees

EGGS: 3 – 6; plain pale glossy blue; tapered oval; 31 x 22 mm

INCUBATION: 17 – 18 days, by female; fledging period 22 – 24 days

DISTRIBUTION
Introduced to Australia, now resident in most cities and towns on the east coast, and spreading. Natural range southern and south-east Asia

STATUS
Introduced; locally abundant

T. Ivison

Adult Yellow Oriole at nest (H. & J. Beste)

Sponsors: Peter & Margaret Buckley

Orioles and Figbirds

Orioles and Figbirds constitute the family Oriolidae, some 25 species of which are distributed throughout Africa and Eurasia to Australasia (excluding New Zealand). Three species occur in Australia and the single species of figbird is endemic. Despite striking similarities in their plumage, the Oriolidae should not be confused with the orioles or blackbirds of the Americas (Icteridae). The two families bear little relation to each other and provide an example of inappropriate nomenclature by early naturalists in Australia.

The biology of the oriolids has not been well explored; their arboreal habits and elusive behaviour make them difficult to observe. Very few species have been studied in depth and some have yet to be fully documented. Two species are now listed as threatened: *Oriolus isabellae* of the Philippines and *O. mellanus* of China.

The name oriole is derived from the Latin *aureolus*, meaning golden; many members of the family have bright yellow or green plumage. The name was first applied to the Golden Oriole, *O. oriolus*, in which the male has spectacular, contrasting, golden yellow and black plumage. However, a few species, such as the Black Oriole, *O. hossi*, and the Maroon Oriole, *O. chlorocephalus*, have neither yellow nor green plumage.

Orioles are moderately sized, slender birds, characterised by a slender, slightly decurved bill, which is often red in colour. Long, tapered wings make their flight swift and undulating. More often heard than seen, they are renowned for their melodious, bubbling calls that are a familiar sound throughout the forests of the Old World. In his book, *Adventures Among Birds*, Hudson describes his first encounter with their call:

> ... the clear whistle of a Golden Oriole, when I listened to it in a strictly guarded wood, where it breeds annually and where I was permitted to spend a day, was more to me than the sight of towns, villages, castles, ruins and cathedrals, and more than adventures among people.

Predominantly birds of the tree tops, Orioles move singly or in pairs, feeding on fruits, berries, insects and their larvae. Birds search laboriously through the foliage of the forest canopy, gleaning insects and many other invertebrates from the surface of leaves and branches; insects are occasionally taken on the wing. Orioles show a particular fondness for caterpillars, removing the unpalatable hairs by vigorously rubbing the prey against a branch before swallowing it. Fruit is an equally important part of the diet and in some areas the species is cited as an agricultural pest. The Golden Oriole has a notorious liking for figs, eagerly devouring these soft fruits as it passes through the Mediterranean region on migration.

Since most orioles live in tropical regions where food is abundant throughout the year, they tend to be sedentary or locally nomadic, moving only in response to food sources. However, the Olive-backed Oriole of Australia shifts from the southernmost parts of its range to winter in the warmer, northern parts of the continent. The only true migrants are the Black-naped Oriole, *O. chinensis*, and the Golden Oriole which travels from its breeding grounds in Europe and western Asia to winter south of the Sahara.

Except for the Golden Oriole, courtship behaviour is poorly documented. In this species females arrive at the breeding grounds some 10 days after the males. Courtship is initiated by the female, which darts at the male, sparking off a series of bill-to-tail chases.

Within the Oriolidae, there is little variation in nesting behaviour. Females usually build most of the nest. The result is a sturdy cup-shaped structure made from grasses and strips of bark, placed among foliage and hung by the rim, in a hammock-like fashion, between two branches. Where a suitable fork is not available, the female makes one by weaving several adjacent branches together. Two to five eggs are laid and incubated by the female. Males confine their activities to collecting odd bits of nesting material and feeding the female at the nest until the naked, helpless young are hatched. Both male and female then assume equal roles in raising the brood.

Figbirds are easily distinguished from typical Orioles by a bare patch of warty skin around each eye. The bill is stout and black with a more pronounced curve. Figbirds are wholly arboreal but, unlike their congeners, are gregarious, often travelling in noisy flocks. They feed predominantly on fruits, and have a particular fondness for figs. Conviviality extends into the breeding season, when both parents contribute to building a nest that crudely resembles that of a typical oriole. Often placed in close proximity, nests are so loosely constructed that the eggs may be seen from underneath. One species occurs in Australia.

T. Ivison

Yellow Oriole

Oriolus flavicinctus

This species was first collected by Captain P.P. King near Port Essington on his circumnavigation of Australia and was described by him in 1826. John Gilbert sent two specimens to John Gould in the 1840s. In 1900 A.J. Campbell noted: "Little is really known about the bird, which has a pleasing plumage of dappled greenish-yellow, set off with orange-tinted eyes and blood-red bill". Almost a century later, there is still little known about the Yellow Oriole. The reasons for our ignorance relate to its restricted distribution, its often uninviting habitat and a degree of neglect on the part of ornithologists.

The Yellow Oriole is a typical member of the family Oriolidae. Wholly arboreal, it spends its time in the forest canopy where it feeds, sings, roosts and nests. Individuals are often elusive; at other times they seem unconcerned about human presence. In either case, the deep yellow plumage and arboreal habits make it somewhat inconspicuous, it being more often heard than seen.

It is primarily a bird of dense rainforests and mangroves. In Australia, it is confined to these habitats in the high tropics, but its range extends northwards to New Guinea and neighbouring islands. The *Atlas of Australian Birds* shows that, in Australia, there are two geographically separate populations, one ranging from the Kimberley to the Top End, the other confined to far northern

Adult at nest; Daintree, Qld; 1976 (J. Purnell)

Sponsors: D.H. & E.A.S. Pearse

Queensland. These vary slightly in colour, those of the Cape York region tending to be a richer shade of yellow. Despite this, and their disjunct distributions, they are considered to belong to the same subspecies. The Yellow Oriole is sedentary throughout its range, wandering only in response to local food sources.

The ranges of the Olive-backed Oriole and the Figbird overlap with that of the Yellow Oriole and there are many reports of these birds feeding in close association with one another. Usually found singly or in pairs, the Yellow Oriole feeds mainly on native fruits supplemented with occasional insects. The birds laboriously search the canopy, plucking and gleaning food as it is encountered. Sometimes they venture into clearings to feast from a single fruiting tree, or move further afield into orchards, parks and gardens. Towards the end of the dry season, when resources are scarce, parties of up to eight birds may be seen, drawn to a common food source.

The monotonous bubbling song of the Yellow Oriole, perhaps its most familiar aspect, haunts the tropical forests of northern Australia. It can be heard throughout the year and at any time of day, even when the temperature is at its highest. Birds call from secluded perches within the dense canopy, sometimes continuously for several hours. One such monologue has been recorded in a mnemonic form by Walter R. Petrie:

Two trips (sub-note in the distance), "Bob Chowder," "Old Bob Chunoo," "You come down one," "Jim, come down one," "I tip Chunoo," "I tip Trotol," "That chap don't know," "Up on top, Jack," "Up on top, Joe," "Jack's on top," "Shut your tongue," "I've just come out," "I've just come home," "Is that the restaraunt?" "Eel fryin'," "Some throw," "He shan't come home," "That's talking," "It's all off," "Keep on shunting," "Keep on choking," "Keep on shoving," "It's ready," "Wish you'd keep it," "Keep on jumping," "Keep on joining," "Its not so hot," "It's a beauty," "This afternoon."

The Yellow Oriole breeds throughout its range, beginning in August or September and usually ceasing by December or January. The cup-shaped nest, made from bark, grasses and other fibres, resembles that of the Olive-backed Oriole. It is suspended by the rim from a tree fork, high above the forest floor, often overhanging water. Normally, a clutch consists of two cream eggs marked with brown and lavender blotches. Young birds are a duller mustard yellow than the adults, and have heavier black markings on the underparts.

T. Ivison

Yellow Oriole

Oriolus flavicinctus

Nest, egg and hatchling; Ingham, Qld (C. & D. Frith/Frithfoto)

OTHER NAMES
Yellow-bellied Oriole, Green Mulberry, Green Oriole

MEASUREMENTS
LENGTH: 260 – 280 mm
WING: 135 – 141 mm
TAIL: 100 – 105 mm
BILL: 27 – 30 mm
TARSUS: 27 – 28 mm
WEIGHT: 85 – 97 g

DESCRIPTION
ADULT
Male: Upperparts deep mustard yellow-green, underparts brighter; head, neck, mantle and chest finely streaked black; wings dusky black, edged pale yellow; dusky green-black tail, edged yellow-green and tipped pale yellow. Eye red. Bill orange-red. Legs dark blackish grey

Female: As male but yellower with clearer streaking

IMMATURE
Duller and paler than adult; more heavily streaked above and below. Legs mid-grey. Tail as adult but terminal buff tips reduced. Bill brown

JUVENAL
Similar to immature but fluffier

IDENTIFICATION:
Medium-sized, slender bird; distinctive mustard yellow-green plumage and contrasting orange-red bill

VARIATION:
2 or 3 subspecies; one in Australia, O. f. flavicinctus.

VOICE:
Usual call a monotonous bubbling "yok-yok-yoddle, yok-yok-yoddle"; sustained soft warbling notes; some harsh alarm/aggressive notes. Mimics other birds

HABITAT:
Tropical rainforest, densely vegetated watercourses, wet eucalypt forests, mangroves, orchards, isolated fruit trees in town parks and gardens

FOOD:
Soft fruits, both native and introduced; beetles, insects and their larvae

HABITS:
Nomadic in response to food sources. Wholly arboreal. Forages almost entirely in the upper and mid canopy. Well camouflaged, more often heard than seen. Usually moves singly or in pairs; small foraging flocks in winter

BREEDING
August – January; probably one clutch per year; female builds most of nest, both sexes contribute to care of the young

NEST: A deep sturdy cup shape constructed from vine tendrils, moss and strips of bark, woven together with spider webs and lined with fine twigs and rootlets; hung by the rim amid foliage in hammock-like fashion in a small tree fork, often overhanging water; 5 – 20 m above ground

EGGS: 2 – 3; pale cream, large spots and blotches shades of brown to dusky-brown, concentrated at larger end; tapered oval; 34 x 23 mm

INCUBATION: By female, incubation period unknown

DISTRIBUTION
Coastal northern Australia and islands. Two populations, one from Broome, Western Australia, to eastern Arnhem Land, Northern Territory; the other from Normanton to Cape York and Cape York to Ingham, Queensland. Also New Guinea

STATUS:
Fairly common within restricted range

T. Ivison

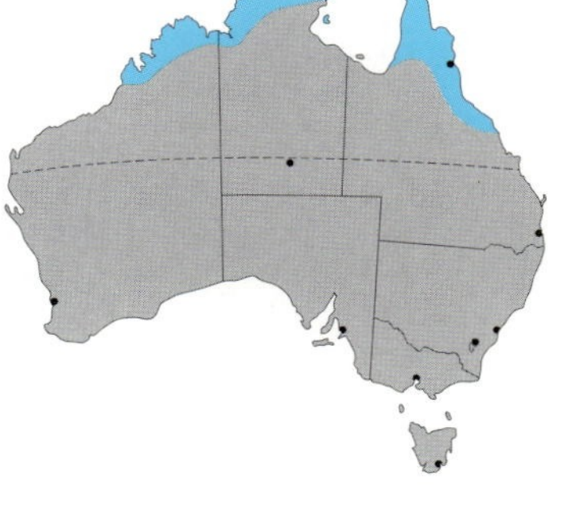

OLIVE-BACKED ORIOLE

Oriolus sagittatus

Of the three members of the Oriolidae represented in Australia, the Olive-backed Oriole has the widest distribution, entering a diverse range of habitats from rainforest and eucalypt woodlands to mallee, as well as orchards, parks and gardens. It is a medium-sized, slender bird, easily distinguished from the female Figbird by its contrasting red bill. As the common name suggests, it is predominantly olive in colour, making it somewhat dull in comparison to other members of the family. Its plumage and arboreal habits make it difficult to study.

Although fairly common throughout its range, it is more often heard than seen. Its presence is usually first revealed by a melodious bubbling, "Ori-ori-ole, ori-ori-ole," uttered persistently from a vantage point in the canopy. However, it is also an excellent mimic, second only to the Superb Lyrebird in versatility. Up to 27 calls have been recorded in its repertoire, including that of the migratory Common Koel, *Eudynamis scolopacea*, which regularly parasitises its nests, as well as some raptors and nest predators such as the Grey Butcherbird, *Cracticus torquatus*.

It also appears to indulge in another form of mimicry. Individuals (sometimes pairs) often forage for short periods among flocks of Figbirds. Their presence seems to go undetected, since the Figbirds do not chase them off, as they do other competitors such as friarbirds and other large honeyeaters. This may be because their plumage is similar to that of the female Figbird, *Sphecotheres virdis*. The Orioles remain quiet and make no attempt to mingle with the flock. No satisfactory explanation has been given for toleration of this direct competition.

Breeding occurs between September and January. The nest is typical of the family, being a deep cup constructed from bark and rootlets, slung from a thin fork in a tree, high in the forest canopy, but not always well concealed. Two to four eggs are laid and incubated entirely by the female. The young are reared by both parents. For the first couple of days they are fed by regurgitation, then on a diet of small insects, caterpillars and other larvae. Soon after fledging, the young become independent and disperse, leaving the adults to nest a second time if food is sufficiently abundant.

Usually seen singly or in pairs, but occasionally in loose companies during autumn and winter, the Olive-backed Oriole spends much time foraging in the forest canopy. It gleans insects and their larvae from the surface of leaves or branches and occasionally takes one on the wing. Native fruits and berries, including mistletoe, are procured from the outer foliage and the species is probably an important disseminator of the seeds of forest trees.

It often enters parks, gardens and soft fruit orchards to gorge on easy pickings. G.A. Keartland, in correspondence with A.J. North, recalled being called to an orchard for:

> ... what was said to be a strange bird, and described as "being all crimson and very wary" ... struck by its peculiar colour, I determined to secure it. After over an hour's chasing, I shot what proved to be one of this species with the whole of its plumage stained with mulberry juice.

In contrast to this and other records of the Olive-backed Oriole being an agricultural pest, there are numerous reports of these birds gathering at caterpillar infestations and foraging to the extent of saving entire crops from imminent devastation.

Three subspecies are currently recognised: *O. s. affinis, O. s. magnirostris* and *O. s. sagittatus*, all of which occur in Australia. The most northern population, *O. s. magnirostris*, has a slightly smaller body and proportionately larger bill than the largest eastern subspecies, *O. s. sagittatus*. Birds of the eastern subspecies occur south of about Brisbane and migrate north in autumn to join the northern populations, some even cross Torres Strait to winter in southern New Guinea. These birds return in August to commence breeding. *The Atlas of Australian Birds* states that in the Top End the reporting rate of this species was much higher in winter than in summer and that reports from south-eastern Australia confirm this by being almost nonexistent during the same period. All subspecies are locally nomadic and wander in response to food sources.

T. Ivison

Adult and chicks; Brisbane, Qld; September 1981 (R. Slater)

Sponsors: Julia, Nicholas, Samuel, Peita, Hugh and Harriet Burton Taylor

OLIVE-BACKED ORIOLE

Oriolus sagittatus

Nest and eggs; Goomboorian, Qld; October 1993 (C. Webster)

OTHER NAMES

Cedarbird, Cedar Pigeon, Greenback, Green Thrush, Oriole

MEASUREMENTS

LENGTH: *250 – 280 mm*
WING: *c 152 mm*
TAIL: *c 103 mm*
BILL: *c 29 mm*
TARSUS: *c 27 mm*
WEIGHT: *95 – 104 g*

DESCRIPTION

ADULT

Male: Upperparts dull olive, streaked black; underparts streaked dark; chin dusky olive green merging to creamy white on breast; wings dusky and edged buff; tail dusky, tipped white. Eye red. Bill orange-red. Legs dark blackish grey

Female: Slightly smaller; clearly streaked above and coarsely streaked below

IMMATURE

Similar to juvenile but plumage sleek; streaks not as heavy, more clearly defined on breast; greenish tinge to eyebrow

JUVENAL

Plumage fluffy; mantle dull green-brown; breast and mantle broadly streaked black; wings brown; wing coverts edged rusty-buff; buff eyebrow. Eye and bill brown

IDENTIFICATION

Medium-sized, slender, olive-green bird with red bill

VARIATION

Three subspecies, all represented in Australia: O. s. magnirostris *in Cape York region;* O. s. affinis *in Northern Territory and west of Norman and Leichhardt Rivers in Queensland;* O. s. sagittatus *in the south*

VOICE

Highly characteristic bubbling "ori-ori-ole, ori-ori-ole"; also scratchy warbles and mimicry

HABITAT

Rainforest, eucalypt woodland, tall scrubs, orchards and parks and gardens with fruit-bearing trees

FOOD

Soft fruits, both native and introduced; beetles, insects and their larvae

HABITS

Nomadic in response to food sources; partly migratory, shifting north in autumn and south again in summer to breed. Diurnal. Wholly arboreal. Forages almost entirely in the upper canopy. Unobtrusive, more often heard than seen

BREEDING

August – January; usually one, occasionally two clutches per year; the female builds most of the nest; both sexes contribute to care of the young

NEST: A deep sturdy cup shape constructed from grasses, moss and strips of bark, woven together with spider webs and lined with fine grasses, hair and sometimes wool; hung by the rim amid foliage in a hammock-like fashion in a small tree fork; 5 – 20 m above ground

EGGS: 2 – 5; cream, spotted and blotched shades of brown to red-brown, concentrated at larger end; tapered oval; 32 x 22 mm

INCUBATION: 17 – 18 days, by female; fledging period 15 – 17 days

DISTRIBUTION

Northern and eastern Australia from south of Broome, Western Australia, across the Top End; Cape York continuing south to south-east of South Australia, inland to about 147°; absent from Tasmania; resident in New Guinea

STATUS

Common nomad in north, regular summer migrant in south

T. Ivison

Figbird

Sphecotheres viridis

The Figbird differs from typical orioles in being very social. It lives in noisy flocks of 20 – 50 individuals; larger flocks may gather at a particularly rich food source. It is largely frugivorous and, as the common name implies, has a particular fondness for figs, both wild and cultivated. Figbirds are common in urban parks where there are mature fig trees, as in Hyde Park in the heart of Sydney. In his book, *Birds of Sydney* Ern Hoskin reports that:

> Since 1946 when the Figbird was discovered in the Royal Botanic Gardens, it has spread rapidly and numbers may be seen in many parts of North Sydney, through to Balmoral and the Eastern Suburbs and city parks. Centennial Park has a breeding population. Its range has extended to Ulladulla on the south coast.

Wild fruits and berries, such as *Pittosporum,* are also taken, along with fruits of introduced plants such as Tobacco Bush and Lantana, and cultivated fruits such as bananas and mangoes. As with most frugivorous species, the Figbird supplements its diet with insects and other invertebrates, which provide essential protein. Flocks converge on isolated trees in parks and gardens, feeding until the tree is virtually stripped of fruit. Figbirds are highly acrobatic and will hang upside down from small branchlets, stretching to procure the outermost fruits.

While feeding, they chatter constantly in parrot-like tones. The song is a low "tu-tu-heer, tu-tu-heer." Figbirds also mimic the calls of other common species (though in whispered tones) interspersed with their own squeaky notes.

A peculiar habit is to take a bath during a rainstorm by swinging from an outer branch. This does not appear to be upsetting for them. As J. N. Yates noted: "Although thoroughly soaked, they did not stop to preen afterwards, but resumed feeding".

The plumage of the Figbird shows distinct sexual dimorphism. The female is similar in appearance to an immature oriole, but is generally browner, with heavier streaking on the chest, and may be readily distinguished from orioles by a bare patch of warty skin around the eye. Males have an olive-green back, a glossy black head with buff-coloured facial skin around the eye; this reddens when the bird becomes excited. Male plumage varies with locality, and in the past this has led to much confusion. All Figbirds are currently considered to be included in one species that encompasses six subspecies, two of which occur in Australia.

The northern form, *S. v. flaviventris*, has bright yellow underparts and the southern form, *S. v. vieilloti*, has grey underparts. These were previously thought to be separate species because of the differences in colour of the males, however the females of all subspecies are similar. The ranges of the two Australian forms overlap in south-eastern Queensland, creating a broad hybrid zone around Townsville. Male hybrids have varying combinations of yellow and grey-olive underparts and most resemble *S. v. salvadorii*, the New Guinean subspecies, having a grey throat combined with bright yellow underparts.

As the breeding season commences in October, the buff-coloured facial skin of the male changes to a bright red. To attract a mate, males seek a prominent perch from which to sing. As a female approaches, a short chase ensues, followed by copulation. Both parents construct the nest which is a loose construction resembling that of orioles, but shallow and unlined, allowing the eggs to be seen from below. Two to three eggs are laid and incubated by both parents. Once the young hatch, the parents are often assisted by a third and sometimes a fourth bird acting as "helpers". These do not necessarily have any relationship to the parents but are young individuals or unsuccessful breeders. The hatchlings, which are naked and helpless, are fed on whole fruits regurgitated by the parents or helpers and held momentarily in the bill before being passed to the young.

Because young birds are particularly confiding, banders have found a novel way to catch them. This involves presenting the young bird with one end of a long stick. The bird will walk onto this and appears not to be disturbed by the movement as the stick is lowered. Birds are returned to the nest in the same manner.

Once young are independent, they disperse to form large flocks that roam widely. Smaller flocks tend to consist of more mature birds and remain sedentary.

T. Ivison

Adult male; Northern race; Cairns, Qld; August 1988 (D. Whitford)

Sponsor: Gould League of N.S.W.

Figbird

Sphecotheres viridis

Adult male; Southern race; Northern N.S.W.; 1989 (J. Purnell)

OTHER NAMES
Green, Grey-breasted, Northern, Southern and Yellow Figbird, Banana-bird, Mulberry-bird

MEASUREMENTS
LENGTH: 280 – 290 mm
WING: 150 – 163 mm
TAIL: 102 – 115 mm
BILL: 22 mm
TARSUS: 25 – 26 mm
WEIGHT: 114 – 145 g

DESCRIPTION
ADULT
Male: All have black head with with reddish facial skin around eye; neck and mantle grey-green merging to olive green on back. Scapulars and wing coverts olive green; flight feathers black, edged grey-green; tail black, outer tail feathers broadly tipped white. Eye dark with reddish tinge. Bill black. Legs pinkish buff

Southern subspecies, S. v. vieilloti*: chin to upper breast mid-grey; lower breast and flanks olive-green; abdomen and undertail coverts white*

Northern subspecies S. v. flaviventris*: underparts, chin to lower breast bright yellow; flanks, abdomen and undertail coverts white*

Hybrids have varying proportions of grey, green and yellow underparts. Eye dark with reddish tinge. Bill black. Legs pinkish buff

Female: Both races similar; upperparts brown, washed olive green and streaked dark brown; underparts creamy white with broad, brown streaks; wings brown; wing coverts edged buff; tail green-brown; outer tail feathers narrowly tipped paler; facial skin around eye mid grey. Eye and bill as male. Legs grey

IMMATURE
Head and neck olive-brown, not mottled; streaks on underparts more sharply defined; secondary coverts off white; skin around eye dark brown. Bill dark brown

JUVENAL
Head and neck brown mottled buff; underparts white with dark brown streaks; secondary coverts edged buff-olive; smaller patch of skin around eye, dark brown. Bill dark brown

IDENTIFICATION
Medium-sized, robust bird; males distinctive; females and immatures may be separated from Olive-backed Orioles by facial skin around eye and black bill

VARIATION
Six subspecies, two in Australia: northern S. v. flaviventris *and southern* S. v. vieilloti, *and a hybrid zone where the two races meet*

VOICE
A variety of harsh squeaky notes; also an emphatic down-slurred whistle "tu-tu-heer, tu-tu-heer". Also mimicry

HABITAT
Rainforest edges, mangroves, eucalypt woodland and urban parks and gardens where food is available

FOOD
Mostly fruit, native and introduced: figs, native cherries, lantana, tobacco bush, cultivated fruits such as bananas and paw-paw; also beetles, insects and their larvae

HABITS
Extremely gregarious; colonial, live in groups of 20 – 50; locally nomadic in response to food supplies. Diurnal

BREEDING
September – January; one or two clutches per year; both sexes contribute to nest building, incubation and care of young; a pair may often have an unrelated "helper" at the nest

NEST: A loose saucer constructed from fine twigs and vine tendrils, fixed by rim in a hammock-like fashion to a tree fork; usually amongst foliage on outer branches; 2 – 15 m above ground; often placed in close proximity to one another

EGGS: usually 3, sometimes 2; pale, muted green, slightly glossy with sparse brown-reddish brown blotches and spots; oval; 32 x 24 mm

INCUBATION: 21 days; fledging period 14 days

DISTRIBUTION
Coastal northern and north-eastern Australia; S. v. flaviventris *Cape York region south to about Cardwell, Queensland; also an isolated population in the Top End and Kimberley Range, Western Australia;* S. v. vieilloti *from about Proserpine, Queensland, south to Illawarra district, New South Wales; hybrid zone around Townsville, Queensland*

STATUS
Common

T. Ivison

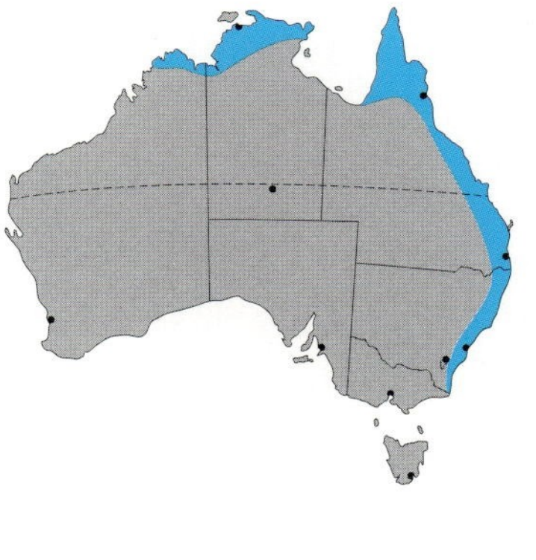

Drongos

Noted for their aggressiveness and acrobatic manoeuvrability, drongos comprise a distinctive family of forest birds, the Dicruridae, with 19 species in Africa, South Asia and one in Australia. All but one species are placed in the genus *Dicrurus*, the exception being a New Guinean species, *Chaetorhynchus papuensis.*

In Australian slang, "drongo" means a fool, but this may relate to the name of an ill-famed racehorse of the early 1920s, which never won a race. Contrary to popular opinion, drongo is not a name that originated in Australia; it is a vernacular name for one of the Northern Madagascan members of the family.

Drongos are medium-sized birds with long tails (approximately the same length as the body) and are generally characterised by iridescent black plumage which imparts a blue, green or purple sheen. Many appear to fly erratically but this movement is really the close and expert aerial pursuit of insect prey. Drongos are also notable for their loud calls and aggressive displays towards raptors and other enemies, real or perceived. Although the tail varies throughout the family, it is mostly forked (the name *Dicrurus* stemming from the Greek "dicroos" meaning forked) and serves as a crucial means of identifying species. Members of the genus *Dicrurus* are unusual among passerines in having only five pairs of tail feathers (retrices). In the racquet-tailed drongos of Asia, barbs are missing along most of the length of the outer tail feathers, resulting in a club-shaped tip. When in flight after prey, the tail "racquets" move in spectacular circular or semi-circular patterns.

Like other aerial feeders, drongos have a large gape and a strong growth of rictal bristles around the base of the bill. The wings are long and swallow-like; employed in conjunction with a long and usually deep-forked tail, they enable them to fly with swiftness and great agility, making them masters of the air.

There is much disagreement about the evolutionary affinities of these birds. Recent studies suggest that they may represent a highly evolved branch of the flycatcher stock, but some authors have suggested a closer link with the orioles, while others relate them to the birds of paradise. Recent comparisons of DNA suggest that there may be a closer link between the drongos, flycatchers and fantails than previously thought, but in this work we retain the traditional classification.

Questions of relationships also turn on the number of tail feathers. While most passerine birds

Adult Spangled Drongo; Clear Mountain, Qld; August 1994 (T. & P. Gardner)

Sponsors: C.S.L. & M.J. Ryman

have six pairs, *Dicrurus* has five, but *Chaetorhynchus* has six. This New Guinean species is widely considered to be related to the Old World flycatchers and it has been referred to as the Bristle-Monarch. It is here regarded as a primitive drongo.

Another feature used in the identification of drongos is the crest, but this can lead to confusion. The Ceylon Crested Drongo, *D. paradiseus lophorhinus*, has a small toothbrush-like crest and an elaborate tail, while the Ceylon Racquet-tailed Drongo, *D. paradiseus ceylonensis*, has a large curved crest and well developed tail racquets. The Ceylon Crested Drongo lives in wet forest areas, whereas the Ceylon Racquet-tailed Drongo lives in dry forest areas. The Spangled Drongo, *Dicrurus hottentotus,* also bears an elaborate crest, but this varies from population to population, being hardly apparent in the Australian population, and has earned the bird the name of Hair-crested Drongo in some areas.

On the whole, drongos are rather noisy, but several species are also excellent mimics. The Greater Racquet-tailed Drongo, *Dicrurus paradiseus*, is particularly skilled, being able to reproduce, among others, the calls of the Grey Jungle-fowl, Cuckoo, Grey Hornbill, Golden Oriole and Crested Serpent Eagle, each being uttered in response to the original caller. As well as vocal mimicry, the Greater Racquet-tailed Drongo has also been observed to mimic the visual display of the Jungle Babbler.

It seems that the primitive (less specialised) drongos are generally smaller and less glossy than other species and have a shallow-forked tail. On these grounds the three African species are regarded as primitive. Two of these, the Shining Drongo, *D. atripennis*, and the Square-tailed Drongo, *D. ludwigii*, are the smallest members of the family. The Asian species are, on the whole, more advanced and include the larger members. Increased body size is approximately correlated with greater tail-length and degree of forking. Another evolutionary trend is a reduction of black colouration and a loss of gloss, but this tends to be more noticeable in birds living in arid environments. For example the Ashy Drongo, *D. leucophaeus*, from India is dark over most of its range but further north in China is a uniform pale grey.

All drongos share similar breeding and nesting habits. The nest is a simple platform of twigs and vine tendrils. It is unlined and generally placed on a horizontal fork, towards the outer foliage of a tree. The nest is vigorously defended by both adults which readily attack humans and the much larger birds of prey, which they can outmanoeuvre with great ease. Their aggressiveness and eager defence of a nesting site encourages less stout-hearted species to nest in the same tree, with the drongos serving as surprisingly tolerant protectors of all.

P. Rowland

Spangled Drongo

Dicrurus hottentotus

Also found in India, China, the Philippines, south-east Asia, Indonesia and New Guinea, the Spangled Drongo is the only representative of the Dicruridae in Australia. This striking bird is common throughout eastern Australia from Cape York Peninsula to north-eastern Victoria and, in northern Australia, from the Kimberley to Arnhem Land.

Its distinctive "fish-like" forked tail and iridescent plumage distinguish the Spangled Drongo from other Australian birds. Sitting conspicuously on a branch or wire, it awaits a suitable prey to fly past, then pursues it in an acrobatic and twisting flight, returning to its perch to eat its victim.

Its presence is frequently betrayed by loud calls, including strange chattering and tearing sounds, and a curious creaking, metallic whistle which has been likened to the twanging of taut fencing wire. Occasionally it mimics the calls of such other birds as the Noisy Miner, *Manorina melanocephala*; Blue-faced Honeyeater, *Entomyzon cyanotis*; Little Wattlebird, *Anthochaera lunulata*; Pied Butcherbird, *Cracticus nigrogularis*; and Australian Magpie, *Gymnorhina tibicen*. Alec Chisholm states:

> I have heard a Drongo imitate cleverly the cry of a Goshawk, and a resident of tropical Queensland actually has credited the bird with using the call of a Butcherbird in order to cause a Magpie-lark to drop the food it was carrying.

Like other drongos, it is bold and aggressive. In his account of the "bird of many moods", E.J. Banfield describes it as a "bully, swaggerer, swashbuckler":

> He has a scolding tongue, and if a hawk hovers over the bloodwoods he tells without hesitation of the evil presence. He is the bully of the wilderness of leaves, bouncing birds vastly his superior in fighting weight and alertness of wing, and chattering his jurisdiction to everything that flies ... The drongo is a bird of many moods. Silent and inert for months together, during the nesting season he is noisy and alert, not only the first to give warning of the presence of a falcon, but the boldest in chivvying from tree to tree this universal enemy... He is then particularly partial to an acrobatic performance, unsurpassed for gracefulness and skill, and significant of the joy of life and liberty and the delirious passion of the moment.

The Spangled Drongo is usually seen singly or in pairs, favouring non-rainforest habitats, such as mangroves and open woodland, but tends to show a preference for lowland rainforest fringes, avoiding the dense interior under the canopy.

In the northern parts of the range breeding occurs from September to March, after which follows an interesting pattern of behaviour. Birds from northern Western Australia and the Northern Territory migrate to Indonesia while those from eastern Australia migrate to New Guinea. However, some birds remain in the same area through the winter and some even move south, turning up – on rare occasions – as far south as Tasmania.

The diet consists mainly of insects plus a little nectar and fruit; it is also known to take small birds in flight. When bushfires break out, Drongos, along with other birds, can be seen sitting nearby, cautiously awaiting insects that are fleeing the flames. It takes most prey on the wing but, in the breeding season, often rips bark from trees in search of insects to feed its nestlings.

It can be a kleptoparasite, often chasing such insectivorous birds as cuckooshrikes, kingfishers, the Magpie Lark, *Grallina cyanoleuca*, and Willie Wagtail, *Rhipidura leucophrys*, in an attempt to make them give up their catch. There is much variation in this behaviour.

In his detailed but somewhat anthropomorphic account of the natural history of the Spangled Drongo, Banfield waxed lyrical on the search for food in the breeding season:

> When the nest on the nethermost branch of the Moreton Bay ash is packed with a hungry brood, his industry is exhilarating. Ordinarily he gets all the food he wants by merely a superficial inspection; but with a family to provide for, he is compelled to fly around, shrewdly examining every likely looking locality. Clinging to the bark of the bloodwood, with tail spread out fan-wise as additional support, he searches every interstice, and ever and anon flies to the Moreton Bay ash and tears off the curling fragments of crisp bark which afford concealment to the smaller beetles, grubs and spiders.
>
> With the loose end of bark in his bill, tugging and fluttering, using his tail as a lever with the tree as a fulcrum, and objurgating in unseemly tones as the bark resists his efforts, the drongo assists the Moreton Bay ash in discarding worn-out epidermis, and the tree reciprocates by offering safe nesting places on its most brittle branches.

It is not shy around humans, often appearing in and around gardens and picnic grounds and allowing reasonably close approach. However, it readily attacks any potential enemy that approaches too close to its nest. Drongos have even been reported to attack snakes that slither past the base of a nest tree. Yet its defensive fervour is not always completely effective, for its nests are sometimes parasitised by the Pallid Cuckoo, *Cuculus pallidus*, or Koel, *Eudynamis scolopacea.*

In 1908, Banfield commented upon the unusual display flight of the Spangled Drongo. This was not reported upon by any other observer until a publication by Longmore and Scoullar:

> On 28 October 1988, at 17:45 hrs, a single bird was heard giving a distinctive song, consisting of five or six single notes uttered in short series. After a brief interlude, the songster appeared, flying over the roof of a neighbouring cottage. It moved above roof level (10 – 14 m) in a westerly direction. Both wings were raised about 45 degrees above the body, which was held almost vertically. The flight ceased its forward momentum, the wings were continued to be held in an outstretched position above the body, which was now dropping forward. With the tail arched well over the body and its head pointing earthwards, the bird commenced to drop vertically and with some rapidity. The calling ceased when the bird was within 3 metres of the ground. The bird then flew off with its normal pattern of flight. After alighting in a nearby tree for less than a minute, it flew away. The display appeared to last only a few seconds and it was within 100 metres of its nesting trees.

This spectacular acrobatic display demonstrated a very successful dual use of the tail. Whether the tail of drongos first evolved as a device for agile flight or for attraction of a mate, it appears to serve both purposes admirably.

P. Rowland

SPANGLED DRONGO

Dicrurus hottentotus

Adult at nest; Cooktown, Qld; 1976 (J. Purnell)

OTHER NAMES
Drongo, Fish-tail, Hair-crested Drongo, Drongo-shrike, King-crow

MEASUREMENTS
LENGTH: *280 – 330 mm (including tail)*
WING: *140 – 169 mm*
TAIL: *107 – 140 mm*
BILL: *27 – 36.3 mm*
TARSUS: *23.2 – 29.5 mm*
WEIGHT: *80 – 107 g*

DESCRIPTION
ADULT
General plumage glossy black, with iridescent blue-green spots (spangles) on head, neck and breast; metallic blue-green reflections from wings and tail. Tail long and curved outwards. Eyes dark red. Bill, legs and feet black. Sexes similar but female slightly smaller. White spotting occasional on scapulars of both sexes

IMMATURE
Head and nape dull black, with iridescent blue spangles; back, rump and wings dull glossy-blue; sides of head and underparts dull black; tail black, slightly glossed with metallic steel-green. Eye brown

JUVENAL
Similar to immature, but plumage blackish brown above and below. Tail less glossy. White tips to undertail coverts and mottled white patches on rest of underparts

IDENTIFICATION
Rather large, conspicuous bird. General plumage black, with iridescent blue-green spotting and overall green sheen. Long flaring forked tail. Bright red eye

VARIATION
Negligible. Some loss of black pigment in plumage of birds from more arid areas (not evident in Australia). Variation in extent of crest formation evident, especially in various island populations

VOICE
Song is a metallic creaking whistle; occasional mimicry. Calls include harsh chattering or cackling, unusual "tearing" sounds

HABITAT
Lowland rainforest fringes, open wooded areas, mangroves, urban parks and gardens

FOOD
Predominantly insects taken on the wing; some fruits and nectar; small birds sometimes taken in flight

HABITS
Sedentary and migratory, northern populations tending to be more migratory. Dispersal in winter is to north and south. Diurnal. Not gregarious, usually alone or in pairs, but small winter flocks have been observed. Arboreal, favouring higher levels. Takes prey in flight. Bold and aggressive. Not shy of humans, allowing close approach

BREEDING
Septembe – March; usually only one clutch per season; both members of pair participate in nest building, incubation and care of young; aggressive defenders of nest

NEST: Simple shallow cup of twigs, vine tendrils and grasses held together with spider web; placed in horizontal fork of tree, toward outer edge, 10 – 20 m above ground

EGGS: 3 – 5; pinkish white or purple-grey, spotted and streaked with reddish-brown or purple; 29 x 21 mm

INCUBATION: Unrecorded; fledging period unrecorded

DISTRIBUTION
Range extends through India, China, the Philippines, south-east Asia, Indonesia and New Guinea. Many island populations. Coastal and near coastal areas of eastern Australia, from Cape York Peninsula to Northern Victoria. Uncommon visitor to Tasmania and South Australia. Northern population from Kimberley to Arnhem Land

STATUS
More common in northern parts of range

P. Rowland

Bowerbirds

The behaviour of male bowerbirds (family Ptilonorhynchidae) is unique in that they build elaborate structures, called bowers, and decorate them with coloured ornaments. From these bowers they display to females. There are nineteen species of bowerbirds; nine of these species are restricted to New Guinea, and eight to Australia. The other two species the Fawn-breasted Bowerbird, *Chlamydera cerviniventris*, and the Spotted Catbird, *Ailuroedus melanotis* occur in both regions. Most bowerbirds inhabit wet forests, but the five species of *Chlamydera* inhabit drier vegetation and two Australian species of the genus are distributed widely throughout inland Australia. In sixteen species of the family, males are polygamous, mostly brightly coloured and courting females at bowers. The cryptically-plumaged females perform all parental care.

There are four basic types of bower: stage or court; mat; avenue and maypole. The male Tooth-billed Bowerbird, *Scenopoeetes dentirostris*, clears an oval or circular display court on the forest floor and decorates it with fresh leaves, the paler undersurfaces turned uppermost.

Archbold's Bowerbird, *Archboldia papuensis*, of the New Guinea cloud-forests, makes a mat of dried ferns on the forest floor and decorates it with heaps of brown snail shells, beetles and blue fruits, draping *Dendrobium* orchid stems on branches over the display area.

Nine species of polygamous bowerbirds of the genera *Chlamydera*, *Sericulus* and *Ptilonorhynchus* build avenue-bowers consisting of two vertical, parallel walls of twigs or grasses erected on a stick platform. The Satin Bowerbird, *Ptilonorhynchus violaceus*, occasionally adds a third wall parallel to the two avenue walls. The Yellow-breasted Bowerbird, *Chalmydera lauterbachi*, of New Guinea builds a unique four-walled bower by adding a short wall at each end of the avenue. Decorations consist of hundreds of greyish stones and some red and greenish berries.

Five bowerbird species build elaborate and intricate maypole bowers around one or more saplings or tree ferns. The exquisite Golden Bowerbird, *Prionodura newtoniana*, Australia's only maypole-builder, constructs a massive bower up to three metres high, consisting of two stick-towers piled up around two saplings with a horizontal display perch between them. The bower is decorated with lichens and flowers. The other four maypole-builders, members of the genus *Amblyornis*, from New Guinea, are known as gardener bowerbirds because their bower platforms consist of a moss lawn decorated with fresh leaves, flowers, fruit, beetles and fungi. These bowers range in complexity from a single cone of sticks around a sapling or tree fern arising from a saucer-shaped moss platform (two species) to elaborate architectural structures resembling roofed huts with a side entrance (two species). The Vogelkop Gardener Bowerbird, *A. inornatus*, constructs a gigantic hut, nearly two metres wide and one to one and a half metres high.

Adult male Golden Bowerbird at bower; Paluma, Qld (C. & D. Frith/Frithfoto)

Sponsors: John & Anne Lucas

Until recently, the Yellow-fronted Gardener Bowerbird, *A. flavifrons*, was known only from three male museum specimens. The home, the bower and the female of this long-lost bowerbird was discovered in 1979 in western New Guinea by an American biologist, Jared Diamond.

Decorations at bowers consist of natural ornaments such as feathers, fruits, flowers, leaves, insects, fungi, bones and pebbles, but may also include human artifacts. The preferred colour of such decorations varies between species. Thus, the Satin Bowerbird mostly collects blue and yellow objects; the Fawn-breasted Bowerbird green; and the Western Spotted and Great Bowerbird mostly pale objects. The mass of decorations at the bower of a Great Bowerbird may amount to 6–12 kilograms.

Another aspect of the behaviour of male bowerbirds is bower-painting. In at least seven of the nine species of avenue-builders, males paint inside bower walls with a mixture of masticated plant material, charcoal and saliva. Some populations of the Vogelkop Gardener Bowerbird also paint their maypole bowers.

In polygamous bowerbirds, the female builds the nest, incubates the eggs and rears the young without assistance. In contrast, the three species of *Ailuroedus* (catbirds) have sexually monomorphic plumage, form monogamous breeding pairs, defend joint territories and build neither bowers nor courts. In the Green Catbird, *A. crassirostris*, and Spotted Catbird, *A. melanotis*, the female of a pair builds the nest and incubates the eggs, while both sexes feed the young. Very little is known of the nesting behaviour of the White-throated Catbird, *A. buccoides*, of New Guinea but, presumably, it is similar to that of the other two catbird species.

Bowerbirds feed predominantly on fruit but also eat leaves, flowers, seeds and invertebrates. They forage at all levels in the vegetation from ground level to canopy.

The relationship between birds of paradise and bowerbirds (and other birds) continues to be debated. For many years it was accepted that the two groups were very closely related and it has been suggested that they are most closely related to the Australian scrub-birds and lyrebirds. However, recent enzyme and molecular data suggest that bowerbirds are more closely related to the corvine (crow) assemblage than to birds of paradise, neither being affiliated to lyrebirds and scrub-birds. Further study is needed to clarify the situation.

The habitat of wet-forest bowerbirds in Australia is secure but some species in New Guinea, especially those with restricted ranges, are threatened by loss of habitat. The range of the Spotted Bowerbird is contracting and it is threatened by extensive clearing of native woodlands in inland New South Wales and Queensland.

R. Donaghey

Fledgling Golden Bowerbird at nest; Paluma, Qld (C. & D. Frith/Frithfoto)

Sponsor: Margaret J. Mashford

Golden Bowerbird

Prionodura newtoniana

The endemic Golden Bowerbird is restricted to north-eastern Queensland in upland wet tropical rainforest. It is Australia's smallest bowerbird but builds the largest bower, a massive "maypole" structure. It was the last of the Australian bowerbirds to be discovered.

The exquisite golden plumage of the adult male contrasts so sharply with the cryptic grey plumage of the female that one zoologist described them as separate species. The first specimen, a bird in female plumage, was discovered by Kendall Broadbent in rainforest near the head of the Tully River in September 1882. The following year, C. W. De Vis, Curator of the Queensland Museum, named it as *Prionodura newtoniana* – *Prionodura* meaning "saw tail" and referring to the partly forked tail, and the specific name commemorating Alfred Newton, an eminent English ornithologist. In 1889, a golden and olive bird was shot by Archibald Meston in the Bellenden Ker Range and named *Corymbicola mestoni* by De Vis. However, a few weeks later, De Vis received more birds collected by Broadbent in the Bellenden Kers together with field observations that the golden and grey birds were different sexes of the same species. Furthermore, they had been seen at bowers and were therefore bowerbirds.

Twenty years elapsed before the nest and eggs were discovered in November 1908 by George Sharp in the Evelyn Tableland rainforest near Atherton. Surprisingly, nests were found in tree crevices that were often close to the ground. In contrast to other Australian bowerbirds, but like the New Guinea gardener bowerbirds, *Amblyornis*, the one to two eggs are creamy white to pure white. In an important monograph on bowerbirds, the Australian zoologist Jock Marshall concluded that on the basis of its bower structure and plain white eggs, *Prionodura*, Australia's only maypole builder, was related most closely to the gardener bowerbirds.

The massive bower, between one and three metres high, often consists of two pyramids of sticks heaped up around the trunks of two vertical saplings about a metre apart. The twin towers of sticks are bridged by a horizontal vine or branch from near ground level up to two metres above ground and together form a massive "U" shape. As first noted by Norman Chaffer, the sticks of the bower are fused together by the growth of a fungus (not by saliva, as suggested by some earlier observers). The centre of the horizontal display perch is kept bare, but the male festoons both ends of it with specific decorations. These include grey-green beard lichen, fresh flowers such as orchids, and the dried seed capsules of *Melicope*. Bowers are built on sloping hillsides throughout the forest. In 1953 Norman Chaffer obtained the first colour photographs of the male at his bower:

> We had only been at the bower for a few minutes when a flash of golden colour heralded the approach of a respondent male. In his bill he held a "bundle" of lichen. Moving back a few yards from the bower, we eagerly awaited developments. Before long the gorgeous creature swooped suddenly to the bower and alighted on the display-perch. And what a wonderful picture he made as momentarily he posed on this prominent position! The whole of the undersurface and the crown and nape were a rich shining gold. Portion of the head, the wings, back and upper surface of the central tail-feathers were olive-brown suffused with a golden sheen. Golden too, was the colour of the iris.

Female at nest; Paluma, Qld (C. & D. Frith/Frithfoto)

Sponsor: Hal Wootten

> As the bird moved around, stray shafts of sunlight glanced on his shining coat and the effect against the sombre shadows of the jungle was electrifying. This was enhanced by his alert and sprightly manner. At times the feathers forming the golden cap were slightly elevated although hardly forming a crest.
>
> Placing the lichen on the side of the display-perch against the main wall, the bird hopped down in front of the bower and, picking up a fallen piece of lichen, returned to his former position. After placing it alongside the first piece, he flew swiftly off as suddenly as he arrived.

Chaffer observed males arranging decorations and recorded a spectacular hovering display at the bower in which a bower-owning male hovers in mid-air with fast-beating wings, body held vertical, head thrust forward, spreading and closing the tail quickly and repeatedly. Fanning the tail wide open exposes the golden yellow outer tail feathers.

Observations made during more than 10 years by Cliff and Dawn Frith of a banded population at Paluma have given new insights into the behaviour and ecology of the Golden Bowerbird. In addition to the twin stick pyramids connected by a perch, they have seen many bowers that consist of a single conical tower, or an irregular mass of sticks, with the bower perch protruding from one side. Bowers owned by individual colour-banded birds have remained the same size and shape for 5 – 10 years.

The Friths confirmed that individual males own, maintain and defend one main bower. However, in hundreds of hours of observation, they have not seen mating at a bower. So, in contrast to the avenue bower-builders, it is possible that mating occurs away from the structure. Like other true bowerbirds, males steal bower decorations from one another. Such theft may enable dominant males to accumulate more decorations and thereby attract more females to the bower.

During the breeding season, a male spends much of the day perched high above its bower. From this high perch, it gives a loud pulsating rattle that serves to advertise the bower and its owner to females and to rival males. The Golden Bowerbird mimics the call of other rainforest birds and the sounds of the forest. A forester noted in 1951 that it reproduced the sounds of a crosscut saw, an axe stroke and a tree-puller. A year later, Alec Chisholm heard a male at its bower utter a scolding note then mimic the voice of the Pied Currawong and the melodious dulcet voice of Bower's Shrike-thrush. It may produce a rapid burst of imitations of a number of other species but Cliff Frith heard it give a perfect imitation of the raucous call of the Sulphur-crested Cockatoo in immediate reply to the calls of a number of these birds as they flew overhead.

An exciting discovery by the Friths was that bower-owning males store bunches of ripe fruits close to the bower. They also saw the males fly from their bowers to eat cached fruits and return to their bowers. This behaviour permits males to maximise the time spent guarding bowers from rival males and waiting for females to visit. Melinda and Stephen Pruett-Jones observed food storage by Macgregor's Bowerbird, a New Guinean maypole-builder.

Little has been published about the nesting habits of the Golden Bowerbird. Observations by the Friths of colour-banded birds at nests have confirmed conclusively that the female builds the nest, incubates the eggs, and raises the young to independence without any help from the male. The nestling diet consists of fruit and insects.

R. Donaghey

Golden Bowerbird

Prionodura newtoniana

Young in nest; Mt Lewis, Qld; December 1989 (D. Whitford)

OTHER NAMES
Golden Gardener, Queensland Gardener, Newton's Bowerbird

MEASUREMENTS
LENGTH: *230 – 250 mm*
WING: *122 – 130 mm*
TAIL: *110 mm*
BILL: *18 – 22 mm*
TARSUS: *30 – 32 mm*
WEIGHT: *Not recorded*

DESCRIPTION
ADULT
Male: Central crown, including short erectile crest, and hindneck glossy golden yellow; rest of upperparts, head, chin and sides of neck yellowish olive; back, wings and central pair of tail feathers olive-brown; underparts and outer tail feathers rich golden yellow. Bill pale olive-brown. Iris yellow. Legs brownish black

Female: Upperparts olive-brown; sides of head and wings dusky brown; tail brown; underparts ashy grey. Iris dull yellow

IMMATURE
Similar to adult female but iris brown. Young males acquiring adult plumage become washed with yellow above and below

JUVENAL
Similar to immature female but more grey on upperparts; head downy grey; sides of head have bare patches and the gape is yellow

IDENTIFICATION
Adult male unmistakable; thrush-sized gold and golden-olive brown bird. Small bill and yellow eye distinguish the brown and grey female from female-plumaged whistlers

VARIATION
No geographical variation but much variation in shade of yellow in individual adult males

VOICE
Rattles, croaks and mimicry

HABITAT
Wet tropical upland rainforest 700 m above sea level and over

FOOD
Predominantly fruit; also insects (beetles and cicadas) and spiders

HABITS
Sedentary. Some dispersal (possibly immature birds) to 600 m in non-breeding season. Arboreal; forages from ground to canopy; difficult to detect amongst rainforest foliage, particularly in non-breeding season. Mostly occurs singly or in small groups. Males and females have overlapping foraging ranges

BREEDING
Delayed sexual maturity in males. Polygynous mating system. Males maintain "maypole" bowers dispersed in forest interior. Females perform all parental care. Nests October – February. One brood per season

NEST: A deep cup of twigs, bark and dry leaves bound with tendrils, lined with fine twigs; built in a tree-trunk crevice or between two vertical tree buttresses, mostly 1 – 3 m above ground

EGGS: Usually two, sometimes one; oval shaped; uniformly pale cream to white; 35 x 26 mm

INCUBATION: 20 – 21 days; fledging period 19 – 20 days

DISTRIBUTION
Patchily distributed in upland rainforest in north-eastern Queensland just south of Cooktown to the Paluma Range north of Townsville; mostly 900 – 1500 m

STATUS
Locally fairly common. Resident. Endemic

R. Donaghey

Satin Bowerbird

Ptilonorhynchus violaceus

The glossy blue-black male Satin Bowerbird is the best known and most photographed member of its family, probably because it lives near coastal cities. The male builds an avenue bower of sticks on the ground and decorates it with ornaments that are coloured blue, violet, yellow or brown. Early naturalists regarded these structures as "playgrounds" and endowed the species with an aesthetic sense to such an extent that when John Gould saw the birds in 1838 he regarded the bowers "as the most wonderful instances of bird-architecture yet discovered". He supposed, correctly, that bowers were used by males to attract females for display and mating.

The first known specimen of the Satin Bowerbird was described by a French ornithologist in 1816, the genus name, *Ptilonorhynchus*, referring to the partly-feathered bill. The specific name refers to the glossy lilac-blue sheen of the plumage of the adult male, an identifying character that is not attained until the sixth or seventh year, prior to which it approximates to the grey-green plumage of the female.

The Satin Bowerbird inhabits rainforest, wet eucalypt forests and woodlands in eastern Australia from Victoria to south-eastern Queensland, with an isolated population in north-eastern Queensland. Its life history has been well documented in field studies by Jock Marshall, Reta Vellenga, Gerald

Immature male displaying; Forestry Reserve, West Pennant Hills, N.S.W.; 1978 (J. Purnell)

Borgia and the myself. An adult male builds, maintains and defends one bower in each breeding season. This is an avenue, built on a circular platform of sticks on the ground and consists of two parallel vertical walls of twigs oriented in a north-south direction. Males decorate the northern end of the platform with a variety of natural and man-made ornaments. Away from human habitation decorations include blue parrot feathers; blue, violet and yellow flowers; yellow leaves and grass; brown snail shells and cicada cases; and snake skins. Closer to settlement, the Satin Bowerbird has a reputation for collecting such artifacts as blue plastic straws, bottle tops, clothes pegs, glass and paper.

The bower is built on the same or a nearby traditional site, year after year by one male or his successor. Bowers are spaced hundreds of metres apart and are dispersed throughout the habitat. Jock Marshall discovered that males raid each other's bowers, wreck nearby bowers and steal decorations. His physiological studies established that bower displays are synchronised with the reproductive cycle. He suspected that males may occasionally be polygamous but proposed that a male usually maintains a brief pair bond with a single female. Studies of colour-banded Satin Bowerbirds in Wallaby Creek woodland and Tooloom Scrub rainforest in New South Wales over five breeding seasons, first by myself then by Gerald Borgia, showed that during winter and early spring, large flocks formed at Wallaby Creek to forage on leaves of herbs. Throughout September and until about mid-October, bower-owning males spent some time at their bowers, then flew long distances to join these flocks. Adult males advertise their bowers by loud repetitive calls delivered from high exposed branches near the bower. They court visiting females at their bowers and also display to immature

Adult male at bower; Paluma, Qld (C. & D. Frith/Frithfoto)

Sponsor: Reader's Digest Services Pty Ltd (Inc. in N.S.W.)

males early in the season but expel these birds during the mating period. Immature males visit the bowers of adult males and, in the owner's absence, "paint" the bower, steal decorations, and display both solitarily and to other visiting immatures and females.

At Wallaby Creek and Tooloom Scrub, mating occurs from early November to late December. At this time, males spend most of their time perched in trees, watching the bower for visitors. They descend to the bower to display to females, to expel raiding males and to perform maintenance activities such as building, decorating and "painting" the inside avenue walls with vegetable matter and saliva.

A female visitor triggers a striking visual and vocal display by the male. He holds an ornament in his bill and performs a display of strutting, bowing and flinging open his wings as he dashes back and forth across the avenue entrance, uttering rhythmic buzzing and rattling notes and quiet mimicry. The female solicits copulation by crouching in the avenue with her primary flight feathers spread and quivering. The male then dashes behind the female and mounts her. After mating, the female stands in the avenue and vigorously flaps her wings. Male hostility toward the female is rare.

Reta Vellenga colour-banded more than 600 birds in the Blue Mountains west of Sydney. In 1967 she observed one adult male at a bower and, for the first time, confirmed polygamy: it mated with five different females in a season.

Borgia, who used cine cameras and infra-red beams to study behaviour of the Satin Bowerbird, found that mating success varied widely among males at the 22 bowers that he monitored continuously over two mating seasons. In 1981, 212 copulations were recorded. One male mated with 33 females in a season and the five most successful males were responsible for 56% of the matings. He found that the number of matings directly correlated with bower quality and the number and type of decorations. Males that displayed more blue feathers, snail shells and yellow leaves obtained more copulations. He also found that blue feathers were stolen more than any other decorations (because these are rare, they are most readily obtained by stealing). The number of parrot feathers at bowers of successful males peaked at the height of the mating period.

A female mates once for each clutch that she lays. She performs all parental care, including nest building, incubation and care of the young. At Wallaby Creek and Tooloom Scrub, I found that females usually build nests in forks in the upper and outer foliage of eucalypts in open woodland, but some are made in mistletoe clumps and against trunks. Nestlings are fed a high protein diet consisting of about 5% fruit and 95% insects, mostly scarab beetles taken from eucalypt foliage near the nest tree.

Nesting failure is high, probably due to predation, although this has not been directly observed. The passive anti-predator behaviour of nesting females suggests that potential predators are large birds such as raptors and kookaburras. When such predators are nearby a female shows very cryptic behaviour in delivering food to nestlings. One such bird was observed approaching the nest cautiously and pausing, motionless, for nearly 50 minutes before a goshawk flew off. She then paused for 80 minutes before visiting the nest. The ability to spend so much time in passive defence is related to the high nutritive quality of the food provided to the nestlings.

The Satin Bowerbird feeds mainly on fruits but its diet also includes leaves, flowers, seeds and animals; those inhabiting woodland eat more leaves and less fruit than those in rainforests. From winter to early spring, it forages on the ground, eating leaves of herbs and a few flowers, mostly white clover. In woodland, winter flocks of up to 200 birds are formed; in rainforest, groups of 2 – 20 birds forage on herbs. Animal food, especially scarab beetles and cicadas, is eaten mostly during the nesting and mating seasons. The birds usually take these insects by sitting motionless, then gleaning them from the foliage, but some are taken in flight. Lemons and pumpkins may be hollowed out on the ground by pecking out the soft contents. Large fruits such as figs are eaten *in situ*. Smaller fruits are taken while the birds are perched upright with the head down, or may even be snatched in flight.

R. Donaghey

Adult female at nest; New Lambton, N.S.W.; November 1993 (G. Weber)

Sponsor: Strathlea Pty Ltd

SATIN BOWERBIRD

Ptilonorhynchus violaceus

Male and female mating in bower (P. Green)

OTHER NAMES
Satinbird

MEASUREMENTS
LENGTH: *280 – 340 mm*
WING: *162 – 181 mm*
TAIL: *94 – 116 mm*
BILL: *28.2 – 36.4 mm*
TARSUS: *44 – 59 mm*
WEIGHT: *165 – 238 g*

DESCRIPTION
ADULT
Male: General plumage deep blue-black with pronounced violet-purple to blue sheen, particularly on neck, breast and rump; feathers of forehead extend forward over base of bill to cover nostrils; flight feathers and tail dull black with some glossy margins and tips; bill greenish yellow with blue-grey base; iris blue; legs dull greenish yellow

Female: Upperparts generally ashy grey-green; rump and uppertail coverts washed with green; primary wing feathers dark brown; greater wing coverts and secondaries dull reddish brown washed with greyish green; ear coverts, cheeks and throat brown with pale buff streaks; rest of underparts yellowish buff, each feather with a dark brown crescent-shaped marking, giving a scaly appearance; tail rufous brown. Bill dark horn, tinged olive. Iris blue. Legs pale grey

IMMATURE
First to third year plumage like adult female but underparts more whitish and bill dark brown. Fourth year male has a green collar on breast and throat; bill changes to greenish-yellow with blue-grey base. Blue-black feathers appear in fifth and sixth year males. Full blue-black plumage of male attained in seventh year

JUVENAL
Similar to first year immature but with brown iris and long wisps of down on top of head

IDENTIFICATION
Robust, with short tail and swooping flight with upturned wing tips. Adult male glossy blue-black with blue iris. Female grey-green above, wings and tail rufous-brown and underparts yellowish buff with brown scalloping

VARIATION
Two subspecies: P. v. minor *from north-eastern Queensland is smaller and female has a greener breast and smaller light spots of underparts than the more widespread southern subspecies,* P. v. violaceus

VOICE
Harsh churring, drawn-out whistle; ringing advertising calls; buzzing, whirring and rattling; mimicry

HABITAT
Rainforest, rainforest edge, wet eucalypt forests and adjacent woodland. Visits orchards, gardens and pastures and crops bordering forest

FOOD
Predominantly fruit throughout the year, but feeds more on leaves in winter and on insects (mostly beetles but also cicadas, moths, stick insects and grasshoppers) in summer

HABITS
Sedentary and mostly solitary in breeding season. May form large foraging flocks in non-breeding season and disperses to orchards, gardens, pastures and crops. Arboreal, foraging from ground to tree canopy. Wary, female-plumaged birds blend in well with vegetation. Moves in twos and small groups when foraging and at bowers. Males and females have large overlapping home ranges. Males defend bower territory and females defend nest tree.

BREEDING
Delayed sexual maturity in males. Polygynous or promiscuous mating system. Males maintain two-walled avenue bowers solitarily dispersed throughout forest or woodland and along the forest edge. Female performs all parental care, nests September – February. Usually one brood per season

NEST: A shallow saucer of twigs lined with dry Eucalyptus *or* Acacia *leaves; built in fork of tree or shrub in outer foliage and against trunk, in mistletoe clump or in cavity in top of stump; often 1 – 35 m above ground*

EGGS: Usually 2, sometimes 3; oval in shape; dark cream to soiled yellow-brown; lightly marked, mainly at brood end, with chestnut-brown to dark brown spots and irregular wavy lines with underlying mauve-grey blotches; 43 x 30 mm

INCUBATION: 21 – 22 days, fledging period 20 – 21 days

DISTRIBUTION
P. v. violaceus *in coastal eastern Australia from Cooroy district, south-eastern Queensland south to Otway Ranges, southern Victoria and inland to Canberra from sea level to about 1000 m.* P. v. minor *restricted to Atherton highlands of north-eastern Queensland from Mt Amos south to Seaview Range and inland to the Herberton Range, mostly above 600 m*

STATUS
Moderately common to locally common

R. Donaghey

REGENT BOWERBIRD

Sericulus chrysocephalus

The deep velvety black and rich gold plumage of the adult male of this species is the most exquisite of all the Australian bowerbirds. The appropriate specific name, meaning "golden-headed", was given by Lewin in 1808 but it was commonly known as the Regent Bird, in honour of the Prince Regent, who subsequently became King George IV. A short-lived synonym was *Oriolus regens*, conferred by the French naturalists Quoy and Gaimard, who visited Sydney on the voyage of the *Uranie* in 1819. It was placed in its present genus, meaning "silky", by Swainson in 1825. Not until the 1860s was it discovered to be a bower builder.

The adult male builds a two-walled avenue bower on the ground and decorates it with brown to yellowish objects, including snail shells, berries and seeds. The brownish plumage of the female, which raises the young without assistance, is markedly different from that of the male. The closest relatives of the Regent Bowerbird are two New Guinean species. One of these, the Adelbert Bowerbird, *Sericulus bakeri*, was not described until 1928 and its bower was not known until quite recently. Otherwise, the Regent Bowerbird is clearly related to the other avenue-builders such as the Satin Bowerbird, *Ptilonorhynchus violaceus,* and the species of *Chlamydera.* Similarities in bower structure are reinforced by behaviour at the bower and the colouration of the eggs.

Adult male displaying (G. Threlfo)

Sponsors: Joan & Keith Cousins

The Regent Bowerbird lives in subtropical and warm temperate rainforest from eastern Queensland to central New South Wales. It favours subtropical coastal and upland rainforest in the northern part of its range and warm temperate rainforest in the south. In some localities, such as the McPherson Ranges of south-eastern Queensland and the adjacent Border Ranges of New South Wales, the Regent and Satin Bowerbirds coexist with the Green Catbird in the same rainforest. In other localities, such as the coastal rainforest at Iluka, near the mouth of the Clarence River in New South Wales, only the Regent Bowerbird occurs.

In the Tooloom Scrub rainforest in New South Wales, the three species are segregated to some extent. The monogamous Green Catbird is dependent on the fruits of strangler fig trees which are dispersed throughout the rainforest interior, and lives in "all-purpose" territories throughout the forest. The two polygamous bowerbird species feed from a variety of fruiting trees and shrubs that tend to occur in clusters and to fruit irregularly, and as a result they forage widely in search of ripe fruit. Male Regent and Satin Bowerbirds place their bowers along the rainforest edge. Female Satin Bowerbirds utilise the adjacent woodland for nesting and foraging, much more so than female Regent Bowerbirds which are more dependent on the rainforest interior and the rainforest edge. Male Regent Bowerbirds, like male Satin Bowerbirds, use the rainforest edge for bower sites and for eating fruits of regrowth species, but the Regent Bowerbird places its bower in denser vegetation. There is least

Female at nest (G. Threlfo)

Sponsor: N.S.W. Field Ornithologists Club

habitat overlap between the Green Catbird and either of the Satin and Regent Bowerbirds. The latter two species overlap more in their habitat requirements than they do with the Green Catbird.

The Regent Bowerbird forages from ground level to the tree canopy. It feeds predominantly on fruit throughout the year, but takes more beetles, cicadas and spiders in the breeding season. In relation to the kinds of fruit they eat, the Satin and Regent Bowerbird have more in common with each other than they do with the Green Catbird. The smaller Regent Bowerbird takes much smaller fruit (3 – 5 mm diameter) than the Satin Bowerbird. The Regent Bowerbird feeds on the small fruits of exotic species such as the Coral Berry, *Rivina humilus*, Inkweed (both in the family Phytolaccaceae) and lantana, all of which colonise the rainforest edge and disturbed areas. Other small fruits of trees, shrubs and vines in the diet include those of *Polyscias*, *Piper*, *Citriobatus*, *Rapanea*, *Celastrus*, *Rhodamnia* and other Myrtaceae. Like the Satin Bowerbird, the Regent Bowerbird pecks at large fruits like figs, *in situ*. Smaller fruits such as *Neolitsea* from the Lauraceae family, and the native quince, *Alectryon*, are eaten whole. Fruits are taken by birds when perched upright, upside down and in flight.

The avenue bower of the Regent Bowerbird is similar to but smaller than that of the larger Satin Bowerbird and is decorated with fewer ornaments. The bower has been described as scanty, but some are neat and well built with denser walls of sticks. Differences in the quality of bower construction probably reflect variation in the age and status of an individual builder.

The Regent Bowerbird's bower is seldom found. Jock Marshall offered several reasons for this. The bower is well concealed; it is nearly always built in the forest interior; the small bower and meagre decorations harmonise with the forest floor; and the display is not noisy. At Tooloom Scrub, 10 bowers of Regent Bowerbirds were found over three breeding seasons. Half of these were avenue bowers and half were platforms of sticks attended by female-plumaged birds. All the platforms and one walled bower were built in the forest interior. Four other walled bowers were built amongst vine tangles and thick vegetation in the rainforest-woodland ecotone. They were all extremely well concealed by a dense layer of vegetation directly above the bower. The scarcity of bowers located led Tom Gilliard to suggest that the Regent Bowerbird rarely builds bowers and is in the process of discarding the practice of bower building. This was strongly refuted by Marshall who observed that captive males regularly built bowers.

The male Regent Bowerbird, unlike male Satin and *Chlamydera* bowerbirds, does not deliver any loud advertising calls near the bower during the breeding season. In contrast to other avenue bower builders, it probably uses its bright plumage near and away from the bower to entice females to the bower. Its bower is usually, but not always, built on a stick platform. Avenue walls measure 25 – 30 centimetres high, 5 – 7.5 centimetres wide, 17.5 – 20 centimetres long and the avenue is 8.75 – 10 centimetres wide. The bower is decorated with a few ornaments such as brown snail shells, brown seeds, fresh purplish-tinted leaves, blue, red or black berries and occasionally yellow flowers and leaves. Unlike the Satin Bowerbird, but like *Chlamydera* species, decorations are placed mostly within the avenue. Like other avenue builders, bower-owning males paint the inside avenue walls, using a yellowish substance or a wet mixture of saliva and green vegetable matter. Sometimes the fresh tender tips of fern fronds are crushed in the bill, mixed with saliva and applied to the inner avenue walls.

Displays by males have not been well documented. At its bower, a displaying male holds a decoration in its bill and performs strutting, head-bobbing, bowing, wing-opening, bill-gaping and backward somersaults to a visiting female or immature male. Throughout its display, the male utters quiet chattering, twittering and mimicry, audible only about a metre or so from the bower. Like the Satin Bowerbird, immature males display on platforms and rudimentary bowers to females and immature males. Mating occurs at the bower. Males steal decorations from each other.

All parental care such as nest building, incubation, brooding and feeding of young is performed by the female parent. Reports of an immature male attending nestlings are erroneous. Nestling Regent Bowerbirds are fed a diet of small fruits and insects such as beetles and moths. An intriguing facet of the Regent's social organisation is that many walled bowers found in the breeding season are demolished within one to two weeks of their discovery. The recent study of the Regent Bowerbird in south-east Queensland by Norbert Lenz confirmed that most walled bowers are temporary and are used only for a few days (always less than two weeks) and rarely stand for the entire breeding season. Further research is needed to clarify whether bower destruction is by neighbouring Regent Bowerbirds or Satin Bowerbirds. At Tooloom Scrub, immature male Regent Bowerbirds pilfer decorations from Satin Bowerbird bowers and Satin Bowerbirds visit Regent Bowerbird bowers.

R. Donaghey

Immature male; Gosford, NSW; September 1977 (G. Weber)

Sponsors: Sally White & Peter Maund

Regent Bowerbird

Sericulus chrysocephalus

Adult male and female mating in bower (P. Green)

OTHER NAMES
Australian Regentbird

MEASUREMENTS
LENGTH: 260 – 280 mm
WING: 120 – 140 mm
TAIL: 100 – 110 mm
BILL: 24 – 26 mm
TARSUS: 35 – 39 mm
WEIGHT: 80 – 138 g

DESCRIPTION
ADULT
Male: Forehead, crown, nape to mantle rich golden yellow, washed with orange-red on forecrown and hindneck; two outermost primaries and their coverts black, remaining flight feathers golden yellow with black tips; secondary coverts and tertials rich golden yellow; other plumage, including tail, black. Bill yellow. Iris pale yellow. Legs grey-black

Female: Crown and nape sooty black; rest of head greyish with feathers edged brown; feathers of cheeks and ear coverts with greyish-white shaft-streaks; mantle and back dull olive-brown; feathers of back with wide whitish concentric markings, small brown centre spots and sooty margins; lower back to uppertail coverts olive-brown; tail brown; wings olive-brown; central throat patch black; underparts dull buff, feathers particularly on upper breast with prominent olive-brown margins; abdomen paler and with less conspicuous olive-brown margins to feathers. Bill black. Iris yellow. Legs dark grey

IMMATURE
Females and first and second year immature males similar to adult female, but with dark brown iris. After the second year, males and females attain yellow iris, males acquire a pale yellow bill. Adult male plumage is attained slowly. Black crown and throat patches expand, back becomes darker and yellow feathers appear on hindneck

JUVENAL
Similar to immature, but paler areas much lighter, nearly white and with softer texture; throat almost naked and lacking dark patch

IDENTIFICATION
Striking velvety black and rich golden yellow of adult male is unmistakable. Females and immatures recognised by brown and whitish scalloping on underparts and upperback, black crown and throat patches and olive-brown wings and tail

VARIATION
Two very similar subspecies: S. c. chrysocephalus *has golden yellow head; neck and upper back of adult male lightly tinged reddish;* S. c. rothschildi *adult male similar to* S. c. chrysocephalus *but golden-yellow of crown and hindneck more strongly tinged with orange-red*

VOICE
Metallic "te-aar". Displaying male at bower utters subdued chattering and mimicry

HABITAT
Wet forests and scrub, particularly rainforest, and rainforest edge. Visits orchards and gardens

FOOD
Predominantly fruit, but also leaves, insects and spiders

HABITS
Sedentary in breeding season. Disperses to orchards and gardens in non-breeding season. Arboreal, foraging from ground to canopy. Shy, difficult to detect in rainforest foliage. Mostly solitary in breeding season. Moves in twos and small groups in fruiting trees and at bowers. Forms larger groups in non-breeding season

BREEDING
Delayed sexual maturity in males, probably a polygynous mating system. Males maintain two-walled avenue bowers solitarily dispersed throughout the forest and along the forest edge. Females probably perform all parental care; nests October – February; probably one brood per season

NEST: A flimsy saucer of twigs, plant stems and rootlets lined with fine twigs and wire-like stems; built amidst dense foliage of climbers and in forks of trees 3.5 – 12 m above ground.

EGGS: Usually 2, sometimes 3; glossy and oval; cream to pale buff with fine brown to black irregular scrawls and streaks and pale mauve-grey blotches, particularly on the broad end; 38 x 26 mm

INCUBATION: About 20 – 21 days; fledging period about 20 – 21 days

DISTRIBUTION
Coastal eastern Australia. S. c. chrysocephalus *occurs from McPherson Ranges, south-eastern Queensland, south to Gosford, New South Wales.* S. c. rothschildi *ranges along the central and southern coast of Queensland from Eungella Ranges near Mackay south to McPherson Ranges, and inland to Bunya Mountains. Occurs on Fraser Island*

STATUS
Generally scarce, but locally common

R. Donaghey

Adult male decorating bower; Bollan, Qld; August 1983 (R. Slater)

Sponsors: Sue & Richard Miles

Spotted Bowerbird

Chlamydera maculata

The endemic Spotted Bowerbird belongs to the only bowerbird genus adapted to drier, sparsely vegetated habitats and grassland habitats. It is aptly named, both sexes having cryptic spotted plumage that blends in with the semi-arid and arid landscape of inland Australia. In contrast, males of rainforest-dwelling bowerbirds, except the Tooth-billed Bowerbird, are brightly coloured.

The Spotted Bowerbird, the second species of *Chlamydera* to be named, was described by John Gould in 1837 in London from specimens collected in New South Wales. Like other *Chlamydera* species and the closely related Satin Bowerbird, the male Spotted Bowerbird builds a twin-walled avenue bower and decorates it with ornaments to attract females for mating. When John Gould travelled to Australia he found several bowers in inland New South Wales and described them as:

> ... outwardly built of twigs, and beautifully lined with tall grasses, so disposed that their heads nearly meet; the decorations are very profuse, and consist of bivalve shells, crania of small mammalia and other bones bleached by exposure to the rays of the sun or from the camp-fires of the natives ... the immense collection of decorative materials are placed in a heap before the entrance of the avenue, the arrangement being the same at both ends. In some of the larger bowers, which had evidently been resorted to for many years, I have seen half a bushel of bones, shells, etc, at each of the entrances. I frequently found these structures at a considerable distance from the rivers, from the borders of which they could alone have procured the shells and small round pebbly stones; their collection and transportation must therefore be a task of great labour.

The male builds an avenue bower in late winter or early Spring and maintains it throughout the display season until December or January. Bowers are built on the ground on a shady site, often under the drooping branches of dense, prickly bushes below taller trees and often near water or habitation. The bower consists of two parallel walls of fine twigs placed vertically in a platform of interlocking twigs. The inside avenue walls are lined with grass stems. One bower was recorded as measuring 30 centimetres long and 13 centimetres thick with 18 centimetres separating the internal walls of the avenue.

Unlike those of the Satin Bowerbird, the Spotted Bowerbird's bower walls have no consistent orientation. Some avenue walls run north-south, others are orientated east-west. The Spotted Bowerbird collects hundreds of ornaments, the most abundant being the bleached white bones of sheep and rabbits, and grey to white pebbles and stones. Other decorations include the white shells of land snails, seed pods, green berries and glass which is pale green, amber, mauve or white.

In contrast to the Satin Bowerbird, the Spotted Bowerbird profusely decorates both ends of the avenue and places many ornaments within it and sometimes on the walls. In bowers in north-western New South Wales, Norman Chaffer noted that certain ornaments were grouped together. The avenue floor contained small objects such as pebbles, glass and bones. Each end of the avenue was decorated mostly with dark grey, water-worn pebbles, lots of pieces of glass and sometimes small bones. Further away from the avenue, either at both ends or in separate heaps at one end, were hundreds of sheep vertebrae. Such grouping and separation of different ornaments has also been described in the Western Bowerbird.

Since European settlement, the Spotted Bowerbird has become legendary for stealing an astonishing variety of ornaments including glass, metals, jewellery, cutlery, coins and other bright objects. Many tales have been told about the more unusual items pilfered and later retrieved, such as jewellery, entomological forceps, vehicle ignition keys and even a bushman's glass eye kept in a cup of water overnight.

The Spotted Bowerbird is a highly accomplished mimic, particularly of mechanical sounds. It was described by zoologist Jock Marshall as "probably the most gifted mocking-bird known". As well as bird calls such as those of the Whistling Eagle and butcherbirds, the Spotted Bowerbird's repertoire of mimicry includes the barking of dogs, twanging of fence-wires, chopping of wood, the crack of a stockwhip, the whirring flight of a crested pigeon and the tinkling sound of picking up and dropping bower ornaments.

Throughout the year, Spotted Bowerbirds can be located by their harsh calls as they forage singly or in small groups. However, at the onset of the display season, bower-owning males deliver ringing, metallic advertising calls from an exposed perch near the bower to attract females. The visit of a female to the bower triggers an intense bout of aggressive display by the male.

John Warham recognised two types of male display at the bower: central and peripheral, neither of which are governed by a fixed sequence. Throughout both displays the male's crest is expanded and raised. Four display postures are given while the male stands among the bones and shells near the avenue walls. Standing in an upright posture, with feathers sleek, the male looks at the female and utters hissing and scolding sounds. With wings slightly raised and the tail horizontal, the male crouches low with open bill, and calls loudly. Turning sideways to the avenue entrance, or facing it, the male picks up an ornament, holds it in his bill, jerks his head up and down and directs his raised lilac-pink neck patch to the female. These two crest-presentation postures are key displays. During central displays the male flings ornaments around, hops from side to side or jumps into the air and advances down the avenue toward the passive female at the other end of the bower. Peripheral displays consist of postures given by the male around, and often off, the bower platform. He runs and hops in circles around the bower with tail cocked up, wings drooped and feathers ruffled, always watching the female. During these displays, the female moves to keep the walls of the avenue between her and the male. Successive copulations of a female at the bower were observed after a 20 minute bout of intense display.

A bower-owning male Spotted Bowerbird displays to all females visiting the bower and mates with as many females as possible throughout the display season. In contrast, the female Spotted Bowerbird performs all parental care, building the nest, incubating the eggs and rearing the young. Cliff and Dawn Frith watched one nest and confirmed that only one female attended the young, with no male assistance. They observed that one female fed two nestlings at a rate of 6.7 times per hour. Fifty-nine per cent of the nestling meals delivered by the female were animal and 41% were fruit. The animal meals mostly consisted of grasshoppers and crickets and also included soft arthropod larvae, mantids, katydids, stick insects, beetles, caterpillars, spiders and a skink. Fruits of the locally common current bush, *Carissa lanceolata*, constituted 57% of the fruit meals.

The sole female parent mimicked the calls of the Whistling Kite, *Haliastur sphenurus*, and Blue-winged Kookaburra, *Dacelo leadii*, during one inspection of the nest with nestling. A Noisy Friarbird, *Philemon corniculatus*, that intruded into the nest tree was chased away vigorously by the female parent. However, when a potential nestling predator, such as the Blue-winged Kookaburra, perched near the nest tree, the female parent Spotted Bowerbird approached the nest and paused motionless until the kookaburra departed. The two young left the nest 20 – 21 days after hatching and were fed solely by the female.

R. Donaghey

Adult female feeding chicks; Near Charters Towers, Qld (C. & D. Frith/Frithfoto)

Sponsor: Peter Miles

SPOTTED BOWERBIRD

Chlamydera maculata

Bower; Near Thallon, Qld; October 1985 (R.E. Viljoen)

OTHER NAMES
Mimic bird

MEASUREMENTS
LENGTH: c 310 mm
WING: 145 – 155 mm
TAIL: 99 – 155 mm
BILL: 23 – 27 mm
TARSUS: 40 – 43 mm
WEIGHT: 135 – 154 g

DESCRIPTION
ADULT
Male: Forehead to nape buff-brown, tinged with silver-grey; dark margins to feathers, giving a scaly appearance; hindneck with erectile tuft of vivid lilac-pink (followed by a grey band in one subspecies); rest of upperparts dusky brown to blackish, each feather tipped with buff to orange-buff, giving a prominent spotted appearance, particularly pronounced on mantle, back, uppertail coverts and wing coverts; tail dusky brown with bold buff to whitish tips; sides of head, chin to upper breast buff to buff-brown with dusky-brown barring and spotting; rest of underparts cream-buff. Iris brown. Bill black. Legs brown. Inside of mouth yellow

Female: Similar to adult male, but lilac-pink patch on hindneck reduced or absent

IMMATURE
Similar to adult, but plumage paler above and below; back more greyish to whitish; spotting on underparts less pronounced; lilac-pink hindneck feathers absent or reduced to a few pink feathers

IDENTIFICATION
A robust, blackish-brown to buff bird, boldly spotted above with rufous to buff; erectile, vivid lilac-pink hindneck

VARIATION
Negligible. Birds from South Australia may have lighter-coloured bill, paler nuchal crest and reddish abdomen

VOICE
Ringing metallic advertising calls; harsh grindings, hissing and clicking. Renowned for mimicry of natural bird songs and mechanical sounds resembling whistles, barking of dogs, ratcheting, squeaking, chopping of wood and cracking of stockwhips

HABITAT
Dry, open grassy woodlands; inland riverine vegetation amongst arid and semi-arid scrubs. Visits homestead gardens and orchards

FOOD
Predominantly fruit, but also nectar, seeds, insects (grasshoppers, moths and larvae, beetles) and spiders

HABITS
Usually encountered singly or in small groups. Forms winter foraging flocks. Forages from ground level to tree canopy. Rather wary, quickly disappears out of sight amongst foliage. Undulating flight on upswept wing tips

BREEDING
Delayed sexual maturity in males; promiscuous mating system; males build avenue bowers, solitarily dispersed; females perform all parental care. Nests August – February, probably one brood per season

NEST: Bulky, loosely built saucer of larger sticks and twigs, lined with fine twigs and grass stems; built in horizontal or upward-sloping fork in foliage of trees or mistletoe clumps; 2 – 20 m above ground

EGGS: 1 – 3 (usually 2); slightly glossy; pale greenish-yellow with light greyish-mauve blotches, overlaid by intertwining hair-like scribbles of brown to brownish black; 38 x 27 mm

INCUBATION: unknown; fledging period 20 – 21 days

DISTRIBUTION
Patchily distributed in interior of eastern Australia from about Charters Towers and the upper Leichhardt River in Queensland, through inland New South Wales to north-western Victoria and adjacent eastern South Australia along Murray River.

STATUS
Endemic. More common in the north of range, in central Queensland, declining in the south. Threatened and rare in north-western Victoria and the southern Murray-Darling System.

R. Donaghey

WESTERN BOWERBIRD

Chlamydera guttata

The first specimen of this beautiful bird was given to John Gould by the explorer F.T. Gregory, who had collected it from north-western Australia, probably in the region of the upper Fortescue River, in 1861. Tom Carter, quoted by Gregory Mathews, writes of a bird collected at Tantabiddy, 16 kilometres south-west of North-West Cape, on 5 February 1892:

> I believe this was only the second time this species was recorded for West Australia, the first being that obtained by Mr T.F. Gregory in early exploration (1837) on the Glenelg and Prince Regent River District (Kimberley, Western Australia) and mentioned in Gould's Handbook, p. 452, as the type.

Matthews went on to add:

> Gregory's bird was not procured until about 1861, being described at once in 1862, and it was suggested by Gould that it was the one which constructed the bowers described by Captain Gray in his Travels, found near the Glenelg and Prince Regent's Rivers. The exact locality where Gregory's bird was collected is not known at present.

Although distinctly smaller and more richly coloured, this striking bowerbird was, until recently, widely considered to be a subspecies of the Spotted Bowerbird, *C. maculata*, found in eastern and central Australia. The association is understandable, since both species share many habits, such as their style of bower-building, although the Western Bowerbird often constructs its avenue-type bower on a raised platform, up to 20 centimetres above the ground.

The Western Bowerbird is well distributed across central and mid-western Australia from north-western South Australia and north-eastern Northern Territory to the Pilbara in the north and south to about Leonora, Western Australia. Some records may also exist of individuals from the Georgina River, western Queensland, although these may be Spotted Bowerbirds. Its distribution is widely considered to be restricted to that of the Wild Fig, *Ficus platypoda*, but in recent years its range has extended beyond that in the south-eastern interior of Western Australia. Although it is usually solitary, groups of up to four birds are commonly seen feeding on fruiting trees around homesteads. Captain S.A. White stated:

> Met with this bird in the Macdonnell Ranges in 1913, also met with this bird in the Everard Ranges in 1914. Several bowers were seen, generally placed under a large native fig tree (Ficus platypoda) in the deep ravines of the range; in one instance quite a barrow load of bleached snail shells were arranged in front of the bower. The birds were very shy and seldom seen.

Although favouring stands of the Wild Fig, the Western Bowerbird is also found near gum-fringed creeks and in dense vegetation (such as *Casuarina* and *Acacia*) near bores, troughs and dams.

As mentioned above, the bower of the male Western Bowerbird is of the "avenue-type", often constructed on a foundation of twigs. Flexible twigs from low bushes are gathered and woven together to form two parallel walls about 10 – 12 centimetres thick and 25 centimetres long. The bower is

Adult male decorating bower; Pilbara, W.A.; circa 1980 (M. & I. Morcombe)

Sponsor: Dr John Newlinds

then adorned with objects of white and green, such as seed pods, rocks, bones and leaves. Gould quotes from Sir George Gray's *Travels*, regarding his encounters with the bowers:

> This very curious sort of "nest", which was frequently found by myself and other individuals of the party, not only along the sea-shore, but in some instances at a great distance of six or seven miles from it, I once conceived must belong to a kangaroo, until I was informed that it was a run or playing-place of a species of Chlamydera. These structures were formed of dead grass and parts of bushes, sunk a slight depth into two parallel furrows in sandy soil, and then nicely arched above. But the most remarkable fact connected with them was that they were always full of broken sea-shells, large heaps of which protruded from each extremity.

Breeding has been recorded in October but is most common from July to September. When a prospective female mate is present at the bower, the male displays to her with a series of ritualised dances and calls in an attempt to convince her to allow him to mate with her. If he fails to dazzle her, she simply leaves in search of a more attractive mate. In display, the male erects the feathers of his pink nuchal crest. Amid various cackles and harsh cries he moves into the centre of the playground area and grabs at items of decoration lying on the floor. His dancing consists of a variety of hopping movements, while continually moving backwards and forwards and from side to side. The display can last more than an hour.

As well as uttering harsh, cackling notes, the Western Bowerbird is an accomplished mimic, often imitating the calls of birds of prey (especially the Whistling Kite, *Haliastur sphenurus*, the Blue-winged Kookaburra, *Dacelo leachii*, and Pied Butcherbird, *Cracticus nigrogularis*. According to Serventy and Whittell, it also emits a cat-like sound that may be its natural call. Thomas P. Austin wrote:

> ... I had not been long waiting when there was a great row in the playground. Just for a few moments I could not think what it could be, as I had not seen the Bower-Bird arrive. I ran up to find a fine bird lying on its back, caught by both legs, and doing its best to get free by biting at the horse-hair with its strong bill, at the same time making the most awful noise. Almost immediately I heard another most extraordinary noise above; looking up, I saw about half a dozen other Bower-Birds on the top of a few dead trees; then suddenly they all started imitating various cat calls.

P. Rowland

Western Bowerbird

Chlamydera guttata

Male displaying to female at bower; Cape Range National Park, Ashburton, W.A.; June 1988 (J. Lochman/Lochman Transparencies)

OTHER NAMES
Guttated Bowerbird, Yellow-spotted Bowerbird

MEASUREMENTS
LENGTH: *24 – 28 mm*
WING: *143 – 147 mm*
TAIL: *88 – 106 mm*
BILL: *20 – 25 mm*
TARSUS: *40 – 45 mm*
WEIGHT: *125 – 145 g*

DESCRIPTION
ADULT
Male: Much darker than Spotted Bowerbird, C. maculata*. Feathers of head brownish-black, each feather with chestnut spot, bordered with bronze; upperparts generally brownish-black (including tail); feathers of mantle, back and rump with a large chestnut-buff spot on tip of each feather; primaries and secondaries margined with pale buff on outer webs, whitish-yellow on inner webs. Hindneck has conspicuous lilac-pink patch of feathers; both head and hindneck have silvery gloss. Underparts predominantly creamy-yellow, becoming more brownish on flanks and thighs. Bill black. Legs and feet greenish-horn. Eye dark brown*

Female: Similar to male but lilac-pink patch of hindneck smaller and duller

IMMATURE
Similar to adult, but plumage paler and greyer and lacking silver gloss on head. Lilac-pink patch on hindneck markedly reduced or absent. Bill brownish

JUVENAL
Similar to immature; lilac-pink patch on hindneck absent; spotting of upperparts less distinct

IDENTIFICATION
Smaller and darker than related Spotted Bowerbird, with which its range overlaps. Generally blackish-brown in plumage, liberally spotted with buff. Underparts, except throat, creamy-yellow, becoming brown on sides. Adult birds have prominent lilac-pink patch on hindneck, larger in males

VARIATION
Negligible. Some north to south variation; northern birds darker, as are those from central Australia. Not known to interbreed with Spotted Bowerbird where ranges overlap

VOICE
Much mimicry. Normal utterance considered to be a few metallic calls and harsh cackling and hissing notes. Also some cat-like calls

HABITAT
Inhabits areas of dense vegetation (such as Casuarina *and* Acacia*) near bores, troughs and dams;* Eucalypt*-fringed creeks; also rocky ranges of the arid and semi-arid zone. Largely, but not exclusively, limited in distribution to that of Wild Fig,* Ficus platypoda*, with which it closely associates. Often visits homesteads and orchards*

FOOD
Mainly fruit (especially those of the Wild Fig); also some flowers, nectar and insects taken

HABITS
Generally shy and wary of humans but will approach homesteads to feed on fruiting trees. Not very gregarious, usually solitary or in parties of 4 or 5 birds

BREEDING
Male builds and maintains elaborate bower of two parallel walls of sticks, curved at the top, and decorated with objects of white and green. Unlike Spotted Bowerbird, bower often elevated on platform of twigs. Breeding season generally July – September, occasionally as late as December

NEST: Not well documented. Probably similar to that of Spotted Bowerbird; consisting of large, untidy saucer of sticks and twigs, lined with finer twigs and needles; placed in horizontal fork of tree, 2 – 10 m above the ground, in foliage of tree

EGGS: Normally 2; oval; pale grey-green, streaked and blotched with brown; 32 x 26 mm

INCUBATION: Unknown; fledging period unknown; probably similar to Spotted Bowerbird

DISTRIBUTION
Throughout central and mid-western Australia from the Musgrave and Everard Ranges, north-western South Australia and the Jervois Range, north-eastern Northern Territory to the Pilbara in the north of Western Australia and south to about Leonora. Two populations may be isolated by the Gibson Desert, Western Australia

STATUS
Endemic. Common in central Australia and in Northern Territory. Probably locally abundant in Western Australia, but generally unobtrusive and shy

P. Rowland

Great Bowerbird

Chlamydera nuchalis

As its name suggests, this is the largest of the bowerbirds. It is widely distributed across tropical northern Australia, frequenting riverine woodlands, open savanna, rainforest vine thickets, eucalypt and *Melaleuca* woodlands, mangroves and urban gardens.

It has the most uniform plumage of all the Australian bowerbirds, both sexes being a cryptic brown-grey to grey that blends in well with their brown and grey-green habitat. The specific name *nuchalis* means "nape" and refers to the rose-lilac plumes on the hindneck of the male, exposed as a fan during display to the female at the bower. Females usually lack a crest, but sometimes have rose-lilac feathers on the nape. It was described in 1830 by the English ornithologists William Jardine and P.J. Selby from specimens collected somewhere in northern Australia by the Sydney naturalist, William Macleay. These were sent to the Linnean Society of London but we do not know whether they came from western or eastern Australia. John Gould restricted the type locality to Western Australia because the type specimens were similar to those collected by the officers of the H.M.S. *Beagle* from north-western Australia.

The Great Bowerbird is common and fairly conspicuous around settlements in northern Australia. It builds the largest avenue bower of any Australian bowerbird, and decorates the bower with conspicuous ornaments. Like other members of the genus in inland Australia, it is often detected first by its harsh, mechanical calls or its low undulating flight as it moves from tree to tree in search of fruit.

It is one of three or four Australian *Chlamydera* species that are separated geographically. It ranges widely across the Top End, north of the range of the Spotted Bowerbird, the two subspecies of which occupy the most arid regions of inland Australia. The Great Bowerbird occupies more humid and sub-humid tropical woodlands with an annual rainfall of 500 – 1500 mm. The range of the eastern subspecies of the Spotted Bowerbird overlaps that of the Great Bowerbird from Mount Isa, Queensland, east to the coast. In areas where both species occur, one is more abundant and more widely distributed. In the Mount Isa district, for example, the Spotted Bowerbird is moderately common, whereas the Great Bowerbird is a rare, irregular visitor. Both species occur in the lower Burdekin River area but there, the Great Bowerbird is common throughout all habitat types and the Spotted Bowerbird is uncommon and restricted to open forest.

Similarly, the range of the Great Bowerbird overlaps that of the Fawn-breasted Bowerbird, but the two species occupy different habitats. The Great Bowerbird is more common and widely distributed throughout Cape York Peninsula, except in the north-east coastal belt where the Fawn-breasted Bowerbird is more common.

An adult male Great Bowerbird builds a large twin-walled avenue bower on the ground and decorates it with ornaments to entice females to visit it for mating. Each adult male holds a traditional bower territory that contains an active bower and, unlike other bowerbirds, the walls of up

Adult male displaying at bower (M. Seyfort)

Sponsors: G. & I. Jackson

to nine old bowers from previous seasons may still be standing in a nearby cluster. Bowers of different individuals are spaced hundreds of metres apart.

The onset of the annual sexual cycle is related to environmental factors such as day length and rainfall. The end of the display season is followed by an annual moult in summer and a new bower is built, usually in May but sometimes much earlier. Egg laying may occur as early as July.

Bowers are built under hanging branches of trees and under bushes, such as prickly *Carissa* species, providing concealment from predators overhead. A typical bower is built on a mat or platform of twigs measuring 1.8 – 2.4 metres by 0.9 – 1.5 metres and about 7.5 centimetres thick. The two upright walls measure 30 – 35 centimetres high and arch over at the top to form a tunnel about 45 centimetres long and 25 centimetres wide; the outer walls are about 55 centimetres apart. The avenue walls are built of twigs, not grasses as in the Spotted Bowerbird. Sometimes a bower-owning male stretches upward and places twigs horizontally on top of the arched avenue walls. An exceptionally long bower found near Townsville had avenue walls 1.2 metres long, apparently due to a second bower having been built as an extension to an existing one. The avenues are oriented more or less on a north-south axis.

Hundreds of bower ornaments decorate the platform at both ends of the avenue and on the avenue floor, the number and type varying with individuals, localities, and the proximity of human settlement. In inland regions, stones, wallaby bones and bleached snail shells predominate; seashells and corals tend to be used near the coast. Most decorations are white, grey, silver or pale green. Natural green objects include leaves, flowers, fruits and plant galls. In contrast to the Spotted Bowerbird, the Great Bowerbird collects bright red objects such as cloth, glass and plastic, especially when dwelling near human habitation. Like the Spotted Bowerbird, it tends to group various types of decorations together: snail shells may be put in one place, apart from kangaroo bones. There are many reports of unusual and shiny human-made objects having been stolen by bowerbirds from bushmen's and mining camps. It has also been confirmed that neighbouring males steal ornaments from each other's bowers.

In common with other avenue-builders, the male engages in bower painting behaviour. John Warham observed a bower-owning male wipe its bill up and down the twigs of the inner avenue walls. The bird's mandibles glistened with saliva and it appeared to have a sticky material in its bill, but no "paint" was seen on the bower twigs. He also watched a male displaying to a female at the bower. The central and peripheral displays were very similar to those performed by the Spotted Bowerbird. In the "upright" posture of the Great Bowerbird, its bill is wide open and the tongue is flicked out frequently. This posture often leads to a "forward stretch" posture in which the male faces the female with outstretched neck, crest slightly expanded and eyes bulging. By rotating the head sideways and down, the displaying male directs his rose-lilac nape crest to the female while usually holding a display object, such as a bone or green ornament, in his bill. These central displays may precede or follow peripheral displays similar to those of the Spotted Bowerbird. Males normally display at the northern end of their bowers, where the larger collections of bones and other ornaments are placed.

The onset of the annual tropical monsoon in northern Australia (and heavy rain in the arid and semi-arid zones) stimulates nesting in Great and Spotted Bowerbirds. Jock Marshall suggested that nesting in response to rain is adaptive in that the onset of rain provides an abundance of insect food for nestlings and enables the female to rear the young alone. Clutch initiation is staggered over several months as a result of variation between individual females, their different ages and re-nesting after loss of eggs or young. The promiscuous male has an extended display season that enables it to display to, and mate with, early and late nesting females.

Cliff and Dawn Frith confirm that only one parent, presumably the female, incubates the eggs and feeds the young. Watching a nest from a nearby hide, they determined that the food provided to the nestlings was predominantly animal (55%), the remainder being fruit. Most (57%) of the animal meals were large grasshoppers which were offered after removal of the legs, wings and head of the insects. Other animals fed to nestlings included caterpillars, crickets, mantids, spiders, beetles, moths, a katydid, a fly, a cockroach, a slug or a snail, and a skink. Most of the fruit meals were figs or pieces thereof.

R. Donaghey

Adult female at nest; Townsville, Qld (C. & D. Frith/Frithfoto)

Sponsor: E.S. Dean

Great Bowerbird

Chlamydera nuchalis

OTHER NAMES
Great Grey Bowerbird, Queensland Bowerbird

MEASUREMENTS
LENGTH: 320 – 350 mm
WING: 155 – 171 mm
TAIL: 124 – 155 mm
BILL: 26 – 31 mm
TARSUS: 44 – 50 mm
WEIGHT: 188 – 233 g

DESCRIPTION
ADULT
Male: Head and neck brown-grey, except hindneck which has a concealed erectile tuft of rose-lilac plumes bordered above and at sides with silvery white tips to feathers; rest of upperparts brownish grey, with prominent pale tips to all feathers; upperwing coverts and flight feathers dusky brown with pale grey outer edges and tips; tail dusky brown with prominent whitish tips and pale margins to each feather; underparts including underwing coverts, soft brown-grey sometimes tinged cream-buff and becoming more whitish on central abdomen; undertail coverts dull creamy white with faint to moderate narrow bars of dusky brown. Bill black; inside of mouth bright yellow to orange. Iris brown. Feet greyish-brown

Female: Like adult male but slightly smaller and hindneck ash-grey or ash-grey with some whitish tips to feathers which usually, but not always, replace the rose-lilac crest on hindneck

IMMATURE
Similar to adult female but underparts (particularly abdomen and flanks) more distinctly barred; broad white band across each feather of back and scapulars. Bill black with paler sepia tips. Legs dull greenish yellow

JUVENAL
Plumage generally soft and fluffy; crown and upperparts dusky-grey to brown with whitish spotting on crown and mantle, some of the nape feathers tinted with rose-lilac; throat to upper breast dark grey; rest of underparts pale grey to whitish with narrow dusky barring; central abdomen whitish. Bill dull black. Iris brown. Feet olive-brown

IDENTIFICATION
A large bowerbird with more or less uniform brown grey underparts. Back, wings, rump and tail dusky brown with greyish outer edges and tips to feathers giving a distinct mottled appearance. Rose-lilac nape patch of male usually concealed, but visible during display

VARIATION
Four subspecies comprising two well-defined groups each consisting of two similar subspecies: C. n. oweni *(Kimberley region) and* C. n. nuchalis *(Top End) constitute one group;* C. n. yorki *(Cape York) and* C. n. orientalis *(south of Cape York) the other.* C. n. oweni *males and females have upperparts and underparts rather uniform and greyish.* C. n. nuchalis *is similar to* C. n. oweni *but somewhat smaller.* C. n. yorki *differs markedly from* C. n. oweni *and* C. n. nuchalis *in having the upperparts more contrasting and more variegated blackish and whitish; also, females differ from males in being less uniform below with more barring.* C. n. orientalis *resembles* C. n. yorki *but is smaller; has generally lighter plumage, particularly underparts*

VOICE
A variety of harsh, grating and scolding calls, hissing, whistles, churring and chattering. Mimicry of mechanical sounds such as crumpling of paper and squeaky gates. Mimicry of natural sounds includes barking of dogs, cackling of fowls and calls of Fork-tailed Kites

HABITAT
Rainforest patches and edges, drier eucalypt forest, riverine vegetation, dry woodland, mangroves and urban gardens

FOOD
Predominantly fruit but diet also includes nectar, seeds, insects and spiders

HABITS
Sedentary in breeding season; more nomadic in the non-breeding season. Usually encountered singly or in small foraging groups. In non-breeding season foraging groups of 30 or more visit orchards and gardens in search of fruit. Forages from ground level to tree canopy. Conspicuous undulating flight. Rather wary when foraging and blends in well with foliage

BREEDING
Delayed sexual maturity in males. Probably promiscuous mating system. Males build terrestrial avenue bowers, solitarily dispersed. Females perform all parental care, nesting August – February. Probably one brood per season

NEST: A bulky shallow cup loosely built of twigs and lined with fine twigs and sometimes a few leaves; built in a fork in the foliage, vine tangles or mistletoe of trees, 1 – 10 m above ground and 50 m or more from the nearest active bower

EGGS: Usually one, sometimes two; round to oval and glossy; greyish-green with pale greenish-grey blotches, overlaid by intertwining hair-like scribbles and spots of olive-brown to dark brown and purple and black, mostly concentrated in a broad band around the centre; 41 x 29 mm

INCUBATION: About 21 days; fledging period 20 – 21 days

DISTRIBUTION
Northern Australia from the Kimberley across the Top End to Cape York Peninsula and northern Queensland, subspecies integrating where their ranges overlap. C. n. oweni *restricted to north-west Australia (West and East Kimberley), south to the Fitzroy River and Hall Creek, Western Australia.* C. n. nuchalis *extends from East Kimberley across the northern Northern Territory, including several islands to the Gulf of Carpentaria and Mornington Island.* C. n. orientalis *occurs in north-western Queensland, except Cape York Peninsula, from Cairns south to the Mackay district and inland west to Mt Isa.* C. n. yorki *is restricted to Cape York Peninsula south to the Cairns-Cooktown region, Queensland*

STATUS
Locally common to fairly common. Endemic. Local populations threatened by illegal shooting and alteration of habitat by fire. It has been reported that local populations appear to have declined in the Atherton Region, north-eastern Queensland, since the 1930s

R. Donaghey

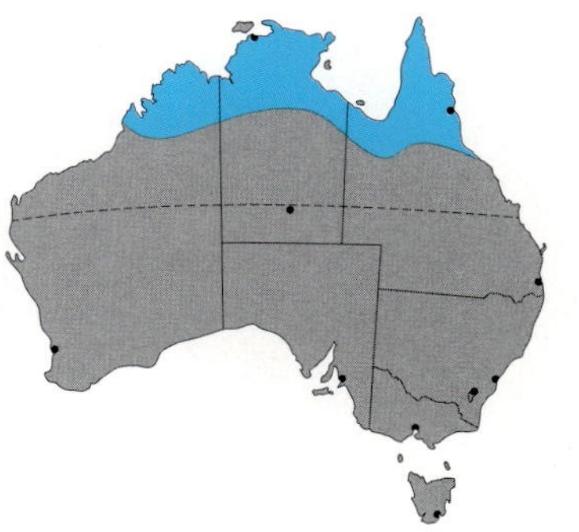

Male displaying to female at bower; Cape York Peninsula, Qld; January 1988 (R. Slater)

Sponsors: Mr & Mrs Cumbrae-Stewart

Fawn-breasted Bowerbird

Chlamydera cerviniventris

Most bowerbirds dwell in rainforest but the Fawn-breasted Bowerbird is one of four or five species of *Chlamydera* that became adapted to drier, open woodlands and grasslands as rainforests shrank during the Pleistocene. It is the only member of the genus occurring in both Australia and New Guinea. Both sexes have similar grey-brown plumage. The species is readily distinguished from other members of the genus by a rich fawn or orange-buff abdomen and the lack of a lilac-rose crest on the back of the neck.

In Australia, it is restricted to Cape York Peninsula, for which reason it was the last bowerbird to be discovered on the mainland. The first specimen was taken in 1849 by John MacGilivray, a naturalist on the survey ship HMS *Rattlesnake,* who found a bower in coastal scrub and shot a male at its bower. He said he had never encountered a more wary bird. Suspecting that it was a new species, he sent a specimen and a bower to London, where it was described by John Gould in 1850.

Bertie Jardine, a pioneer naturalist of Cape York Peninsula, wrote that near the Jardine River it:

> ... frequents scrubs growing in soft sandy soil in the neighbourhood of the sea, and low brush bordering creeks and rivers. Usually they are shy and retiring, and are met with in pairs, but sometimes they may be seen in small companies from four to eight in number as they traverse the open parts of the forest in search of food.

Like other members of the genus, it has a variety of harsh calls, whistles and soft mimicry that help birdwatchers to locate this wary bird. When foraging, it has a distinctive, swooping undulating flight as it moves between patches of forest and woodland.

The adult male builds a substantial, two-walled avenue bower that is ornately decorated. Individual males probably construct a new bower each year on traditional bower territories. Most are built in June or July and are attended for about eight months. Bowers are spaced about 500 metres apart and are built on the ground within clumps of vegetation or near the forest edge. Most are easily located by the vocalisations of displaying males, or by their advertising calls near a bower. In the foothills of the McIlwraith Range one September, I watched a bower-owning male as it delivered a repetitive, metallic advertising song from the high, bare branch of a dead tree. Advertising calls uttered in the display season serve to attract females to the bower for mating.

The twin vertical walls of a bower are erected on a mat of sticks. Each wall is about 90 centimetres long, 60 centimetres high and 10 centimetres thick. To build the vertical walls, the male holds each twig in his bill, turns its head sideways and thrusts the twig into the platform. Distinctive features of the bower are that the avenue or passageway runs more or less east-west and that the platform is very thick, especially at the western end. Decoration and arrangement differ from other members of the genus. The principal decoration comprises a hundred or so individual green fruits. Bunches of green berries and fruits mostly decorate the western platform but also adorn the floor of the avenue and are hung in bunches from the tops of each avenue wall. One pioneer naturalist noted that the bowers were:

> ... adorned chiefly with green fruits broken into fragments, leaves of a silver-grey colour and the immature flower shoots of a Melaleuca, with a few pieces of land mollusc shells.

Fresh berries are brought to the bower almost daily in the display season. Withered fruits are discarded and may accumulate in a pile off the bower platform.

Avenue bower builders paint the inside of the avenue walls. Bower painting by the Fawn-breasted Bowerbird was first observed by Bill Peckover near Port Moresby in Papua New Guinea. A male bird mandibulated green vegetable matter, then wiped it on the avenue walls by pushing the bill between the twigs. Freshly painted twigs are green but dry to a dull grey. The lower twigs are stained a rich red-brown. Males maintain and decorate their bowers and advertise themselves by visual and vocal displays that attract many females to the bower for mating. A visit to a bower by a female triggers off a display by the resident male. A female visitor may perch above the bower, but usually descends to the east end of the avenue.

In many members of the genus *Chlamydera*, the male displays by exhibiting the crest on the back of its neck. The Fawn-breasted Bowerbird lacks such a crest but displays by holding a spray of green berries or a single large fruit in its bill while twisting its head, presenting its nape to the female. Tom Gilliard suggests that this presentation is a relic of the behaviour of a once-crested ancestor. Copulation at the bower has been observed only in New Guinea, where a displaying male mated twice in succession with a visiting female. After each copulation, the female engaged in vigorous wing-flapping. Breeding appears to continue throughout the year in New Guinea. In Cape York Peninsula, nesting probably extends from September to February. Females build the nest and incubate the eggs

Male displaying to female at bower; Iron Range, Qld (C. & D. Frith/Frithfoto)

alone. Nestlings are fed mostly insects (including caterpillars) and fruit. Males are probably promiscuous, maintaining their bowers for eight to nine months of the year.

The Great Bowerbird also inhabits Cape York Peninsula. Its range overlaps with that of the Fawn-breasted Bowerbird in the eastern coastal fringes but the species are segregated by habitat. The Great Bowerbird frequents the drier, more open forest and woodlands, whereas the Fawn-breasted Bowerbird occupies the coastal lowland fringes of eucalypt and paperbark woodlands, mangroves, riverine vegetation and rainforest edges.

Males of the two endemic *Chlamydera* species, the Great and Spotted Bowerbird, possess a rose-lilac erectile nape patch and use it in display to females. The two predominantly New Guinean species, the Fawn-breasted Bowerbird and Lauterbach's Bowerbird, both lack a crest. Tom Gilliard noted that in bowerbirds there is an inverse relationship between bright plumage and the complexity and ornamentation of the bower. The drab, crestless Fawn-breasted Bowerbird builds the most elaborate, ornately decorated of the two-walled avenue bowers. The two Australian crested bowerbirds build less ornate bowers. Gilliard speculated that the crestless bowerbirds lost this structure through a transfer of the forces of sexual selection from bright plumage to display objects.

Interestingly, in Cape York Peninsula, where two bowerbird species occur, the presence of the lilac-rose crest is much more frequent in the eastern population of the Great Bowerbird than in the western population. Gilliard suggests that, since the eastern Great Bowerbird interacts with the crestless Fawn-breasted Bowerbird, the crest may act as an isolating mechanism.

R. Donaghey

Male on raised platform at bower; Portland Roads, Qld (C. & D. Frith/Frithfoto)

Sponsors: Peter & Isabel Davidson

Fawn-breasted Bowerbird

Chlamydera cerviniventris

Adult male; Cape York Peninsula, Qld; January 1988 (R. Slater)

OTHER NAMES

None

MEASUREMENTS

LENGTH: 290 mm
WING: 144 – 156 mm
TAIL: 110 mm
BILL: 30 – 32 mm
TARSUS: 39 – 42 mm
WEIGHT: Not recorded

DESCRIPTION

ADULT

Male: Forehead to hindneck pale grey-brown; buff-white shaft-streaking and edging to feathers of forehead, forecrown, lores and sides of head and neck; rest of upperparts dusky grey-brown, feathers with buff-white shaft-streaks, central spots and tips giving a pronounced mottled appearance; flight feathers dusky brown with prominent buff-white tips and pale narrow margins; chin to upperbreast buff-white with faint brownish grey edging to sides of feathers producing a streaked appearance; breast buff, brown margins to feathers becoming broader toward sides of breast; rest of underparts tawny buff, brighter on abdomen; undertail coverts pale brown-grey tinged with buff; tail dusky grey-brown. Bill black; inside of mouth orange-yellow. Iris dark brown. Legs greyish brown

Female: Similar to adult male

IMMATURE

Like adult, but forehead more extensively streaked greyish; broad buff-white central streaks to some feathers of upper back

JUVENAL

Like adult, but hindneck naked; plumage generally soft and fluffy; pale streaking of upperparts more extensive; throat and neck greyish without bold margins; rest of underparts pale greyish fawn with traces of dusky barring

IDENTIFICATION

A robust, crestless bird, grey-brown and spotted above with streaked throat and orange-buff abdomen and stout black bill. Distinguished from other Chlamydera *species by lack of crest and presence of orange-buff abdomen*

VARIATION

Negligible

VOICE

A variety of harsh, loud and soft churring, hissing, rasping, rattles and whistles. Mimicry by males at bowers includes natural bird calls and mechanical sounds. Nesting female mimics bird calls

HABITAT

Lowland coastal tropical eucalypt and paperbark woodland, rainforest regrowth and mangroves but specially the clearings and margins of rainforest

FOOD

Predominantly fruit of trees, shrubs and vines, and insects

HABITS

Sedentary during breeding season. Apparently forms small foraging flocks in post-breeding season. Forages from ground level to the tree canopy. Occurs singly, in twos and in small groups. A noisy, wary bird, difficult to locate

BREEDING

Promiscuous mating system. Adult males build and decorate terrestrial avenue bowers to attract females for mating. Females perform all parental care. Females nest May – March (virtually all year round)

NEST: An open cup loosely built of sticks and fine twigs and lined with fine rootlets; built in a fork in a shrub or tree 1 – 10 m above ground, as close as 20 m to an active bower

EGGS: Usually 1 or 2; oval-shaped and slightly glossy; pale cream tinged with olive and marked with intercrossing hair-like lines and scribbles of brown to brownish-purple to black, often encircling the eggs; 41 x 28 mm

INCUBATION

About 21 days; fledging period 21 days

DISTRIBUTION

Widely distributed along the entire coastal lowlands of Papua New Guinea and extending west into Irian Jaya on the north and south coasts. Disjunct, restricted distribution in the Vogelkop, Irian Jaya. Restricted range in north-eastern Cape York Peninsula, Queensland, south to Silver Plains

STATUS

Locally common to moderately common

R. Donaghey

TOOTH-BILLED BOWERBIRD

Scenopoeetes dentirostris

Like the Golden Bowerbird, the endemic Tooth-billed Bowerbird has a restricted range in north-eastern Queensland where it inhabits wet tropical rainforest. The promiscuous male clears a distinctive display court on the forest floor and decorates it with fresh leaves. It also utters an extraordinary range of vocalisations.

Discovered by Robert Johnstone in 1874 on the Seaview Range north of Townsville, it was named by J.P. Ramsay in the following year. The name *Scenopoeetes* ("stagemaker") refers to the display court; *dentirostris,* ("tooth-bill") refers to the serrations on the bill of both sexes. Since males use their bills to saw through the stalks of leaves for use as court decorations, it had been suggested that this was its sole adaptation. However, individuals have been seen to eat the margins of leaves in the rainforest canopy during winter, and the stomachs of birds collected in this season contained only leaf material. It seems that the serrated bill is used primarily for tearing and eating leaves – particularly in winter when fruits are least abundant – and secondarily by males in the breeding season, to obtain leaves for decoration of their courts.

Immature male showing notched bill; Paluma, Qld (C. & D. Frith/Frithfoto)

Sponsor: Bruce Walker for the Sporting Shooters' Association of Australia

Many Australian male bowerbirds have bright plumage that enhances their courtship display; the Tooth-billed Bowerbird is atypical in that both sexes have the same cryptic plumage. Males attend their display courts during the breeding season, from September to January, but are most active and vocal at their courts from October to December. Around September, an individual male clears a space, or display court, on the forest floor to make an oval or circular area 0.9 to 2.4 metres in diameter. Each court is decorated with fresh leaves, the paler undersurfaces of which are turned uppermost, creating a striking visual contrast with the dark forest floor. The number of leaves on a court varies from about 40 to a little more than 100. Leaves of particular tree species are selected and most contain leaves of only three to four species of rainforest trees although a few may contain only one kind and others up to 13 species. Leaves used for decoration are mostly 30 to 38 centimetres in length, longer than the bird itself.

The male compensates for its lack of bright plumage with its impressive vocalisations. Throughout the breeding season, males perch above their courts and the forest resonates with their loud, prolonged singing bouts. Males sing and counter-sing against each other to advertise their presence at their courts and to attract mates. Vocalisations include mimicry and are unrivalled among those of other forest-dwellers in their penetrating quality.

Sidney Jackson, who collected and photographed the nest and eggs of the Tooth-billed Bowerbird in 1908, reported that it mimics the songs of other forest-dwellers and weaves this mimicry into its own repertoire. A male on its singing perch above the court throws back its head, fluffs out its whitish throat feathers, opens its bill wide and pours out a constant bout of song for 45 minutes or so before pausing and repeating the performance. Stanley Breeden described its song and mimicry:

> He now imitates a range of other birds: the fluting notes of a black butcherbird, the rat-tat-tat of a Lewin's honeyeater, the scratchy rollicking song of a bridled honeyeater, calls of the golden whistler, the subdued whisperings of a scrubwren. Not only does he mimic the sounds but also their varying volumes. The effect is such that if you just listen to the bird sing, without seeing him, you have the impression that the different songs come from different places; that you hear a number of birds, some close and others further away. But all the sounds are made by just this one bird, sitting in the one place. At other times I have heard tooth-bills mimic a grey goshawk, a grey-headed robin, a crested hawk, a drongo, a figbird, a black-faced flycatcher and once the screech of a flying-fox.
>
> Apart from the rather clear and melodious sounds, he sometimes produces a harsh strange noise; a wheezing kind of screech followed by the penetrating crackling sounds made by certain cicadas. But even that is not the end of his vocal gymnastics. At first I think a Lewin's honeyeater has joined him in a duet, but I soon realise that the tooth-bill utters the wheezing screeches and the honeyeater's song himself and both at the same time. The tooth-bill has this uncanny ability because its syrinx (the part of its throat that produces the sounds) is divided, enabling the bird to sing with two voices at once.

An adult male spends much of the day during the display season perched above his court, singing to attract females. The elements and sequence of the courtship displayed by a male to a female were unravelled recently by Cliff and Dawn Frith who showed that each court contains the trunk of at least one special "display tree" the base of which is kept clear of leaves. The courtship display has two main components, first vocal then visual. When a female approaches a singing male, he descends to the court and hides from her behind the display tree, while continuing to sing. Once concealed, he

Adult male calling over his court; Paluma, Qld (C. & D. Frith/Frithfoto)

Sponsor: Yutta Lewis

changes his song to soft vocal mimicry. As the female nears the display tree, he darts out, gapes open his bill, puffs out his throat feathers to reveal a white "beard", hops sideways flicking his wings and vigorously leaps toward the female, flapping the wings. The Friths watched many displays but have yet to see mating on the court. However, when the stuffed skin of a female was placed on the court, a male performed the above described vocal and visual display: when near the mounted skin, he flew toward it and attempted copulation.

What is the relationship of the Tooth-billed Bowerbird to other bowerbirds and catbirds? Although egg colour is similar in *Scenopoeetes* and *Ailuroedus*, the catbirds have been engulfed in *Ailuroedus* on account of an alleged observation of a captive green catbird laying leaves on the ground. The courtship display and mating system of *Scenopoeetes* and *Amblyornis*, the New Guinea "maypole" bower builders, is similar, suggesting a close relationship. The male tooth-bill hides from the female behind a display tree and utters subsong mimicry and the males of at least two *Amblyornis* species hide behind the base of their maypole bower and give mimicry. Recent molecular evidence suggests that *Scenopoeetes* is more closely related to the "maypole" bower-builders than to the catbirds.

The first nest and eggs of the Tooth-billed Bowerbird were collected by George Sharp in November 1908 in the Evelyn Scrubs in north-eastern Queenlsand. Sid Jackson, a collector and zoologist for H.L. White, located nine nests in December 1908 by getting Aborigines to climb and search likely nest trees. The small nests were incredibly difficult to find because they were hidden in high, dense vine thickets.

The parental care and mating system of the Tooth-billed Bowerbird has long remained obscure. Earlier observers assumed, without evidence, that the species was monogamous, with both parents feeding the young. However, Cliff and Dawn Frith, who watched two nests, found that in the nest with eggs, one individual (presumably a female) was alone responsible for brooding. In the nest which contained nestlings, one individual (again presumably a female) provided all the food. Sole parental care by the female is to be expected for a species in which the male spends much of the day singing and guarding its display court.

R. Donaghey

Tooth-billed Bowerbird

Scenopoeetes dentirostris

Adult female at nest; Paluma, Qld (C. & D. Frith/Frithfoto)

OTHER NAMES
Stagemaker, Tooth-billed Catbird

MEASUREMENTS
LENGTH: 260 – 280 mm
WING: 147 – 157 mm
TAIL: 100 – 105 mm
BILL: 25 – 28 mm
TARSUS: 32 – 37 mm
WEIGHT: 143 – 198 g

DESCRIPTION
ADULT
Head and rest of upperparts olive-brown; sides of lower throat buffy-white forming a whitish throat patch; underparts buff-white, feathers broadly edged with olive-brown, giving a bold streaked appearance. Bill black and notched. Iris dark brown. Legs greenish-grey

IMMATURE
Similar to adult, but bill brown. Iris pale brownish-grey. Legs deep olive green

JUVENAL
Similar to immature, but downy grey head; neck bare; buff underparts mottled and streaked grey. Bill grey. Iris grey. Gape yellow; mouth pale orange. Legs light grey

IDENTIFICATION
A stocky bird; olive-brown above with underparts buffy-white, distinctly streaked brown; with a stout notched bill

VARIATION
Negligible

VOICE
A loud and varied repertoire of wheezing harsh screeches, scratchings, loud melodious whistles, soft whisperings and loud "chucks" interwoven with mimicry

HABITAT
Upland wet tropical rainforest

FOOD
Predominantly fruit, but also leaves, seeds, insects, other arthropods and snails. Leaves form the bulk of the diet in winter

HABITS
Sedentary. Arboreal, mostly forages in the mid to upper levels of the forest. Mostly seen singly, or in twos or small groups. Inconspicuous and wary in the non-breeding season, but much more conspicuous and less wary in the breeding season

BREEDING
Presumably a polygamous mating system. Males clear and decorate terrestrial display courts dispersed in dense clusters throughout the forest. Females perform all parental care. Female nests November – January. One brood per season

NEST: A frail, shallow cup of thin vine tendrils and dry sticks built in dense foliage of saplings or vine tangles from 5 – 30 m above ground

EGGS: Usually 2, sometimes 1; oval shaped; a uniform rich dark cream to cream-brown; 42 x 29 mm

INCUBATION: Unknown; fledging period unknown

DISTRIBUTION
Ranges widely in upland rainforest in north-eastern Queensland from Mount Amos, south of Cooktown to Mount Elliot, south of Townsville at 600 – 1400 m above sea level

STATUS
Common

R. Donaghey

Sponsor: Mrs G. Dechert

Spotted Catbird

Ailuroedus melanotis

The common name of this bowerbird refers to its "cat-like" song and spotted plumage. John Gould and other early naturalists likened its voice to the wailing of a domestic cat. The generic name, *Ailuroedus*, means "cat-singer" or "cat-voiced" and the species name, *melanotis*, means "black-eared", hence the alternative common name, Black-eared Catbird.

The Spotted Catbird and northern populations of the Green Catbird are more vocal and musical and have a greater repertoire of songs than the southern populations of the Green Catbird. John Gould wrote that the Green Catbird has a:

> ... loud, harsh, and extraordinary note ... which differs so much from that of all other birds, that having been once heard it can never be mistaken. In comparing it to the nightly concerts of the domestic cat, I conceive that I am conveying to my readers a more perfect idea of the note of this species than could be given by pages of description. This concert is performed either by a pair or several individuals, and nothing more is required than for the hearer to shut his eyes to the neighbouring foliage to fancy himself surrounded by London grimalkins [alley cats] of house-top celebrity.

The Spotted Catbird is very vocal and conspicuous at the onset of the breeding season, when neighbouring males sing and counter-sing against each other while establishing and maintaining their territories. Studying the Spotted Catbird intensively in a rainforest, one becomes fond of its song, which is much more musical and varied than cat-wailing. Males have a repertoire of songs and match their song types in response to that of a neighbour. Singing continues throughout the day, not only at dawn and dusk. As territories are re-established each spring, neighbouring breeding pairs engage in song bouts and there is much singing and chasing as neighbouring pairs pursue each another and chase off intruders that are trying to establish new territories.

The Spotted Catbird inhabits rainforest in north-eastern Queensland and Cape York Peninsula and is widespread in New Guinea. The first Spotted Catbird was discovered by Alfred Russel Wallace, the distinguished naturalist and evolutionist, near sea level in the Aru Islands in 1857. The following year, it was named by G.R. Gray as a bowerbird, *Ptilonorhynchus melanotis*, but was subsequently moved to a new genus, *Ailuroedus*. The first Australian specimen was collected in the Cardwell – Rockingham Bay area of north-eastern Queensland in 1874 and was named by E.P. Ramsay as *A. m. maculosus*. Tom Gilliard, an American ornithologist, recognised eight other subspecies distributed throughout much of the mid-montane forests of New Guinea, the lowlands of southern New Guinea, the Aru Islands, and Misol, in the western Papuan Islands.

The Spotted Catbird is treated by the Royal Australasian Ornithologists Union as a distinct species. Unfortunately, many popular bird books have lumped the two Australian catbird species together as *A. crassirostris*, the Green Catbird. Consequently, biological information on the two species has been pooled and attributed to the wrong species. Tom Gilliard has argued strongly that although the Green and Spotted Catbirds are undoubtedly closely related, they demonstrate pronounced differences in morphology. The lack of blackish ear coverts and the uniform bright green plumage of the larger Green Catbird throughout its range distinguish it from the Spotted Catbird. He observed that:

> The Spotted Catbird is a remarkably uniform polytypic species considering the pressures of geographical isolation to which it has been exposed. It is composed of nine thinly differentiated races, one on Cape York Peninsula, one in the Aru Islands, one on Misol Island and the rest on the mainland of New Guinea. All have the ear coverts blackish, the head to upper back brown to black with pale feather centres (spotting).

As early as 1858, John Gould recognised that the Spotted Catbird was a bowerbird and suggested that it "might construct a bower similar to that of the Satin Bower-bird". Other early naturalists suspected that catbirds might build a rudimentary bower or display court but, despite searching, none

Adult at nest; Paluma, Qld (C. & D. Frith/Frithfoto)

Sponsor: R.S.P.C.A., N.S.W.

was ever found. Observations of a captive Green Catbird placing leaves on the ground fuelled speculation that both species of catbirds may make display courts, but recent long-term field observations of Spotted Catbirds by Cliff and Dawn Frith at Paluma, north Queensland, reveal that the Spotted Catbird does not construct bowers or display courts as do the polygamous bowerbirds; in field observations totalling thousands of hours, they have found no sign of leaf-laying.

Observations by the Friths of a leg-banded population at Paluma show that breeding Spotted Catbirds are truly monogamous. The female of a pair builds the nest, incubates the eggs and broods the nestlings. Both parents feed the nestlings and fledglings. Males defend the territory and the nest site. Both species of catbird perform a distraction display to lure predators away from the nest. In a nest that contained one recently hatched young, R.S. Miller noted that the brooding bird, a female, "flushed from the nest and fluttered along the ground across the roadway and into the scrub on the other side".

Spotted Catbird pairs occupy rainforest territories throughout the year. Adults have few predators and a high survival rate, but one observer had a rare view of a Grey Goshawk taking an adult Spotted Catbird. After the breeding season, birds moult and are fairly quiet, right throughout winter. Foraging birds are difficult to detect because their green plumage blends with the rainforest foliage. In winter, the Spotted Catbird rarely sings but utters high-pitched squeaks or "chip" calls that serve to maintain contact between members of a pair that are foraging separately. The chip call, wailing songs, strong flight and energetic bouncing from limb to limb assist in their ability to locate each other.

R.S. Miller, who studied Spotted Catbirds around Cairns in 1937 wrote that:

> The seeds of palms, quandong, wild figs, etc, are eaten with avidity. One of the favourite berries is the fruit of the stinging-tree, whilst the finger-cherry, which is reliably stated to cause blindness to human beings, is eaten readily, and apparently, harmlessly, by these birds.

In addition to fruit and berries, the Spotted Catbird catches beetles and various forms of insect life from the rotten and moss-laden tree-trunks, and, as well, eats new buds and shoots of ferns, etc. The Friths have shown that the diet is predominantly fruit, especially figs, but also includes buds, shoots, flowers, seeds, arthropods, lizards and nestling birds. Taking samples of meals that parents fed to chicks, they showed that the nestling diet was both fruit and animal, including other nestling birds.

An intriguing feature of the nests of the Green and Spotted Catbird is that, below the stick lining, there are pieces of decayed wood. Sidney Jackson described this situation in 1921 and gave a plausible explanation of its function. He was surprised to find:

> ... about one inch below the thin stick lining, a layer of wet, decayed wood, some pieces measuring over two inches long by a half inch thick. This wet wood filled two 2 pound jam tins, and, after drying in the sun thoroughly, was with difficulty pressed into one of these tins. Probably this wet layer is instrumental in promoting heat [insulation] when the female is sitting on the nest.

R. Donaghey

Spotted Catbird

Ailuroedus melanotis

Fledglings; Paluma, Qld (C. & D. Frith/Frithfoto)

OTHER NAMES
Black-eared Catbird

MEASUREMENTS
LENGTH: 260 – 290 mm
WING: 146 – 169 mm
TAIL: 110 mm
BILL: 33 – 38 mm
TARSUS: 44 – 47 mm
WEIGHT: 143 – 195 g

DESCRIPTION

ADULT
Head to upper back dark blackish-brown, strongly spotted with buff on upper back; ear coverts brown-black; rest of upperparts, including tail, emerald green; tail, secondaries and wing coverts tipped white; chin black; throat blackish, spotted buff-white, rest of underparts buffy white with olive green to brown edges; olive green feathers streaked creamy white to give a spotted appearance. Bill light grey. Iris red. Legs pale grey. Sexes similar but female slightly smaller

IMMATURE
Similar to adult but darker and duller. Iris mid-brown

JUVENAL
Like immature, but fluffy down still on head

IDENTIFICATION
Robust green, dark-headed, black-eared bird with stout bill, strong legs and red iris

VARIATION
About 10 subspecies, two in Australia: A. m. maculosus *occurs in north-east Queensland;* A. m. joanae *in eastern Cape York Peninsula. These have similar plumage with little variation. The eight subspecies in New Guinea vary in the degree of brown to black of the upperparts and the buff to green of the underparts.*

VOICE
Cat-like wailing songs, baby cries, varied repertoire of songs, high pitched squeaks or chip calls

HABITAT
Wet tropical rainforest, rainforest edge and adjacent forest. Visits gardens and orchards

FOOD
Predominantly fruit, especially figs. Diet also includes buds, shoots, flowers, seeds, insects, nestling birds and lizards

HABITS
Sedentary. Breeding pairs territorial and occupy larger home ranges in non-breeding season. Mostly occurs singly or in pairs during breeding season but may form small foraging groups in non-breeding season. Arboreal, forages from ground level to upper canopy. Very active and wary. Blends with foliage

BREEDING
Nesting season September – March. One brood per season. Monogamous breeding pairs. Female constructs nest, incubates eggs, broods and feeds young. Male defends territory and shares equally in feeding the young

NEST: Bulky cup of twigs and broad, dried leaves bound with stems of vines and lined with fine twigs; pieces of decaying wood below the twig lining; built in forks in densely foliaged canopy of small rainforest trees and palms often 2 – 5 m above ground

EGGS: Usually 2, sometimes 3; oval-shaped, glossy, uniform cream to pale olive brown; 40 x 28 mm

INCUBATION: 23 – 24 days; fledging period about 21 days

DISTRIBUTION
Two Australian subspecies separated geographically: north-east Queensland subspecies, A. m. maculosus, *extends from Cooktown south to Mt Spec in Paluma Range; Cape York Peninsula population,* A. m. joanae, *extends along the east coast between the Claudie and Rocky Rivers. Coast and tablelands from near sea level to above 900 m*

STATUS
Common resident of north-east Queensland. The population in northern Cape York Peninsula is less common than the north-east Queensland population

R. Donaghey

Sponsors: Sir John & Lady Proud

Adult; Atherton Tableland, Qld (M. & I. Morcombe)

Sponsored in memoriam: Dr Judith Ross, M.B., F.F.A.R.A.C.S.

Green Catbird

Ailuroedus crassirostris

This robust bird has a pale stout bill, bright green wings and tail, dull green underparts with distinct white spots and a red iris. It bounds up branches on strong legs. In contrast to most other bowerbirds and birds of paradise, it is monogamous and the sexes are similarly coloured. The Green Catbird dwells in lowland and upland rainforest from south-eastern New South Wales to south-eastern Queensland, a range which does not overlap that of the Spotted Catbird. Green Catbirds usually move about singly or in pairs which are territorial and show great fidelity to the same territory, year after year.

This species was the first of the Australian catbirds to be described but its discoverer and the location from where it was collected are not known. The original specimen was named by a Swedish ornithologist in 1815 as a shrike, *Lanius crassirostris*, in reference to its thick bill. It was later placed in the bowerbird genus, *Ptilonorhynchus*, and it was not until 1851 that the genus *Ailuroedus* ("cat-voices") was erected to accommodate it. The name refers to its wailing song.

During the cooler winter months, the Green Catbird is rather quiet, but members of a foraging pair maintain contact with single or double squeaks or chip notes. In spring, at the onset of the breeding season, males are very vocal as they establish territories by singing and by vigorously chasing intruders from their space. Their wailing songs can be heard throughout the day as competitors sing and counter-sing against each other, changing their songs to match those of their rivals. Males in north-eastern New South Wales have a much more varied repertoire than their southern relatives. Members of a pair sing a duet against neighbouring pairs, particularly during the early phases of the breeding cycle. The female song, given while swinging her head from side to side, consists of rapid whirring and slurring notes and ends with a sharp "tsit".

The pair bond is maintained by song duets and other vocalisations and by breeding behaviour and joint defence of territory or nest. The male feeds his mate throughout the year, the female begging for food by squeaking and wing-quivering. Figs are the principal item provided by the male, but during the pre-laying period males provide their mates with insects and their larvae.

Intensive study of colour-banded birds enabled me to unravel some of the complex relationships between diet and the breeding system. The Green Catbird feeds mainly upon fruits but also eats shoots, flowers, seeds, insects and occasionally, small vertebrates. In the upland subtropical rainforest of Tooloom Scrub, north-eastern New South Wales, massive strangler figs tower above the forest canopy and the fruits of these trees are a major element of the diet of adults and nestlings. Most breeding territories contain one or two such trees, which regularly produce an abundance of ripe fruit for several months during the catbird's breeding season. Catbirds forage high in the canopy, in association with other bowerbirds and frugivorous birds such as pigeons.

Unlike many other frugivores that eat large fruits *in situ*, the Green Catbird uses its stout bill to detach a fig from its stem, then carries it to a feeding place below the canopy – often a broad limb, a crevice in a tree, or a birdnest fern. The seeds of the fig are thereby dispersed on the trunks and limbs of other rainforest trees and eventually form an interlocking network that envelops these hosts.

A Green Catbird usually perches upright when plucking a fig but it may occasionally do so in a head-down posture or in flight. When feeding on flowers, particularly epiphytic orchids, it may sometimes hang upside-down. Animal food, which constitutes about 10% of the diet, is mostly arthropods gleaned from branches and foliage. Tree-frogs and nestlings of small songbirds are occasionally eaten.

Like riflebirds, the Green Catbird drinks from, and bathes in, pools of water in crevices of rainforest trees. When such pools are unavailable, it bathes by clinging onto dripping wet foliage and shuffling and shaking its plumage.

Early naturalists suspected that it might construct rudimentary bowers or clear display courts, but this behaviour has never been verified despite thousands of hours of field observations at Tooloom Scrub in New South Wales. No leaf-laying has been observed at display courts. Breeding pairs defend multi-purpose territories in which they raise offspring, obtain food and find shelter from enemies. The female builds the nest, incubates the eggs and broods the nestlings, spending about 20 minutes per hour in this activity over the 20 day nestling period. Nestlings are fed almost equally by both parents at a combined rate of about five trips per hour. Figs constitute about half of the food provided, but the nestling diet also includes beetles, cicadas, millipedes and nestling birds. Breeding catbirds cache fruit near their nests in birdnest ferns, tree forks and depressions. Brooding females retrieve food from the caches, permitting them to spend more time near the nest. Faecal sacs from the nestlings are eaten by the parents. The young (usually two, sometimes three) fledge at about 20 days but remain dependent for a further 80 or so days.

In the Tooloom Scrub, about 65% of nestings were successful in each of three breeding seasons that were studied. There appear to be few nest predators. The green plumage of brooding females blends well with the bright green rainforest foliage and most nests, placed in dense foliage, are well concealed from above. Potential reptilian predators include the nocturnal Carpet Python, *Morelia spilota*, and the diurnal Lace Monitor, *Varanus varius*. Parents perform a distraction display against a monitor approaching a nest tree, but predation by monitors has not been recorded. A known avian predator is the Grey Goshawk, *Accipiter novaehollandiae*, which hunts for prey below the forest canopy and has been seen to take young from the nest. Adults appear to have a high survival rate. The regular, abundant crops of strangler fig fruits and the random dispersion of fig trees has helped promote monogamy, territorial behaviour and biparental care of nestlings.

R. Donaghey

Green Catbird

Ailuroedus crassirostris

Adult at nest; Lamington National Park, Qld; February 1992 (R. Brown)

OTHER NAMES
None

MEASUREMENTS
LENGTH: 295 – 340 mm
WING: 155 – 171 mm
TAIL: 120 – 131 mm
BILL: 26 – 33.5 mm
TARSUS: 43 – 50 mm
WEIGHT: 194 – 224 g

DESCRIPTION

ADULT
Male: Crown and sides of head yellowish green to olive, feathers tipped black; nape to upper back olive green, feathers with greenish white centre streaks or spots; rest of upperparts bright green; secondaries and secondary coverts with white tips; central tail feathers green; rest of tail feathers bright green with white tips; feathers at sides of neck and upper breast often tipped buff-white to give an irregular spotting effect; malar region olive to yellowish green with indistinct black barring; throat and posterior of ear coverts greyish washed with green, shafts of feathers with greenish white streaks; breast and sides of abdomen dull green, feathers with broad greenish white spots and streaks; abdomen and flanks greenish yellow, feathers with yellowish-white central streaks; undertail coverts pale greenish yellow. Bill horn-coloured. Iris brownish red. Legs greyish brown. Adult female similar to males

IMMATURE
Similar to adult, but plumage duller. Bill more greyish. Iris reddish brown

JUVENAL
Similar to immature, but with fluffy grey down on head

IDENTIFICATION
Robust, green, with distinct white-spotted underparts and nape, pale stout bill, strong legs and reddish iris

VARIATION
None recorded

VOICE
Cat-like wailing songs, baby cries, varied repertoire of songs, high pitched squeaks or chip calls

HABITAT
Favours subtropical and warm temperate rainforest. Also inhabits rainforest edge and adjacent eucalypt forest or woodland. Visits gardens and orchards

FOOD
Predominantly fruit, especially figs. Diet also includes buds, shoots, flowers, seeds, insects, millipedes, frogs and nestling birds

HABITS
Sedentary. Breeding pairs are territorial, occupying larger home ranges in non-breeding season. Mostly moves singly or in pairs, but may form small foraging groups in non-breeding season. Arboreal, foraging from ground level to upper canopy. Very active and wary. Blends with foliage

BREEDING
Nesting season September – February. One brood per season. Monogamous. Female constructs nest, incubates eggs, broods and feeds young. Male defends territory and shares equally in feeding young

NEST: Bulky cup of twigs and large, broad dried leaves, bound with stems of climbers and lined with many fine twigs and sometimes rootlets. Layer of leaves forms base of nest; many leaves are placed on edge around the sides; pieces of decaying wood below the twig lining; mostly built in fork near the top of slender understorey tree with densely foliaged crown, in stinging trees and trees with thorns, in bird nest ferns and in crown of tree fern at heights of 1 – 18 m

EGGS: Usually 2, sometimes 3; oval-shaped; uniform rich cream; 44 x 31 mm

INCUBATION: 23 – 24 days; fledging period 21 days

DISTRIBUTION
From Mount Dromedary, south of Narooma, south-eastern New South Wales, north to the Cooloola and Kingaroy districts, south-eastern Queensland from near sea level to above 1000 m

STATUS
Locally common. Endemic

R. Donaghey

Adult female Paradise Riflebird; Lamington National Park, Qld; December 1975 (T. & P. Gardner)

Sponsor: Nell Filewood

Birds of Paradise

Of the 43 species of birds of paradise, all but six are confined to New Guinea. Two of the remaining number are found in the Moluccas, while four are located on the eastern coast of Australia (two of which are also found in New Guinea). In comparison to the richly coloured and highly ornamented species found in New Guinea, the Australian species are relatively drab, although the glossy plumage of the male riflebirds of the Genus *Ptiloris* are a beautiful sight, and their courtship display is spectacular. The Trumpet Manucode, *Manucodia keraudrenii*, which is found in the rainforests of northern Cape York Peninsula and New Guinea, is so named for its call – a single loud note resembling that of an organ blast. The sound is produced from an unusually extended windpipe coiled between the skin and the muscles of the chest.

Birds of paradise vary considerably in size and shape. The King Bird of Paradise, *Cicinnurus regius*, and Wilson's Bird of Paradise, *Diphyllodes respublica*, are the smallest, with body lengths of 15 centimetres, while the Black Sicklebill, *Epimachus fastosus*, measuring some 70 centimetres (excluding central tail feathers), is the largest.

The first records of these remarkable birds date back to 1521, when the crews of Ferdinand Magellan's ships docked at Tidore, in the Moluccas, several months after Magellan himself had been killed during a battle with natives of the Philippines. The Spaniards were trading goods for spices but during their stay at Tidore, were given specimens of the Lesser Bird of Paradise, *Paradisaea minor*, as gifts for the King of Spain from the King of Bacchian (today called Batjan), a remote volcanic island in the northern Moluccas. Only one of the fleet, the *Victoria*, commanded by El Cano, returned to Spain in September 1522, thus becoming the first ship to circumnavigate the world. The only comprehensive record of the voyage was kept by a young Venetian nobleman, Antonio Pigafetta, who stated that the King of Bacchian sent as presents for the King of Spain some spices, a slave and:

> ... two very beautiful dead birds which are as thick as stock-doves with small head and long beak, and legs a palm in length and as thin as a feather. They have no wings, but have instead long feathers of diverse colours like large plumes. The tail is as long as that of a stock-dove, and all the other feathers except the wings are of a tawny colour, and they never fly except when there is wind. We were told that these birds came from the earthly paradise and were called Bolon diuata, that is to say, birds of God.

A few years later, Portuguese explorers returned with skins of the Greater Bird of Paradise, *Paradisaea apoda*, probably originating from the Aru Islands. The wings and legs had been removed and the skins were prepared to emphasise the colourful plumes. From these specimens arose the myth that these legless birds lived always high in the sky, the female incubating her eggs in a hollow on the back of the male, neither birds ever coming to land. In this way arose the concept of "birds of paradise".

This myth was widely accepted until late in the sixteenth century when Dutch navigators learned that the birds lived in trees, and in 1605 Carolus Clusius, a professor at Leiden University, used this information to destroy the legend of legless birds. More scanty information trickled back to Europe over the next couple of centuries, but it was not until 1824 that the first naturalist actually saw a bird of paradise in the wild. On 26 July 1824, a french corvette, *La Coquille*, arrived at Dorey Bay (today called Manokwari), on the Vogelkop Peninsula. On board was Rene P. Lesson who wrote of his first observations of a Lesser Bird of Paradise:

> The sight of this first bird of paradise was out of this world. The flintlocks in our hands did not move, we were so amazed. Whilst we were walking very carefully on a wild pig trail through dense scrub, suddenly in a slight curve a bird of paradise flew gracefully over our heads; like a meteor it trailed a light when flying through the scrub. With its display feathers close to its sides, this bird looked like an egret.

Following this visit, many expeditions were made to New Guinea and the Moluccas and, as the areas of New Guinea were brought under colonial management, collectors ventured farther afield, gathering comprehensive information on the distribution and habits of these birds.

The second naturalist to see birds of paradise in the wild was Alfred Russel Wallace, who is probably best remembered as the co-originator of what is known as Darwin's theory of evolution. In March and May of 1857 Wallace spent six weeks in the Aru Islands, where he observed male Greater Birds of Paradise, *Paradisaea apoda*, in display. In 1869, he visited Batjan, where he discovered Wallace's Standardwing, *Samioptera wallacei*, so called because of its long white flag-like wing plumes. Following this sighting, it eluded observation for nearly 60 years, and then it was only seen a few times until 1953, when it disappeared again. It was not until 1983 that a British ornithologist, David Bishop, rediscovered it and made the first detailed observations of this extraordinary bird of paradise. Even now very little is known of its habits or behaviour, and no nests have ever been found.

The Ribbon-tailed Bird of Paradise *Astrapia mayeri* was the last bird of paradise to be discovered, in 1939.

P. Rowland

Paradise Riflebird

Ptiloris paradiseus

This endemic species is the most southerly of the birds of paradise, living in rainforest from south-eastern Queensland to northern coastal New South Wales. It is noticeably larger than the closely related Victoria's Riflebird of the rainforests of north-east Queensland. The adult male is velvet black with metallic oil-green crown, nape and upper breast shield, iridescent blue-green central tail feathers and a purple sheen to the abdomen and upper parts. The plumage sparkles green, blue or purple, depending on the bird's orientation to the sunlight and enhances its courtship display; even humans are entranced by this unforgettable experience. In promiscuous male birds of paradise, as in bowerbirds, sexual maturity is delayed. Males acquire adult plumage later than females but we do not know at what age a male Paradise Riflebird attains adult plumage – it may be between three and seven years.

The plumage of the female is olive-brown and chestnut-brown above and buff-white barred dark brown below. She builds the nest, incubates the eggs and feeds the nestlings. Her cryptic plumage tends to make her inconspicuous but there is probably a high rate of predation of eggs and young.

The generic name, *Ptiloris*, was given to this bird in 1825 by the British ornithologist William Swainson, in reference to feathers that extend over the base of the mandible. *Paradiseus* refers to its alliance with the better known birds of paradise of New Guinea. It was the first member of that family to be discovered in Australia. Why it should be called a "riflebird" is less certain. The French naturalists F. Lesson and P. Garnot of the French exploring ship *La Coquille*, which visited Sydney in 1824, wrote (rather unconvincingly) that it was called "rifleman" because the first recorded specimen had been shot by a soldier in the garrison at Port Macquarie. Certainly, male and female specimens were sent from Port Macquarie to the Edinburgh Museum, but these were more aptly labelled "velvet birds". Less abstrusely, it has been suggested that the plumage resembled the "well-known uniform of the rifle regiments of the British Army".

It tends to be sedentary in rainforest and at rainforest edges, but immatures and even adults may disperse into adjacent forest and woodland, particularly outside the breeding season. The most northerly population is isolated in the Calliope Range, east of Biloela, Queensland. Otherwise it ranges more or less continuously, mostly in upland forest, from the Bunya Mountains in south-east Queensland to the upper Williams River, north of Dungog in central eastern New South Wales. Much of the original habitat of the Paradise Riflebird – particularly lowland rainforest – has been cleared for agriculture and, today, many of the remaining protected rainforest remnants are severely reduced and fragmented. However, large tracts of rainforest are reserved in the Border Ranges and the contiguous Macpherson Range, Queensland.

The Paradise Riflebird feeds predominantly on insects and other arthropods by probing with its decurved bill under bark, into tree crevices and epiphytes such as staghorns, elkhorns and birdnest ferns. Individuals usually forage from ground level to the uppermost canopy. The species is nowhere common, but foraging birds can be located by the rustling of their wings, their sporadic, strident calls and their "treecreeper" foraging behaviour. In 1918, J.P. Ramsay described the foraging behaviour of a female Paradise Riflebird in the forests of the upper Clarence River:

> Clinging like a Climacteris to the side of a branch or trunk, she would pick a few times at some loose bark or rotten timber, and then lay her ear against it listening for any movements of insects within. She would peck and listen a few times, move on a foot or so, repeat the performance and then suddenly pick and tear furiously, while bark and rotten wood showered down, until the unfortunate beetle or centipede was discovered and devoured. The birds showed great strength of legs and claws and clung to the under side of a branch for long periods; one was timed for over 40 minutes hanging head downwards on a broken branch apparently devouring ants or their larvae.

Sidney Jackson's observations in 1919 in the upland rainforests of the Macpherson Range are also pertinent:

> It was interesting to watch these birds climbing the trunks of the large Beech (Fagus moorei) and other trees, and levering off the big scales of bark with their long curved bills when hunting for beetles ... they also examine all grub holes and when "probing" these holes in old dead scrub trees they often place their head flat up against the tree, no doubt "listening" to hear if any grub or beetle etc, moved inside. We found that they usually did most of their hunting on the east side of the trunk of such trees, and on that side they would have all the rotten bark removed.

In addition to insects and other arthropods, the Paradise Riflebird also eats fruit. At Tooloom Scrub in New South Wales, during January and February, one to three riflebirds were seen feeding on the fruits of the native pepper, *Piper novae-hollandiae*. J.P. Ramsay observed the Paradise Riflebird feeding in association with the Satin Bowerbird on the fruits of the Bastard or Scentless Rosewood, *Synoum glandulosum*, and the Native Tamarind (presumably *Diploglottis australis*).

In upland rainforest, there is often little surface water, particularly in winter and spring. Rainwater accumulates in tree cavities and in pockets between the buttresses of trees – it is here that the Paradise Riflebird drinks and bathes, as described by S. Jackson:

> We counted nine "dips" that the lovely bird had in his little bath, and, although we were fully 30 feet away, we could plainly hear him splashing, and each time he had a "dip" he flew up on to a vine close by, about four feet from the ground. We got a splendid view of him each time, cleaning and combing his gorgeous feathers with his long curved bill.

Adult males are vocal and conspicuous on their display perches during the breeding season. Breeding females are cryptic and their nests are rarely found. In north-eastern New South Wales females build nests at extraordinary heights above ground and consequently, little is known about the nesting behaviour. The nest is a cup-shaped structure of vine stems and leaves, decorated with cast-off snake skins. It is similar to that of the Victoria's Riflebird but more bulky. Two nests found by J.P. Ramsay in November in the forests of the upper Clarence River were situated in mistletoe clumps in the tops of eucalypts 35 metres and 39 metres above ground. One tree was at the rainforest edge, the other on a steep wooded hillside a kilometre from the nearest rainforest patch. S. Jackson found a nest at a height of 21 metres in a mass of vines. All three nests had been deserted after being lined and had either been abandoned before egg-laying or after eggs were predated during early incubation.

Adult males, dispersed solitarily throughout the forest, intensify their vocalisations in October at the start of the display season. Each adult male acquires a territory in the rainforest that includes more than one display site from which it delivers unmusical advertising calls to attract females for mating. Display sites are usually high up in trees and consist of an exposed dead branch protruding from the top of a very large fig tree or a thick, moss-covered horizontal limb. When a female alights on a display perch, the male is transformed into a shimmering spectacle of green, blue and purple as the early morning sunlight strikes his plumage. He raises his tail and opens and fans out his wings in front of his body. The head is thrown back while the bill is pointed up and opened to expose the lime-green gape. In October at Binna Burra, southern Queensland, naturalists observed a female hop toward a displaying male. The male embraced her with wings so outstretched that the tips met behind her head, forming a collar. The male clapped his wing tips together, making a sound audible 60 metres away on a calm day. As the female advanced and the male retreated along the display perch, wing-clapping intensified. On one occasion, 17 slow wing claps and 20 faster ones preceded mating.

R. Donaghey

Adult male displaying; Lamington National Park, Qld; December 1990 (R. Brown)

Sponsors: Sir John & Lady Proud

Paradise Riflebird

Ptiloris paradiseus

Adult male; Lamington National Park, Qld; December 1990 (R. Brown)

OTHER NAMES
Riflebird

MEASUREMENTS
LENGTH: c 280 mm
WING: 142 – 158 mm
TAIL: c 100 mm
BILL: 44 – 53 mm
TARSUS: 32 – 38 mm
WEIGHT: Not recorded

DESCRIPTION
ADULT
Male: Forehead, crown and nape metallic oil-green, glossed with blue; rest of upperparts velvety black with pink-purple sheen; feathers surrounding nostrils, lores, chin and sides of head and neck velvety black glossed with purple; throat and upper breast metallic oil-green glossed with blue, feathers of sides of upper breast forming an erectile fan; abdomen feathers and flanks black, edged oil-green; undertail coverts black; tail black with shorter central feathers iridescent blue-green. Bill black. Iris brown. Legs grey-black

Female: Upperparts, except wings and tail, olive-brown; broad, dull white band above eye; throat and cheeks buff-white; underparts buff-white, the feathers with subterminal bars of dark brown; wings and tail dark chestnut-brown. Bill grey-black. Iris brown. Legs grey-black

IMMATURE
Similar to adult female

IDENTIFICATION
Adult male velvety black with decurved bill, short tail and iridescent green crown, throat, upper breast and central tail. Female olive-brown above except for chestnut-brown wings and tail; buff-white, barred dark brown, below

VARIATION
None recorded

VOICE
Strident, guttural "yaa-aass-yaas-aass"

HABITAT
Subtropical and temperate, mostly upland, rainforest and adjacent eucalypt forest and woodland

FOOD
Mostly arthropods (insects, spiders, centipedes) and fruit

HABITS
Sedentary in breeding season. Some individuals disperse from rainforest into adjacent woodland and forest, mainly in non-breeding season. Arboreal, foraging (mostly singly) for arthropods from near ground level to high in canopy. Gathers in twos and small groups in fruiting trees and vines

BREEDING
Delayed sexual maturity in males. Probably a polygynous mating system. Displaying adult males solitarily dispersed throughout forest. Females probably perform all parental care, nesting October – January. Probably one brood per season

NEST: A shallow cup of vine tendrils, fine twigs and dead leaves lined with fine vine tendrils and adorned with cast-off snake skins, moss and orchids; built in dense foliage of tree, in dense vine tangle or in mistletoe clump at height of 6 – 39 m, mostly above 10 m

EGGS: 2, occasionally 1; oval-shaped and glossy; rich reddish-cream with longitudinal steaks of reddish-brown, purple and grey; 36 x 24 mm

INCUBATION: Unrecorded; fledging period unrecorded

DISTRIBUTION
Upper Williams River, north of Dungog, central-eastern New South Wales, north, more or less continuously, to Bunya Mountains, south-eastern Queensland. Most northerly population isolated in the Calliope Range, east of Biloela, Queensland

STATUS
Locally fairly common to uncommon

R. Donaghey

Victoria's Riflebird

Ptiloris victoriae

Victoria's Riflebird inhabits rainforest on the coast and mountains of north-eastern Queensland. It was named in 1849 in honour of Queen Victoria by John Gould, based upon specimens taken by his collector John McGillivray in 1848 from the Barnard Islands near Cardwell. The male has a gorgeous, iridescent plumage and an elaborate courtship display, and is probably promiscuous. The female has cryptic brown plumage and almost certainly performs all parental care.

It is noticeably smaller than the Paradise Riflebird. The male is larger than the female but the female has a longer and slightly more decurved bill. Adult male plumage is similar in both species. Males are velvety black with iridescent blue-green on the head, throat, upper breast and central tail feathers. The plumage of the female is similar to that of the female Paradise Riflebird, but paler, more grey, and less brown above; the underparts are more ochreous and buff, less whitish and less heavily marked with brown.

Victoria's Riflebird is usually encountered in rainforest. For much of the year, particularly in the non-breeding season, female-plumaged birds are more often seen than males. It is a rather quiet foraging bird that can be detected by the rustling of its wings as it flies from tree to tree in search of food, and by its sporadic, strident calls. Like other riflebirds, it forages like a treecreeper up tree trunks and along branches in search of insects, spiders and centipedes. The decurved bill is used to pry

Adult male displaying (J. Sharland)

Sponsor: Alex W. Auldist

off large pieces of bark, to chisel into dead branches and to probe into tree crevices and rotten logs and stumps on the forest floor. Birds foraging high in the canopy can often be detected by a shower of small debris raining down. The diet also includes native and cultivated fruit. In summer and autumn, when fruits are most abundant, male and female riflebirds are often conspicuous as they gather together in association with bowerbirds and other fruit eaters. One observer noted that the Victoria's Riflebird fed on the red fruits of the Alexandra Palm and chased off smaller fruit-eating birds such as the Olive-backed Oriole and Lewin's Honeyeater.

During the breeding season, adult males are more vocal and conspicuous on their display perches whereas breeding females are very cryptic and their nests are not often found. Little is known of nesting in this species. Almost certainly, the female is responsible for nest building, incubation, feeding and brooding the nestlings, and care of the young to independence.

The cup-shaped nest is placed in dense foliage, in the crown of a tree in a vine thicket, the crown of a palm, or in an epiphytic fern. It is constructed of stems, vine tendrils and a few dead leaves and is lined with various materials including fine vine tendrils, fern roots and stems and *Casuarina* needles. The rim of the nest, like that of the Paradise Riflebird, is often decorated with a cast-off snake skin and green fern fronds. By watching a female Victoria's Riflebird preening on a perch and then following it in flight Sidney Jackson found its nest in a birdnest fern and wrote in 1909:

> The nest was situated about 25 feet from the ground, was not very large, and had been rather loosely constructed of dead leaves, rootlets, etc, and closely resembled that of the Bower Shrike-Thrush ... save that it was much less compact. It measured about seven inches across, and the cup part was about two and a half inches deep; the lining was similar to that of the nest of the New South Wales species ... and comprised fern-roots and thin yellow fern-stems. There was no cast-off or "sloughed" snake-skin about the fern growth, but on the nest itself there was a small portion of one, and under the eggs I found several large belly scales belonging to a cast skin. The nest was also decorated with green sprays of a climbing fern [Polypodium] , the leaf of which is small and thick and similar to that of the fern used in the same way by the New South Wales species.

The strange habit of using cast-off snake skins as nest material is practised by the two species of riflebird that are endemic to Australia. No other bird in Australia or New Guinea is known to share this behaviour on a regular basis. Sloughed snake skins may be used simply as nest decoration but it is possible that they scare off potential nest robbers. This hypothesis has not been experimentally tested, nor have the rate and causes of nest predation been studied in any species of riflebird.

We do not know how long a male riflebird takes to attain adult plumage. It may be as little as four years or as long as seven years. Males are most vocal and conspicuous at the onset of the breeding season. As they acquire territories, they become dispersed throughout the forest. Each territorial male has special display sites from which it gives advertising calls and displays to attract females. Favoured display sites include dead, short, horizontal limbs among the upper branches of tall trees, the topmost branch of a dead tree that projects above the forest canopy, and the tops of dead stumps (often only six metres above ground). The rays of the early morning sun on these high perches changes the black plumage of males to a shimmering iridescence of green, blue and purple. At the height of the display season females visit a male's display site and trigger off spectacular behaviour, as described by Stanley Breeden:

> The male faces her. He raises his shining tail and fans his wings till their tips meet under his chin. He spreads the dark golden feathers of his underside into a kind of skirt. The iridescent feathers on his head flash blue and green as he ripples them in the sunlight. Swaying from side to side and moving up and down in slow knee bends in a mesmeric dance, he slaps his wing-tips together. He opens his beak wide to show the yellow of its inside. The female hops along the branch towards this brilliant and passionate display. She inspects the male closely for a few moments then flies off. The male is literally deflated. He folds his wings and sleeks his feathers.

An adult male may also display to an immature male. One such display was seen on top of a six metre stump. The birds faced each other with their breasts almost touching. Then each bird, in turn, slowly stretched and extended its opened wings forward, threw back its head and opened the bill wide. The tails were spread and the two birds swayed from side to side and up and down with their wing tips meeting out from the breast. The display stopped abruptly and both birds paused before resuming the performance. A male Victoria's Riflebird may also give a solitary display, perhaps as "overflow" behaviour or to advertise its presence to females. The display season lasts from October to December, permitting males to mate with late-nesting females and also re-nesting females that have lost eggs or young through predation.

R. Donaghey

Adult female; Kuranda, Qld; June 1990 (D. Whitford)

Sponsor: Dr Susan Stevens

Victoria's Riflebird

Ptiloris victoriae

Nest and eggs; Atherton Tableland, Qld (C. & D. Frith/Frithfoto)

OTHER NAMES

Queen Victoria's Riflebird, Lesser Riflebird

MEASUREMENTS

LENGTH: *230 – 250 mm*
WING: *120 – 138 mm*
TAIL: *80 mm*
BILL: *34 – 42 mm*
TARSUS: *32 – 39 mm*
WEIGHT: *77 – 107 g*

DESCRIPTION

ADULT

Male: Crown and nape metallic blue-green; rest of upperparts velvety black glossed with purple; chin, sides of head and lower breast velvety black tinged with violet and greenish-bronze; throat and upper breast metallic oil-green glossed with blue; feathers on sides of upper breast forming an erectile fan; abdomen feathers blackish, edged oil-green; tail black with shorter central feathers iridescent blue-green. Bill black. Iris dark brown. Legs grey-black

Female: Upperparts brownish-grey; pale buff stripe above the eye; throat and cheeks buff-white; underparts buff to ochreous with dusky brown bars and spots; wings cinnamon-brown; tail rufous brown. Bill grey-black. Iris dark greyish-brown. Legs grey-black

IMMATURE

Male like adult female in first and second years, with a few iridescent green feathers in second and third years

JUVENAL

Similar to immature female

IDENTIFICATION

Adult male a blackish riflebird with decurved bill, short tail and iridescent green crown, throat and central tail. Female very different from male, brownish-grey above and deep buff barred brown below

VARIATION

No geographical variation in size, but females and young males from near the southern limit of range approach Paradise Riflebird in being more heavily marked; females slightly paler than those from the northern limit of range

VOICE

A loud, rasping "ya-as"

HABITAT

Wet tropical, mostly upland, rainforest, gallery rainforest, swamp woodland and adjacent open woodland. Visits gardens and orchards

FOOD

Fruit and arthropods (insects, spiders, centipedes)

HABITS

Sedentary. Some individuals disperse from rainforest into adjacent woodland, mainly in the non-breeding season. Arboreal, foraging (usually singly) for arthropods in the mid – high canopy. Gathers in pairs and small groups in fruiting trees and vines

BREEDING

Delayed sexual maturity in males. Probably polygynous mating system. Solitary displaying adult males dispersed throughout forest. Females probably perform all parental care. Female nests October – January. Probably one brood per season

NEST: A shallow cup of stems, vine tendrils and a few dead leaves lined with fine vine tendrils, adorned with snake skin and built in dense foliage at heights of 2 – 9 m above ground

EGGS: 2; round-oval in shape; buff pink streaked with reddish-brown and purplish-brown; 32 x 23 mm

INCUBATION: Unknown; fledging period unknown

DISTRIBUTION

Coast and tablelands of north-east Queensland from Mt Amos near Cooktown, south to Mt Spec, northern Paluma Range. Also present on North Barnard, Dunk, Goold and Hinchinbrook Islands

STATUS

Locally fairly common. Endemic

R. Donaghey

Magnificent Riflebird

Ptiloris magnificus

The adult male of this species has a velvety plumage with iridescent green nape and breast shield and black flank plumes. The cryptic plumage of the female is brown above and dull white below, strongly barred with brown. Three distinct subspecies are recognised, two of which have extensive ranges in the lowlands and foothills of New Guinea and the Aru Islands. The Australian subspecies, *P. m. alberti*, is restricted to tropical rainforest on Cape York Peninsula.

The Magnificent Riflebird differs from the two other Australian riflebirds in size, plumage, voice, display, nest and eggs. The adult male is easily distinguished from the smaller Paradise Riflebird and Victoria's Riflebird by its larger size, a more expansive greenish-blue breast shield extending to the chin and lower down on the breast, and its filamentous flank plumes. The adult female is more red-brown above and whiter below than in the other two species.

Adult female feeding chicks; Iron Range, Qld (C. & D. Frith/Frithfoto)

Sponsor: Lloyds Bank NZA Limited

It can be located in dense forest by its loud, resonant distinctive whistles, rustling wings, and foraging behaviour. Outside the breeding season it utters two sharp whistles, but W. McGillivray noted that during the nesting season this call alters to three sharp whistles: "wheeoo" and a long "who-o-o". These calls are very different from the harsh "ya-as" of the other Australian riflebirds.

It forages like a treecreeper, up tree trunks and on branches and limbs. The decurved bill is used with the feet to prise bark from tree trunks and the base of large branches in search of insects and their larvae, being employed like a chisel to probe into tree crevices. The diet is predominantly of arthropods (insects, spiders and centipedes) but includes much small fruit. The outer primary feathers of the wings are modified to produce a distinctive rustling sound which can be heard during a male's display and in flight, as foraging birds move from tree to tree.

The Magnificent Riflebird was described in 1819 by the French ornithologist L.J.P. Vieillot, from a specimen collected from the Vogelkop in western New Guinea. The first Australian specimen was collected by John McGillivray from Cape York Peninsula and named by D.G. Elliot in 1871.

Adult male; Cape york, Qld; 1984 (M. & I. Morcombe)

Sponsor: Marie Kinnersly

Adult males have a spectacular courtship display. In the breeding season, they are dispersed throughout the forest, each holding a territory that includes several display sites from which they give advertising calls and displays that attract potential mates. Each male attempts to mate with as many females as possible during the display season. In Cape York Peninsula, favoured display sites are exposed bare branches in the topmost part of the tallest trees. In New Guinea, display sites include the uppermost branches of tall emergent trees, but the display site is usually a horizontal or gently sloping perch in the forest substage, about one to four metres above ground. Males give advertising calls from display sites and also while they are foraging. C.H.B. Grant observed the Magnificent Riflebird in the dense forests of Cape York Peninsula and recorded in 1915 that the advertising call "consists of a loud whistle resembling 'wheeoo' repeated three times and ending abruptly in a note like 'who-o-o'". The call was given two or three times at intervals of from two to five minutes. Grant made a further observation:

> If a female be near, the male frequently perches on a conspicuous dead twig in a crouching attitude, rapidly opening and closing his wings, the feathers of which, by their peculiar form and texture produce a loud rustling noise, which ... may be heard at a distance of a hundred yards.

The male performs a spectacular courtship display when a female visits the display site, spreading open both wings, stretching the head back and swinging it from side to side to expose the metallic iridescent throat and breast shield. He rises on his perch, elevates the wings, lowers the body and then snaps back the wings to their full extent. This produces a rustling sound in time with the swaying head. The display is repeated again and again. No vocalisations accompany the display and, unlike Victoria's Riflebird, the bill is not opened to expose the yellow gape.

In New Guinea, Brian Coates observed a display between two adult males which faced each other. One performed a display similar to that described above, pointing the bill up, spreading the wings and turning its head towards one wing then towards the other; this behaviour was repeated several times. The second male crouched with head held low. Then the roles were reversed, the second male displaying several times before both flew off.

Similar confrontations have been observed in New Guinea between female-plumaged birds. Birds point the head and bill upwards, stretch up and open the wings. However, in one display the extended wings were bent forward so that the tips almost touched each other. In another variation involving two birds in female plumage, one elevated the tail to a horizontal position, opened the wings and bent them backwards so that they nearly met behind the back.

Very little is known of nesting but it is almost certain that the female builds the nest, incubates the eggs, feeds and broods the nestlings and cares for the young until independence. Fifty nests were examined by H.G. Barnard in the Lockerbie scrubs at the tip of Cape York Peninsula in 1910 and 1911. None of these had snake skin decorations (as in Victoria's Riflebird). Most were hidden in a clump of pandanus at the base of its long leaves, 1 – 10 metres above ground. Males were never seen at nests.

R. Donaghey

Magnificent Riflebird

Ptiloris magnificus

Adult female (D. & M. Trounson)

OTHER NAMES

Albert Riflebird, Prince Albert Riflebird

MEASUREMENTS

LENGTH: *280 – 330 mm*
WING: *148 – 190 mm*
TAIL: *100 mm*
BILL: *50 – 60 mm*
TARSUS: *41 – 47 mm*
WEIGHT: *Not recorded*

DESCRIPTION

ADULT

Male: Crown and nape metallic oil-green; remainder of upperparts, including forehead, lores and sides of head, velvety black with rich violet-purple sheen; wing and tail bluish-black with central pair of tail feathers iridescent greenish-blue; chin, throat and upper breast iridescent blue-green bordered below by a pectoral band of velvety black, then another with traces of red; side breast feathers form an erectile fan; rest of underparts, including long, hair-like flank plumes, velvety black with dark pink-purple sheen. Bill black. Iris dark brown. Legs grey black. Inside of mouth greenish yellow

Female: Upperparts cinnamon brown including broad stripe from lores through eye to sides of neck; a narrow dull white streak over eye extending down side of neck; wings and tail rufous brown; throat dull white finely marked with brown; rest of underparts dull white with a profusion of fine blackish-brown barring. Bill brown. Iris dark brown. Legs grey. Inside of mouth greenish-yellow

IMMATURE

Immature male similar to adult female for 2 years then changes to adult. Immature female like adult female

JUVENAL

Like immature but sooty brown; bars on underparts more widely spaced

IDENTIFICATION

Adult male: A spectacular, curve-billed, velvety black bird of paradise with iridescent green nape and breast shield, black flank plumes and short black tail

Adult female: Brown above and dull white strongly barred blackish-brown below, with decurved bill

VARIATION

Three fairly well defined subspecies, one confined to western New Guinea, one to eastern New Guinea and one to northern Queensland. The male of the Australian subspecies, P. m. alberti, *is distinguished from the two New Guinea forms by its lower breast and abdomen which are a dusky black with little or no pinkish-purple suffusion*

VOICE

Two to three loud, sharp whistles, "wheeoo", followed by a long "who-o-o". Female-plumaged birds give more subdued whistles. Also harsh, rasping growls

HABITAT

Tropical rainforest, monsoon forest, gallery rainforest

FOOD

Fruits, seeds and arthropods (insects, spiders, centipedes)

HABITS

Sedentary. Arboreal, mostly forages in mid-upper canopy usually singly or in pairs

BREEDING

Delayed sexual maturity in males. Probably polygynous or promiscuous mating system. Displaying adult males solitarily dispersed throughout forest. Females probably perform all parental care. Female nests October – February. Probably one brood per season

NEST: Scanty, loosely constructed cup of dried leaves and vine tendrils lined with fine vine tendrils; built in crown of pandanus palm or small dense leafy trees (occasionally the top of a broken-off tree trunk); at heights of 1 – 12 m

EGGS: Two; glossy ovate; dull white to cream with longitudinal broad streaks of brownish-olive or rufous brown with underlying streaks of grey; 34 x 24 mm

INCUBATION: Unknown; fledging period unknown

DISTRIBUTION

From northern Cape York Peninsula south along the east coast to Rocky River and on the west coast to about Weipa, Queensland. Also on offshore Albany Island

STATUS

Common to fairly common resident

R. Donaghey

Trumpet Manucode

Manucodia keraudrenii

Many people are unaware that two birds of paradise inhabit Queensland's Cape York Peninsula. The Riflebird is well known but the lone Australian representative of a second group of birds of paradise, the Trumpet Manucode or Trumpetbird, is far less familiar. It is an obscure, dark species, in terms of its colour and in our knowledge of the behaviour of the Australian population, which originated from an incursion out of New Guinea possibly as recently as the last Ice Age, some 10,000 years ago. At this time there was a land connection between Australia and New Guinea, permitting several species of birds to move southward and establish breeding populations in Australia. Other species that took advantage of this opportunity were the Eclectus Parrot, *Eclectus roratus*, the Red-cheeked Parrot, *Geoffroyus geoffroyi*, and the Green-backed Honeyeater, *Glycichaera fallax.*

Little has been published about possible movements of the Trumpet Manucode, either locally or as regular migrations. In 1914 it was reported that numbers first arrived in the Somerset area during August, gathered in flocks in April, then departed to an unknown destination, possibly New Guinea. Such movement has been reported again as, recently as 1977.

Australian populations of the Trumpet Manucode, identified as *M. k. gouldi*, were first recognised by a British ornithologist as a distinctive group as early as 1859, relevant specimens having been taken to England by John MacGillivray, the naturalist on the HMS *Rattlesnake*. Characters cited as characteristic of the subspecies include a general steel-green (not purplish blue) plumage coloration. All New Guinea populations have some degree of purple on the body or wings and close observation reveals only a trace of this in Australian birds.

The Trumpet Manucode favours mature tropical rainforest such as can be found at the Lockerbie Scrubs near Cape York Peninsula's tip. Other distributional records are from monsoonal scrubs and

Adult at nest; Iron Range, Qld; December 1990 (C. & D. Frith/Frithfoto)

riparian situations. It also occurs further south at Iron Range and the McIlwraith Range. Reports from south of Coen, usually about the Atherton Tablelands, have so far proved to be erroneous or unsubstantiated. It is often confused with the similarly shaped and coloured Spangled Drongo, *Dicrurus bracteatus*, but it can readily be distinguished by its rounded tail, in contrast to the strongly forked tail of the drongo. Another identification feature is elongation of the feathering behind the eyes of the manucode, appearing like ear tufts. However, the Drongo, while displaying, raises a superciliary crest reminiscent of this structure.

Fruits of many rainforest trees provide primary sustenance for the Trumpet Manucode but it also eats invertebrates. In both Australia and New Guinea it frequently feeds in company with other frugivorous species.

Vocalisations are characteristic and memorable. Within the chest area of the male, overlying the breast muscles, is a very long, coiled trachea that enables the male to utter a variety of loud resonant reverberations, to which the "trumpet" part of its name refers. J.A. Thorpe remarked that:

> ... males utter very loud and deep guttural notes, unlike that of any other bird I am acquainted with, and it astonishes me that a comparatively small bird could make so much noise.

Mating displays are elaborate, with the courting male often chasing his prospective mate over a wide area. During this procedure, the male calls and engages in displays that involve repeated bowing, raising the ear tufts, and fluffing the feathers while slightly raising the wings and lowering the tail. A cryptic species, the manucode is usually seen in pairs or individually. Sometimes birds assemble in small groups. Whether these consist of adults and their young, are feeding aggregations, or pre-movement associations, is unknown.

In the nineteenth century, J.A. Thorpe reported that birds could be attracted to an observer by mimicry of their calls, but others have had little or no success. One recent instance concerned a single calling individual in a small area of monsoonal forest located behind a beach. The bird avoided the watcher by constantly changing its location, yet periodically uttered its distinctive call, enticing the watcher to continue his observation. In 1910, William MacGillivray had similar problems near Somerset:

> Next day we are out before breakfast to observe the Manucode (*Phonygama gouldii*). We see and listen to him calling from the top of a high tree in some thick scrub – no easy matter, as the openings in the scrub by which one can view the tree-tops are few, and even at this early hour the atmosphere is humid, like that of a glass-house.

Nests are usually constructed either in tall forest trees or in small leafy trees from October to January, the hotter and wetter months. Some records imply that nesting is not restricted to forested areas. The nest is an open cup of vine tendrils constructed by both parents, although, as reported as early as 1914, the female undertakes most of the work. Two eggs complete a clutch, which is incubated by both parents.

Records of a possible close association with the Black Butcherbird, *Cracticus quoyi*, deserve further investigation.

N.W. Longmore

TRUMPET MANUCODE

Manucodia keraudrenii

Adult; Cape York Peninsula, Qld; November 1990 (C. & D. Frith/Frithfoto)

OTHER NAMES
Trumpet bird

MEASUREMENTS
LENGTH: *270 – 320 mm*
WING: *142 – 160 mm*
TAIL: *120 – 140 mm*
BILL: *25 – 33 mm*
TARSUS: *31 – 33 mm*
WEIGHT: *Not recorded*

DESCRIPTION
ADULT
Generally black with greenish blue gloss; head and neck "ear" feathers elongated to form a plume; greenish-blue sheen to head, mantle, back and underparts; violet sheen to wings and tail. Bill black. Iris red. Legs black

IMMATURE
Similar to adult but with little or no gloss to feathers of underparts and back; wings and tail glossed. Iris brown

JUVENAL
Like immature

IDENTIFICATION
A glossy blue-black, starling-like bird with oily-green sheen, long nape plume, long rounded tail, short bill and red iris

VARIATION
Eight subspecies extensively distributed in lowlands and foothills of southern New Guinea island and satellite islands from the Vogelkop west to east Papuan islands. Subspecies vary in size and the extent of violet and green gloss. The Australian subspecies, M. k. gouldi, *is restricted to Cape York Peninsula and offshore islands; resembles nominate subspecies but sheen to general plumage is more uniformly bluish-green and less purple, especially on wings and tail*

VOICE
Loud, resonant trumpet-like squawk by male. Other notes include rolling call, long drawn-out call, mournful call and a croak

HABITAT
Lowland and foothill tropical rainforest

FOOD
Fruits, especially figs, also insects and other arthropods taken from foliage

HABITS
Sedentary. Arboreal, mostly foraging in mid-upper canopy of forest, singly or in pairs. Easier to detect in breeding season

BREEDING
Nesting season October – February, probably one brood per season. Monogamous breeding pairs. Both parents share feeding and brooding of young. Role of male in nest building and incubation uncertain

NEST: A shallow cup or saucer, sparsely made of interwoven tendrils and stems and lined with fine tendrils; built in fork in the top outer foliage or the inner upper foliage, 6 – 25 m above ground

EGGS: Two; oval-shaped; slightly glossy; pale pink to buff-pink, blotched and spotted with brown, brownish-purple and grey; 37 x 25 mm

INCUBATION: Unknown; fledging period unknown

DISTRIBUTION
Eastern Cape York Peninsula from the tip south to Rocky River and offshore islands. Recently recorded on western side of Cape York Peninsula

STATUS
Fairly common resident

R. Donaghey

MAGPIE-LARKS

Confined to the Australasian region, Magpie-larks comprise the small, distinct family Grallinidae, the relationships of which are obscure. Until recently, this family included the White-winged Chough and Apostlebird, on the basis that all build mud-nests. However, there are many other, clearly unrelated, species that also build mud-nests. Substantial differences in behaviour have led to separation of the Chough and Apostlebird into the family Corcoracidae, leaving the little-known Torrent-lark of New Guinea and the Australian Magpie-lark as the only grallinids.

Magpie-larks are related neither to magpies nor to larks, the common name of the Australian species referring to its pied (black and white) plumage and supposedly lark-like feeding habits. Grallinids are medium-sized, ground dwelling birds, conspicuously clad in sleek, glossy black and white plumage, with distinct differences in patterning between male, female and juvenile.

Usually seen singly or in pairs, both species appear to be monogamous and, since pairs are sedentary and remain in contact throughout the year, there is no elaborate courtship display. The nest is simple in structure and unlike those of corcoracids, has no drainage pad. It is cup-shaped and made from droplets of mud mixed with short grasses and rootlets for reinforcement. These are pressed together and smoothed over with the bill. The nest of the Magpie-lark is usually cemented to a horizontal tree branch, that of the Torrent Lark being constructed on a stream-side bush. It is usually situated under foliage that provides some protection from predators. The two species differ markedly in behaviour. The Torrent-lark is generally shy and elusive, seldom offering more than a fleeting glimpse; the Australian Magpie-lark is bold and somewhat pugnacious.

Locally, the Torrent-lark is known as "Jigayain", a name derived from its call; or "Kotleg", a word also used to describe the black and white pebbles in a river bed. New Guineans are somewhat wary of this bird, believing that it is the "essence of witchcraft" and that there was no witchcraft before these birds appeared.

Reliant on mud for nest building, grallinids are seldom found far from surface water. Both species are largely insectivorous but worms, snails, crustaceans and small vertebrates are also taken. In New Guinea the Torrent-lark inhabits heavily forested, fast flowing, rocky streams in mountains at altitudes of 450 – 2300 metres. Here, it forages actively along the edges of gravelly banks and from the surface of large rocks and boulders, even those in mid-stream. The more adaptable Australian

Magpie-lark is found throughout Australia and parts of southern New Guinea, where it prefers open country and is quite at home foraging around the edge of a farm dam or park pond.

It is interesting to note that both species have associations with other black and white birds. The Torrent-lark is commonly found foraging near the smaller, black and white River Flycatcher, *Monarchella mulleriana*. The Magpie-lark often nests in close proximity to Willy Wagtails, *Rhipidura leucophrys*. The implications of these associations are unknown.

T. Ivison

Adult female Australian Magpie-lark at nest; Brisbane, Qld; September 1981 (R. Slater)

Australian Magpie-lark

Grallina cyanoleuca

The Australian Magpie-lark, also known as the Pee-wee, Mudlark or Murray Magpie, is bold in attire and attitude. Familiar by sight and sound in both town and country, it is as much a part of our culture as the Laughing Kookaburra or Australian Magpie, but has never acquired emblematic significance. It is widespread and common throughout mainland Australia wherever there is surface water, but not in dense rainforest. Although it is only an occasional, non-breeding visitor to Tasmania, it was first recorded from a specimen collected by a farmer near Hobart in 1888. Pee-wee is an onomatopoeic representation of its call but this varies throughout its range: birds in Western Australia sound quite different from those on the east coast. Some descriptive Aboriginal names include: By-yoo-gool-ye-de, Dil-a-but, Kulyaberuk and Jillinberri. The name Mudlark refers to its habit of strolling across muddy ground, from which it obtains material for its nest.

The characteristic piping, five note call of the Magpie-lark is often given in duet. It is uttered by both sexes and used by pairs to maintain contact and to proclaim territorial boundaries. With each utterance, the calling bird raises its wings. Either bird may start the duet, the other responding so quickly that the notes sound as though they are continuous. A series of high-pitched notes constitute an alarm signal and pairs utter a lower-pitched contact call when flying together.

The Magpie-lark prefers open areas, where it spends much time on the ground. It walks with a bouncing terrestrial gait across paddocks and grasslands, or wades around the muddy margins of rivers, streams, farm dams and park ponds, where it forages on seeds and a wide variety of both terrestrial and aquatic invertebrates. Insects are occasionally snatched on the wing and there is a record of a bird tackling a small, insectivorous bat in flight. In pastoral areas, a considerable number of freshwater snails, *Lymnea tomentosa*, are eaten. These are an intermediate host for a liver-fluke, which is a serious parasite of sheep and cattle. When consumed, the life cycle of the parasite is halted, thus making the Magpie-lark a welcome resident at any farm dam. In the hope of controlling numbers of these snails, attempts were made to introduce the Magpie-lark to New Zealand, the Hawaiian Islands and Fiji, but were unsuccessful.

The Magpie-lark is monogamous and usually seen in pairs. Established pairs are sedentary and throughout the year both vigorously defend a territory of 8 – 10 hectares. It is a competent flier and quick to confront intruders on the wing, even when these are much larger birds. Males become particularly aggressive in the breeding season and will even attack their own reflection if seen in a window, car hub-cap or other reflective surface. However, they do not exhibit the hostility towards humans that is shown by nesting Magpies. In fact, in urban areas where they have become accustomed to humans, they can be very confiding, often venturing close to a person or even to vehicular traffic, to snatch a prey item.

Both members of a pair share in the construction of the characteristic mud-nest, a sturdy structure made from droplets of mud reinforced with short grasses and rootlets. These are pressed together and smoothed over with the bill. A nest takes several days to complete, since each layer needs to harden before the next can be added. It is usually conspicuously placed on a horizontal limb but in areas where trees are scarce, alternative sites are used. Telegraph poles are popular and G.H. Morton, in correspondence with A.J. Campbell, noted that "One built its nest on the rail of a swing gate; another upon the top of a post; whilst a third selected some iron hooks suspended in an outshed". Several may be seen in close proximity, being the nests of successive seasons. There are reports of more than two birds helping at a nest, but these are unusual and are probably the result of offspring remaining in the parental territory. Active nests are often parasitised by the Common Koel, *Eudynamis scolopacea*. Abandoned nests are frequently re-lined and used by other species such as woodswallows, cuckooshrikes and the Willie Wagtails.

Three to five eggs constitute the usual clutch. Incubation is shared by both parents, as is the feeding of the nestlings. Young hatch with a coat of soft, grey down, unlike those of the mud-nest building Corcoracidae, which are naked. Development is rapid and the young fledge in a distinctive, intermediate version of the parents' pied plumage. Young birds are easily identified. Whereas adult males have black above and below the bill, adult females have white and white respectively; juveniles are blackish above and white below. Young are fully independent a little less than a month after fledging and leave the parental territory. Breeding resumes and as many as four broods (usually two) may be reared in a good season.

Young birds hold no territory and join with other non-breeding individuals to form nomadic flocks of 20 to several hundred birds. These flocks roam widely. They move north in winter – some reaching Cape York Peninsula – and south again in summer. Birds within a flock stay close together and large communal roosts are formed at night.

The Australian Magpie-lark is a species that has been able to adapt and, in some instances, prosper amidst urban and rural development. Although often feeding on roads and verges, it is rarely recorded as a victim of traffic. It faces no known threat and it seems likely that it will persist as long as Australia can provide mud.

T. Ivison

AUSTRALIAN MAGPIE-LARK

Grallina cyanoleuca

Adult male; Round Hill, N.S.W.; June 1987 (G. Chapman)

OTHER NAMES
Peewee, Little Magpie, Murray Magpie, Mudlark, Peewit, Pugwall, Soldier, Tilyabit.

MEASUREMENTS
LENGTH: *260 – 300 mm*
WING: *173 – 185 mm*
TAIL: *118 – 125 mm*
BILL: *26.4 – 28.5 mm*
TARSUS: *42.3 – 43.5 mm*
WEIGHT: *72 – 87.5 g*

DESCRIPTION
ADULT
Distinctively patterned black and white

Male: Face black, prominent white eyebrow and small white bar below eye; black throat, joining black bib on upper chest; broad white panel on sides of neck; upperparts black; wings black with broad white band from base of inner primaries, through shoulder, joining lower breast; underparts white; tail white with broad black subterminal band, pointing up in centre. Bill cream with black tip. Legs black. Eye pale yellow

FEMALE: Similar to male; face and throat white; no white eyebrow. Broad black band from crown, through eye, joining black bib

IMMATURE
Similar to adult

JUVENAL
First 3 months, black forehead, white throat, white eyebrow. Eye dusky brown. Bill dark grey

IDENTIFICATION
Striking black and white plumage. Familiar piping "pee-wee" call

VARIATION
None

VOICE
Loud piping "pee-wee, pee-wee," often sung as a duet. Strident "pee, pee, pee" in alarm. Slight variations to call throughout range

HABITAT
Diverse; typically near water. Avoids dense forest and true deserts

FOOD
Mainly small insects and their larvae plus other invertebrates such as earthworms and fresh water snails found in pastoral and muddy areas. Occasional small vertebrates taken

HABITS
Established pairs. Sedentary and highly territorial. Nomadic flocks of immatures and unmated adults; flocks drift north in winter and south in summer. Flight is strong, even and direct. Usually seen singly or in pairs. Common in urban areas

BREEDING
Throughout year, mainly August – December; usually 2, but up to 4 broods per year. Both sexes contribute to nest building and incubation

NEST: Bowl-shaped; made from mud reinforced with fibre, lined with feathers, fur and grass; plastered to a horizontal branch or other suitable structure such as a telegraph pole or windmill; anywhere between 2 – 20 m above ground. Similar to nest of Apostlebird but shorter

EGGS: 3 – 5; colour varies from pale pink to dull white; blotched and spotted reddish-brown and dusky purple; tapered oval; about 29 – 11 mm

INCUBATION: 17 – 18 days; fledging 18 – 23 days

DISTRIBUTION
Throughout mainland Australia except in dense forest and dry desert. Rare vagrant to Tasmania. Also southern New Guinea and Timor

STATUS
Has adapted well to human-made environments and is abundant throughout mainland Australia

T. Ivison

MUD-NEST BUILDERS

Unique to eastern Australia, the mud-nest builders, family Corcoracidae, comprise only two monospecific genera: the White-winged Chough, *Corcorax melanorhamphos* and the Apostlebird, *Struthidea cinerea.* Choughs seem to prefer the wetter, more timbered areas and the Apostlebirds drier, open areas. Where their ranges overlap, the two species mingle harmoniously.

Although the two birds differ in outward appearance, both build intricate mud-nests that are similar in structure and both lead remarkably similar, close-knit communal lives. They are very gregarious birds and live in family groups or clans which enthusiastically share in most activities. They forage together, preen each other, roost side by side and cooperate in nest building, incubation and care of the young.

The plumage of neither species shows any sexual dimorphism and is soft and dull in comparison to that of the sleek magpie-larks. The White-winged Chough is about the size of a Currawong but sooty black, with large, white wing patches. The bill is long, slender and decurved, whereas the Apostlebird is small and plump with dull grey plumage and a short, stout bill that is somewhat finch-like in appearance.

Groups usually consist of one breeding pair and their offspring of several years, ranging from four to 20, but usually five to 10. Groups are sedentary and defend a small territory in the breeding season, but have a large home range. Out of the breeding season, groups roam throughout their home range, moving to another section once food supplies become exhausted. Home ranges are usually too large to be defended and often overlap with those of either species.

Apostlebirds and Choughs often feed in close association, with little or no competition. Both are omnivorous, but the differences in bill structure reflect different diets. The Chough is largely insectivorous and rakes through the litter with its bill in search of beetles and grubs. The Apostlebird is predominantly a seed eater and will follow Choughs, sorting through the up-turned litter for smaller items.

Corcoracids are competent fliers, but flight is used only for travel and escape. They are largely terrestrial with long, strong legs used for walking, running and occasionally hopping. Both species forage almost entirely on the ground but the feet are never used to scratch through the litter.

Although there is only one breeding pair in a group, all members help with other aspects of the breeding cycle. Both species build a mud-nest, strengthened with bark and grasses. First, a layer of mud is cemented to a horizontal limb, this acts as a drainage pad as well as a point of fixation of the nest. From here, a new ring of mud is added around the pad at each building session to form a rim, each contribution of mud smoothed over with the bill before another is added. A nest usually takes about a week to construct, but if wet weather intervenes, several weeks may be needed to complete the structure, since each layer needs to set before the next can be added. Nest building in the Corcoracidae seems to be partly a learned process. First-year birds tend to accompany older, more experienced birds to watch the complex nest building process for the first time. Occasionally assisting with mud collection, they are rarely trusted to help with construction. Although constructed differently, the nest of the Apostlebird is similar in appearance to that of the Magpie-lark, *Grallina leucomela*, whereas that of the Chough is almost twice as large.

Once the nest is completed and the eggs are laid, all but the first-year birds share in incubation. Young birds leave the territory during this time, but return to assist with the feeding of the chicks, which are naked and helpless. All members of the group collect food for the chicks and, in their enthusiasm, it is not uncommon for a queue to form at the nest.

Young leave the nest before being able to fly properly. They stay close together, being supervised and fed by older birds. It is at this time they are most vulnerable to introduced predators such as the feral cat and fox. Once able to fly, young birds follow the group, but are still fed by the adults for another few weeks while they develop an ability to forage for themselves.

Studies have shown that pairs or groups numbering less than four have very little breeding success and that, the larger the group, the more young produced, which in turn increases the group number. The foraging and nesting skills needed for survival have to be perfected and birds are slow to mature; there is debate as to whether this is a cause or result of cooperative breeding.

T. Ivison

White-winged Chough

Corcorax melanorhamphos

The common name of this bird is one of many examples of an inappropriate appellation based on superficial similarity. The Chough of Europe is a member of the crow family, Corvidae, and has entirely black plumage but brightly coloured bill and legs. It is solitary and specialised for a diet of ants and worms on rocky coasts. The bird known in Australia as the White-winged Chough has little in common with this species apart from a somewhat crow-like appearance. It has an overall sooty plumage with large white wing patches; its bill and legs are black. It is communal and makes a nest of mud, unlike the "true" Chough, which builds an untidy nest of sticks. Interestingly, John Gould declined to allot the Australian bird to any known family and referred to it as the White-winged Corcorax. It is unfortunate that Australians did not follow his example or, perhaps, adopt the Aboriginal name, Waybung, which Gould recorded from New South Wales.

For most of the day, groups of Choughs forage on foot, in loose arrowhead formations, covering considerably more ground than an individual could. The group moves across open areas, including pastures and bare understories of introduced pine plantations, travelling as much as 5 kilometres in a

Adult displaying; A.C.T.; November 1993 (C. Andrew Henley/Larus)

Sponsor: D.L.J. & G. Solomon

day. Members of a group keep in constant contact with a low, piping whistle. Tails bob and heads wave to and fro as the litter is constantly tossed by their bills, which are also used to probe into damp patches in search of grubs and worms, or to overturn small logs, stones or cow-pats. Scarab larvae are a favoured prey but a wide variety of invertebrates are taken, along with occasional small vertebrates. Grain is also eaten and consumption of this increases in winter as insects become scarce. There is a record of a group of Choughs feeding on freshwater mussels, holding one mussel with the foot and striking it with another held in the bill.

White-winged Choughs rest as a group beneath a shady bush or perched in a tree. At this time, they mutually preen each other. They occasionally bathe, but prefer "dust baths". Dusting is a frequent communal affair in which all members of the group regularly participate. A form of feather maintenance, dusting is thought to help rid birds of ecto parasites. Most dusting birds utilise existing dust by squatting and flicking the wings, feet and body so that the dust is driven into the plumage, but Choughs are unique in "making" dust by digging a series of narrow ruts into suitable, dry soils. Then, while standing, a bird applies dust to its erect body feathers by the bill-full. In the evenings, the group moves together to a communal roosting site.

Choughs are highly excitable birds. When disturbed for any reason, an individual will display. This can happen when one bird is slow to leave its post at the nest or when greeting another individual or group. The wings and tail are fanned out and, simultaneously, move up and down. Most of the initial study of this behaviour was the work of Ian Rowley who described this as the "wing-wave, tail-wag" display. The skin around the eye retracts and the "white" turns crimson as blood vessels dilate. When performed by the whole group, birds form a scrum and face the attacker, hissing loudly. Similarly, if one group member is threatened, others come to its aid by screeching loudly and "dive-bombing" the attacker.

Choughs usually nest between July and August. However, in dry seasons, rain may trigger mud-gathering, and nesting has been recorded as early as July. Much excitement accompanies nest building. Birds busy themselves collecting bark and grasses which are mixed with mud then carried to the nest site in the bill. All birds participate in the activity, but first-year birds tend to accompany older, more experienced members of the group.

At the nest site, amid much displaying and calling, birds queue for a chance to contribute to its construction. Each layer is smoothed with a scissor-like action of the bill. Building usually takes place in the morning, allowing the mud to set through the heat of the day. Nests take weeks (sometimes a month or more) to complete, as building depends on good weather conditions. Too much or too little rain can cause delays in construction. The finished product is a very sturdy bowl-shaped structure. The nests are so hardy that it is not always necessary to build a new one every year, instead, old ones are patched up and re-lined. In dry seasons, when mud is not readily available, cattle or emu dung can be used to revamp old structures.

More than one female may lay in the same nest, bringing the clutch to as many as nine eggs (usually three or four). Regardless of the number of eggs laid, rarely more than one or two chicks survive to fledging. Incubation is a communal affair in which all group members participate. The

19 days of incubation are shared by all members of the group, except the first-year birds, which wander away from the territory at this time but return to participate in caring for the young. By the time the young have hatched, all members of the group seem to be fully occupied in collecting adult and larval insects, and small vertebrates, which they carry crosswise in the bill, back to the nest. Individuals often have to wait for a chance to feed their offerings to the chicks.

Young leave the nest before they are able to fly properly, about a month after hatching. However, their legs are strong and they are able to scramble into low bushes. For the first few days after leaving the nest, the young remain close to cover and adults continue to feed them. Gradually the chicks start to move around with the group and to develop the techniques of foraging, but are fed for nearly a month after leaving the nest. Once the young are able to fend for themselves, the whole group leaves the territory to wander over its home range. At dusk, the group returns to a communal roosting site.

T. Ivison

Adult and chicks; Waikerie, S.A.; September 1992 (R. Brown)

Sponsor: R.J.C. Christian

WHITE-WINGED CHOUGH

Corcorax melanorhamphos

Adults in flight; July, 1985 (T. Howard)

OTHER NAMES
Black Jay, Black Magpie

MEASUREMENTS
LENGTH: 430 – 470 mm
WING: 250 mm
TAIL: 200 – 225 mm
BILL: 48 – 51 mm
TARSUS: 57.5 mm
WEIGHT: 320 – 347 mm

DESCRIPTION
ADULT
Sexes similar. Body black with large white wing patches, prominent in flight. Eye red with orange inner ring. Bill and legs black

IMMATURE
As adult, with eye turning from brown to orange to red over four years

JUVENAL
Similar to adult but body sooty black and fluffy

IDENTIFICATION
Usually in groups of 5 – 20. Thin, slightly decurved bill. Large white wing patch

VARIATION
None

VOICE
Mellow pipes and whistles. Aggressive throaty clicks. Alarm call a loud grating scream

HABITAT
Open forest with leaf litter and little understorey, such as eucalypt woodlands, scrublands, tall mallee and mulga. Also pine plantations and farmlands

FOOD
Mainly insects and their larvae. Some small vertebrates including frogs and mice. Also seeds and some vegetable matter

HABITS
Communal living in family groups of 5 – 20 that share in all aspects of daily life. Sedentary, groups rarely wander far from home range. Largely terrestrial

BREEDING
August – December, sometimes earlier after rain in arid regions. One brood per female, large groups may produce 2 – 3 broods per season. All group members contribute to nest building and incubation

NEST: Large bowl-shaped structure; made from mud reinforced with fibre, in dry seasons mud may be substituted with cattle or emu dung; plastered to a horizontal branch, usually between 2 – 13 m above ground

EGGS: 3 – 5, sometimes 6 – 10 if more than one female lays; cream with few large brown blotches, underlying lavender blotches; oval-shaped; 40 x 30 mm

INCUBATION: 19 days; fledging 28 days

DISTRIBUTION
Confined to eastern and southern Australia

STATUS
Locally common but restricted to fragments of suitable habitat

T. Ivison

Apostlebird

Struthidea cinerea

The Apostlebird is so called because it was originally thought to live in groups of twelve. In fact, it lives in family groups of one breeding male, two or three females and their young, comprising five to 20 individuals. The species is also known as "Seven Sisters" and "Happy Family".

It is endemic to Australia, where there are two populations, regarded by some as subspecies. The larger population is distributed throughout the inland eastern mainland and appears to have expanded in recent times in the northern part of its range, while declining in the south-east. The other population is isolated in the Northern Territory and appears to be expanding.

The Apostlebird inhabits lightly timbered country such as brigalow, mulga scrub and cypress pine, where there is ample ground cover to provide feeding opportunities and trees to provide nesting sites, shade and refuge from predators. An additional habitat requirement is access to fresh water in the form of a river, stream or dam, which is necessary not only for building the mud-nest but also for bathing. Like most seed-eating species, the Apostlebird also needs to drink regularly, sometimes several times a day in hot weather.

Gregarious by nature, groups of Apostlebirds can become very confiding. At an isolated service station in the Northern Territory which provides a small swimming pool, hundreds of birds have learned to wait around the parking lot for incoming trucks and cars. As soon as its engine is turned off, the birds descend on a vehicle, paying little notice to people, as they scavenge insects from the radiator grille or windscreen.

In rural areas, Apostlebirds will gather near homesteads with bird-feeders and accept any food scraps that are offered. K. Langloh Parker gives a very apt description of these birds around his home:

> Such a noisy, self-assertive crowd. Very home-spun looking birds, dim grey, with light brown touches on the wings, boot-button eyes, and raucous voices which, as they are always in flocks, are rather deafening.

Birds share noisily in all aspects of daily life and maintain contact by constantly chattering amongst themselves and to other species, giving rise to another of their common names, "CWA-Bird" (in reference to the Country Women's Association). They often indulge in what appear to be energetic games such as "follow the leader" around the bases of trees and "tug-of-war" with a leaf or twig, leaving it in tatters. If a group is faced with an attacker, all members quickly come to the defence, screeching loudly and dive-bombing the intruder.

Like the related White-winged Chough, the Apostlebird feeds almost entirely on the ground. It is predominantly granivorous and uses its stout bill in a finch-like manner to pick up even the smallest of seeds from the ground. Seeds are crushed in the bill before being swallowed. One foot is often employed to hold down grasses and larger insects while feeding, but like the Chough, the feet are never used to scrape through leaf litter. Small insects are taken opportunistically from the ground, or from under bark while clambering up trees.

Family groups have a large home range of 10 – 15 hectares, often overlapping with those of other groups of Apostlebirds or Choughs. A small territory is defended within this area during the breeding season of September to February. In Inverell, a group of White-winged Choughs started nest building in a neighbouring territory in July. Apostlebirds were present and closely supervised the process, rearranging nest material and warning the Choughs of intruders. There is also an unusual record of a lone Apostlebird attending to the nest of a pair of Magpie-larks. As the Magpie-larks started to build, the Apostlebird removed the material and used it to refurbish an old structure. This was accepted by the pair and the Apostlebird remained to help with incubation and rearing of the young.

The courtship display of the Apostlebird begins with a dominant male offering some food or a twig to a female. This is held in the bill as the birds walk side by side, quivering. Quivering intensifies, the offering is dropped, and the pair stand facing each other with feathers erect for some time before copulation takes place.

Once the nest site has been chosen, all members of the group contribute to building the mud-nest. Mud is always nearby and the nest is built from grasses and bark, permeated with mud. Construction is similar to that of the White-winged Chough, but the finished product is a much smaller cup. It resembles that of the Magpie-lark but is slightly taller. The nest is lined with grass, feathers and wool. Two to five (usually four) eggs are laid. Occasionally more than one female lays in the nest, incubation and care of nestlings being shared by all members of the group. Young fledge about 15 days after hatching, before being able to fly properly. However, their legs are strong, enabling them to claw their way up to a low perch at night. Once able to fly and feed themselves, they leave the nesting territory of the group and begin to wander over the home range.

T. Ivison

Adults allopreening; Blackdown Tableland, Qld; May 1990 (G. Chapman)

APOSTLEBIRD

Struthidea cinerea

Adult at nest; Tocanmain Station, N.S.W.; September 1965 (G. Chapman)

OTHER NAMES
CWA-bird, Grey Jumper, Happy Family, Lousy Jack, Seven Sisters, Twelve Apostles

MEASUREMENTS
LENGTH: 290 – 230 mm
WING: 146 – 151 mm
TAIL: 148 – 152 mm
BILL: 23.9 – 25.3 mm
TARSUS: 39.3 – 40.6 mm
WEIGHT: 119 – 137 g

DESCRIPTION
ADULT
Sexes similar. Body uniform ash grey. Head, back and breast feathers tipped pale grey; wings dusty brown, edged paler brown; tail dusty black. Bill and legs black. Eye brown in female, grey in male

IMMATURE
As adult. Males have smoky brown eye changing to grey in second year

JUVENAL
Similar to adult female but plumage a soft fluffy grey

IDENTIFICATION
Medium-sized grey bird, usually seen in groups of 5 – 20; gregarious; constantly chattering

VARIATION
None

VOICE
Many harsh scratchy notes including a loud discordant "ch-kew, ch-kew" and chattering "chereer, chereer"

HABITAT
Prefers drier, open forest, mulga scrub, mallee, native pine and brigalow as well as farmlands. Usually near water

FOOD
Primarily seeds and vegetable matter, also insects, their larvae and other invertebrates

HABITS
Communal, lives in family groups of 5 – 20 that share in all aspects of daily life. Sedentary, groups rarely wander far from their home range. Largely terrestrial. Often associated with the White-winged Chough

BREEDING
August – February; sometimes earlier after rain in arid regions. One brood per female, but large groups may produce 2 – 3 broods per season. All group members partake in nest building and incubation.

NEST: Bowl-shaped structure; made from mud reinforced with fibre; lined with fine grasses; plastered to a horizontal branch, usually between 2 – 17 m above ground. Similar to nest of Australian Magpie-lark, but taller

EGGS: 3 – 4, sometimes 5 – 8 if more than one female lays; pale blue-white, sparingly blotched brown or dark grey; tapered oval; 29 x 22 mm

INCUBATION: 18 days; fledging 13 – 18 days

DISTRIBUTION
Eastern Australia, inland of the Great Dividing Range. Isolated population in Central Northern Territory

STATUS
Locally common

T. Ivison

Mixed Woodswallows; Thylungra, Qld; December 1988 (G. Chapman)

Sponsor: South Australian Museum

Woodswallows

The Woodswallows comprise a distinct, largely Australasian family that is probably most closely related to the butcherbirds, magpies and currawongs. It includes 10 species in the genus *Artamus*, of which six occur in Australia, four being endemic. They are uniform in shape, varying mainly in plumage, which is subtle in colouring. Most species are endowed with subdued browns and metallic greys; some have simple black and white markings; and two sport comparatively colourful shades of rust on their chests. The pale blue-grey bills have a dark tip. They are accomplished fliers, their broad pointed wings giving them a flight silhouette similar to that of the Common Starling, *Sturnus vulgaris*.

Woodswallows are unique among passerines in lacking preen glands and possessing powder down, special feathers that grow continuously and produce tips that break down into a fine, water-resistant powder. This powder is distributed through the plumage as the bird preens, giving it a distinctive "soft" look and musky odour.

Also distinctively, woodswallows are the only passerines to habitually soar. They are largely aerial feeders, often seen flying very high. They soar, glide and swoop through and with air currents, catching insects in a manner similar to that of the true swallows of the family Hirundinidae, to which they are not closely related. Occasionally, large prey such as cicadas are caught in the feet, other prey are snatched from trees. Sometimes, where insects such as grasshoppers are in abundance, birds descend to the ground to feed. Although primarily insect eaters, these versatile birds also possess brush-tipped tongues that are used to collect nectar from blossoms and flowers.

Woodswallows generally inhabit open forests, scrubs and cleared areas. Most are fairly sedentary but some Australian species show a definite north-south trend in their movements. The only truly nomadic species are the White-browed and Masked Woodswallows, which form huge mixed flocks that travel widely, pausing only to breed when conditions are appropriate. They are very sociable and, when not on the wing, often perch closely side by side on a bare branch, telegraph wire or other conspicuous perch. Such perches are vantage points from which birds chatter constantly, grooming themselves and each other. Individuals sally out after passing insects and have to jostle for a position upon their return.

A peculiar and little understood habit of woodswallows is cluster-roosting. This usually occurs at night when up to 100 birds huddle together, clinging to the surface of some sheltered spot such as a tree crotch or cavity, rather like a swarm of bees. The reason for this behaviour is not understood but it may be a form of thermoregulation. Birds also cluster on rainy and overcast days, and in the arid interior clustering occurs on even the hottest of days.

Ian Rowley noted that many species of communal birds do not perform elaborate courtship displays, these being notably more common in species where the male mates with more than one female, as in bowerbirds or in species that renew pair bonds each mating season. Perhaps because of their sociability, courtship among woodswallows is vigorous but short-lived. Courtship feeding may precede courtship display. One bird, usually the male, offers food to the other, which then assumes a begging stance to receive it. This display may be initiated by either bird. One rotates and partly spreads its tail and quivering wings; the other bird soon joins in and the display can last up to a minute before copulation. Once copulation takes place, nest building commences almost immediately.

The nest is flimsy. Both parents construct it from small twigs, grasses and rootlets to form an untidy cup-shape. Nests are usually placed in loose colonies, individual nests being placed in almost any available tree crotch, stump hollow or human-made substitute, such as a fence-post. The Little Woodswallow, *Artamus minor*, is an exception in that it also nests in rock crevices.

The usual clutch consists of two to four eggs, which are white, spotted and blotched with varying shades of reddish-brown. The nesting process is carried out with such efficiency that in arid regions the nest may be built and eggs laid within 12 days after heavy rains. The eggs are incubated by both members of a pair and young hatch within 12 – 16 days. Both parents rear the chicks, sometimes assisted by a helper or third unrelated bird. Young fledge in a mottled plumage before they are able to fly properly. They are camouflaged and have well developed claws that enable them to clamber around the substrate near the nest until they are able to fly. Birds seldom mature before two years of age but some species that inhabit arid areas, such as the Masked Woodswallow, *A. personatus*, may be stimulated to breed as young as six months, if conditions are favourable.

T.Ivison

White-breasted Woodswallow

Artamus leucorhynchus

Predominantly a bird of tropical woodland and mangroves, the White-breasted Woodswallow ranges over the north-eastern half of the continent. Seldom far from water, its population extends inland along timbered watercourses, and reaches as far as the coastal mangroves of northern Western Australia. There are other similarly plumaged species in New Guinea, the Bismarck Archipelago, eastern Indonesia, India and across south-east Asia. Throughout the range there are 11 subspecies, one of which occurs in Australia.

Overall, the species is considered to be sedentary but there is evidence that in Australia the southern and southern inland populations are strongly migratory, with a definite northward movement in winter. There is also a suggestion that some northern birds move south during the same period.

The behaviour of the White-breasted Woodswallow is similar to that of the other woodswallows. It spends much time soaring on wind currents in search of prey and is mainly an aerial feeder, catching insects while on the wing. It appears to be more social than other woodswallows, being frequently seen in flocks of 50 – 100 individuals, sometimes 200 or more. Between bouts of feeding, and at night, individuals cluster together.

The White-breasted Woodswallow is the only member of its family to lack a white tip on the tail. There are conflicting reports of its courtship behaviour. The courtship displays described for the New Guinean subspecies by Bill Peckover and Win Filewood in 1976 are that:

> One bird (the female) perches back onto another bird (the male) and fans and twists the tail while bowing slightly forward and spreading the wings in a slow rowing or overarm movement.

Ian Rowley states that the White-breasted Woodswallow displays only by moving its wings. However, on a recent trip to Cairns I observed these birds in large numbers and noticed that while a bird was resting on an exposed perch during the day, its tail was sometimes rotated but never did I observe it fanned. The tail was most commonly rotated when a bird first landed or when another bird joined a resting bird.

The nest is typical of all woodswallows but this species often uses old nests of the Australian Magpie-lark. John Gould noted that:

> Most of those I found breeding on the Mokai (north of the Liverpool Plains, in northern New South Wales) had possessed themselves of the forsaken nest of Grallina melanoleuca (Grallina cyanoleuca), which they had rendered warm and of the proper size by slightly lining it with grasses, fibrous roots, and the narrow leaves of Eucalypti.

T. Ivison

Artamus leucorhynchus

Group; Bush Bay, W.A.; May 1975 (G. Chapman)

OTHER NAMES:
Swallow Shrike, White-rumped Woodswallow

MEASUREMENTS
LENGTH: 170 – 180 mm
WING: 129 mm
TAIL: 59 mm
BILL: 17 mm
TARSUS: 16 mm
WEIGHT: 36 – 46 g

DESCRIPTION
ADULT
Sexes similar. Upperparts dark blue-grey; head and neck dark blue-grey, throat slightly paler; underparts and underwing white; tail dark blue-grey. Eye dark brown. Bill pale blue-grey, tipped black. Leg dusky black

IMMATURE
As adult

JUVENAL
Wings as adult; slight terminal white tip to tail feathers; rump and underparts white; upperparts grey speckled buff

IDENTIFICATION
Only Australian woodswallow without white tip on tail

VARIATION
Eleven subspecies in Australia

VOICE
Brassy chirps "pirrt, pirrt"; loud chattering; quiet twittering; some mimicry

HABITAT
Eucalypt forest, woodlands and mangroves. Usually close to water

FOOD
Insects, mainly taken on the wing, also from ground or gleaned from tree canopy; nectar from flowering plants

HABITS
Nomadic; southern populations partly migratory, shifting north in autumn and south in spring. Move in flocks of 10 – 50, sometimes as many as 100 or more. Flocks cluster together to roost both day and night

BREEDING
August-January; also after rain in arid regions; 1, occasionally 2 clutches per season; both sexes contribute to nest building, incubation and care of young

NEST: Substantial shallow bowl; made from thick grases, rootlets and some twigs, lined with fine grass; placed in upright tree fork, stump hollow or inside nest of the Magpie-lark; 4 – 30 m above ground

EGGS: 3 – 4; off white, spotted and blotched mid-brown to purple-brown, becoming denser at larger end; tapered oval; 23 x 17 mm

INCUBATION: 13 – 15 days

DISTRIBUTION
Northern coastal Western Australia across Kimberley, most of the Northern Territory, Queensland and New South Wales (absent from southern coastal New South Wales), Western Victoria and north-eastern South Australia. Extralimital range: New Guinea to Fiji and the Philippines

STATUS
Fairly common throughout range; dependent on water in drier regions

T. Ivison

Sponsor: Hunter Wetlands Trust

Masked Woodswallow

Artamus personatus

The name of the Masked Woodswallow relates to the distinctive black facial mask of the male. In the female, this is a dusky slate-grey and less pronounced. The habits of this species are almost identical to those of its white-browed counterpart and, where the ranges of the two overlap, they are often found in close association, frequently in mixed flocks. It is an accomplished flier, spending hours aloft, cruising the skies and catching insects on the wing.

It is very social, and when not aloft, individuals spend much time perched side by side on a conspicuous perch, preening each other and chattering. From these vantage points, individuals fly out to snatch insects as they pass by. It is very nomadic, shifting northward in winter and southward to breed in spring. Vast distances are travelled in large flocks that are sometimes mixed with those of the White-browed Woodswallow. Where the two species mingle, the latter predominates in the west and the former in the east. The distribution of the Masked Woodswallow extends to the coast of Western Australia, much further west than that of the White-browed Woodswallow.

Masked Woodswallows stay in one place only long enough to breed, when conditions are favourable. On such occasions breeding is carried out with a fevered rush of activity. Nests may be built and eggs laid within a week of the flock's arrival and may be deserted just as quickly if conditions deteriorate. Nesting is colonial and, in mixed colonies, there is a surprisingly low rate of interbreeding between these two highly social species. The fact that interbreeding can occur suggests that these two species – the only ones with marked sexual dimorphism – are each other's closest relatives. They are also the only Australian woodswallows with identical tail patterns.

Young fledge in a dappled grey and brown plumage and are easily distinguished from the adult birds. At this stage, they are unable to fly properly and must cling tightly to branches with their well developed claws. They remain close to the nest for about a week while being fed by the parents. Once the young birds have mastered the art of flight, the group moves on and continues to roam over the continent.

T. Ivison

Masked Woodswallow

Artamus personatus

Adult male at nest (P. Munchenberg)

OTHER NAMES
Blue Martin, Bluey, Bush Martin, Cherry-bird, Skimmer, Summerbird, Woodmartin

MEASUREMENTS
LENGTH: *190 – 200 mm*
WING: *115 – 125 mm*
TAIL: *77 – 81 mm*
BILL: *17 – 19 mm*
TARSUS: *19 – 21 mm*
WEIGHT: *32 – 39 g*

DESCRIPTION
ADULT
Male: face and throat black with narrow white margin. Upperparts entirely mid-grey; tail mid-grey with white terminal band; underparts light grey. Eye dark brown. Bill blue-grey; tip black. Legs dusky black

Female: As male but face and throat sooty; underparts faint pinkish tinge

IMMATURE
As adult

JUVENAL
Body grey-brown, speckled buff to off-white; wings as adult; tail as adult, but white terminal tip greatly reduced

IDENTIFICATION
Grey woodswallow, whitish below with large black facial mask

VARIATION
None

VOICE
Musical, descending chirrup; soft chattering; some mimicry

HABITAT
Diverse; eucalypt forest and woodlands to heath and spinifex. Also rural areas, orchards and towns

FOOD
Insects, mainly caught on the wing, also taken from ground or gleaned from tree canopy; nectar from flowering plants

HABITS
Colonial; travels widely in large flocks, often in company of White-browed Woodswallow. Nomadic, some migratory, moving north in Autumn and south in spring. Birds cluster together to roost day and night

BREEDING
August – December; though earlier after rain in arid regions; 1, sometimes 2 clutches per season; both sexes contribute to nest building, incubation and care of young

NEST: A shallow basket, loosely made from twigs, grasses and rootlets; similar to that of the White-browed Woodswallow; placed in a tree fork, bush, stump hollow or fence post; 1 – 6 m above ground

EGGS: 2 – 3; dull off-white with brown and dusky-grey blotches; tapered oval; 22 x 77 mm

INCUBATION: 12 days; fledging period not recorded

DISTRIBUTION
Widespread throughout mainland Australia; absent from south-western corner, Cape York and Tasmania

STATUS
Common in western two-thirds of continent, less so in east; irregular east of the Great Dividing Range

T. Ivison

White-browed Woodswallow

Artamus superciliosus

The most colourful of the woodswallows, the male of this species has a metallic grey back and head with a conspicuous white eyebrow and a rich chestnut-coloured breast. The female is similar in appearance but, although much more colourful than other species, is notably duller than the male. It is endemic to Australia and inhabits woodlands and cleared areas. Ranging over the eastern half of the

Adult female; Hattah, Vic; October 1991 (T. & P. Gardner)

Sponsor: Robert J. Davidson

continent, it is very nomadic. White-browed Woodswallows usually travel in large, swarming flocks, often in association with the Masked Woodswallow. The proportion of each species in one flock appears to vary geographically, with the White-browed Woodswallow dominating the east, its masked relative in the west.

In the breeding season, huge flocks may descend on an area – usually one that has had recent rain or is abundant in insect life – and nest building begins almost immediately, having been completed and eggs laid within days of a flock's arrival. However, nests are just as quickly abandoned if conditions fail to remain suitable. The White-browed Woodswallow breeds in loose colonies, often mixed with those of the Masked Woodswallow. Nests are placed almost anywhere. H. L. White, in correspondence with A. J. North, noted that:

> The choice of nesting site appears to be a matter of perfect indifference, the most exposed positions being frequently selected. In 1907 swarms of birds visited us and nests were met with almost anywhere, stumps, mortice or knot holes in posts, shrubs and fruit trees in gardens, in fact wherever it was possible to place a nest.

Because of the nomadic nature of the species, breeding colonies are rarely situated at the same site in consecutive years. In some areas it is locally known as the "Four Year Bird" but it is doubtful that it could even keep to this schedule.

Despite their differences in appearance, it seems that the closest relative of the White-browed Woodswallow is probably the Masked Woodswallow. They are the only two species with pronounced sexual dimorphism. The wing and tail patterns are identical, whereas all other species are distinctively marked. Hybridisation between the two species is known.

Like all woodswallows, the White-browed Woodswallow feeds mainly on the wing. Flocks may stay aloft for hours on end, soaring, swooping and gliding as they catch insects. When travelling, a flock flies at very high altitudes and often may only be identified by their high-pitched contact calls and the height at which they fly.

T. Ivison

WHITE-BROWED WOODSWALLOW

Artamus superciliosus

Adult male; Thylungra, Qld; December 1988 (G. Chapman)

OTHER NAMES
Blue Martin, Skimmer, Summerbird, White-eyebrowed Woodswallow

MEASUREMENTS
LENGTH: *190 – 200 mm*
WING: *118 – 128 mm*
TAIL: *76 – 83 mm*
BILL: *22 – 25 mm*
TARSUS: *19 – 23 mm*
WEIGHT: *32 – 38 g*

DESCRIPTION
ADULT
Sexes similar though female notably duller than male; upperparts deep metallic grey, conspicuous white brow over eye; underwing and undertail greyish-white; wings grey-brown edged buff; upper breast metallic grey; rest of underparts rich chestnut. Eye dark brown. Bill blue-grey; tip black. Legs dusky black

IMMATURE
Similar to adult female

JUVENAL
Body grey, speckled buff off-white; eyebrow absent; wings as adult; tail as adult but terminal white tip greatly reduced

IDENTIFICATION
Most colourful of the Australian woodswallows(other species subtle greys and browns); in flight, chestnut breast contrasts with pale underwings

VARIATION
Negligible

VOICE
Musical "chep, chep"; soft chattering; harsh scoldings; some mimicry

HABITAT
Diverse; eucalypt forest and woodland to heath and spinifex. Also rural areas, orchards and towns

FOOD
Insects, mainly caught on the wing, occasionally taken from ground or gleaned from tree canopy; nectar from flowering plants

HABITS
Colonial; travels widely in large flocks, often in company with Masked Woodswallow. Nomadic; some migratory tendencies, moving north in autumn and south in spring. Birds cluster together when roosting, both day and night

BREEDING
August – December; 1, sometimes 2 clutches per season; both sexes contribute to nest building, incubation and care of young

NEST: shallow basket, loosely made from twigs, grasses and rootlets; similar to that of Masked Woodswallow; placed in any available crevice such as a tree fork, stump hollow or fence post; 1 – 6 m above ground

EGGS: 2 – 3; dull off-white with brown and dusky-grey blotches; tapered oval; 22 x 17 mm; similar to those of Masked Woodswallow

INCUBATION: 16 days; fledging period about 15 days

DISTRIBUTION
Widespread throughout the eastern half of the continent; irregular in western Australia; vagrant northern Tasmania; absent from Cape York Peninsula

STATUS
Common. Locally abundant

T. Ivison

Black-faced Woodswallow

Artamus cinereus

The Black-faced Woodswallow inhabits open country, including eucalypt woodlands, scrubs, spinifex and sandhills and the arid interior, often far from water. A.J. North found it in treeless plains utilising human-made objects for perches:

> It is a common species, and is met with in pairs or small flocks almost everywhere, and frequently many miles from timber. Wire fences and telegraph wires are favourite perching places, from which it often sallies forth to capture some passing insect and return again to the same place.

It has a continuous range across most of Australia, being absent only from the extreme northern and south-western tips, the coastal south-east and Tasmania. Currently, three subspecies are recognised, separated by the colour of the vent region. *Artamus c. cinereus* occurs in Timor; the black-vented form, *A. c. melanops*, inhabits dry country west of the Great Dividing Range; and *A. c. albiventris*, the white-vented form, prefers the more humid areas east of the Great Dividing Range. When the two Australian subspecies meet they interbreed and form a distinct hybrid zone along the spur of the Great Dividing Range, the Gregory Range and the eastern Carpentarian grasslands. Hybrids have varying combinations of black and white on the vent.

Unlike other woodswallows, which are either migratory or nomadic, the Black-faced Woodswallow is only partly nomadic. It lives in flocks of about 6 – 20 birds which are locally nomadic and, in most areas, sedentary. Continually in search of food, a flock roams over the same area from year to year, even in the arid interior where they remain during periods of severe drought. However, there does appear to be some seasonal north-south movement among northern populations, where it is common in winter and scarce in summer.

Like other woodswallows, flocks of the Black-faced Woodswallow roost in clusters, huddling tightly at night in a tree hollow or crotch. It feeds in the same manner as other woodswallows, mainly taking insects on the wing and also by gleaning them from leaves or horizontal branches. Some prey is taken from the ground and there is an extraordinary record of a Black-faced Woodswallow with "eyes bigger than its belly" which was found dead, with half a skink protruding from its bill. The bird presumably suffocated in its attempt to swallow the lizard.

T. Ivison

BLACK-FACED WOODSWALLOW

Artamus cinereus

Adult; Murchinson, W.A.; circa 1980 (M. & I. Morcombe)

OTHER NAMES
Grey, Grey-breasted, White-bellied or White-vented Woodswallow

MEASUREMENTS
LENGTH: 180 – 190 mm
WING: 119 – 126 mm
TAIL: 72 – 76 mm
BILL: 15 – 19 mm
TARSUS: 18 – 21 mm
WEIGHT: 32 – 40 g

DESCRIPTION
ADULT
Sexes similar. Upperparts smoky-grey; rump black; wings blue-grey, buff edging to feathers; black eye patch extending from base of bill to just behind eye; chin black; throat, chest and upper belly pale smoky-grey; vent variable (black in A. c. melanops, *white in* A. c. albiventris, *hybrids with varying combinations of black and white); tail black, broad white top on all but central tail feathers. Eye dark brown. Bill blue-grey. Tip black. Legs dusky black*

IMMATURE
Like adult

JUVENAL
Body grey-brown speckled buff to off-white; wings as adult; tail as adult but with terminal tip greatly reduced

IDENTIFICATION
Dark grey woodswallow with small amount of black on face (as opposed to the black mask of the Masked Woodswallow). Black, white or black and white vent

VARIATION
Three subspecies, two in Australia: the black-vented form, A. c. melanops, *occurs west of the Great Dividing Range; the white-vented form,* A. c. albiventris, *occurs east of the Great Dividing Range. Hybrids occur along the spur of the range*

VOICE
Quiet sweet notes; scratchy "chap, chap" or "chiff, chiff"; harsher alarm call. Some mimicry

HABITAT
Diverse; generally open areas, from farmland and open woodland to spinifex and treeless plains; often far from water

FOOD
Insects, mainly caught on the wing, occasionally taken from ground or gleaned from tree canopy; nectar from flowering plants

HABITS
Colonial; usually in flocks of 6 – 20; partly nomadic and some seasonal north-south movement in northern part of the continent; cluster roosting, both day and night

BREEDING
August – December, though may be earlier after rain in arid regions; both sexes contribute to nest building, incubation and care of young

NEST: Shallow bowl-shape, similar to that of the Masked Woodswallow but slightly tighter weave; made from twigs, grasses and rootlets; usually placed in tree crotch or crevice, well concealed by foliage, occasionally more open positions; 1 – 5 m above ground

EGGS: 3 – 4; white to off-white spotted and blotched red-brown and purple-grey; tapered oval; 22 x 17 mm

INCUBATION: 14 – 16 days, fledging period about 18 days

DISTRIBUTION
Widespread throughout mainland Australia; absent from coastal South Australia, New South Wales and southern Queensland; the top of Cape York Peninsula and Tasmania. Extralimital range to Timor and New Guinea

STATUS
Common

T. Ivison

DUSKY WOODSWALLOW

Artamus cyanopterus

The endemic Dusky Woodswallow is a typical member of the Artamidae. It is uniformly dusky brown merging to a dark blue-grey on the wings and can be separated from the somewhat similar Little Woodswallow by its larger size and a diagnostic white stripe on the outer wing, clearly visible when the bird is in flight.

Like other woodswallows, it is predominantly an aerial feeder, expertly catching insects on the wing either while in flight or by darting from a vantage point to snatch at them. It has been reported to be a nuisance around apiaries, taking considerable numbers of bees as they move to and from their hives.

The Dusky Woodswallow occurs in two separate populations, currently regarded as subspecies. *A. c. cyanopterus* occurs along the east and south-eastern coasts of the mainland and Tasmania; *A. c. perthi* is restricted to the extreme south-western tip of Western Australia. The latter population is sedentary, bounded by hot, arid land to the north of its range. Birds from the south-east are migratory and move north in autumn, and are practically absent from Tasmania in winter. There is one record of nocturnal movement when three birds, later identified as Dusky Woodswallows, collided with a lighthouse on Montague Island, New South Wales, during strong winds in April 1963. Birds return south in spring and may breed in the same area year after year.

It usually lives in groups of 10 – 30 birds which cluster together to roost at night in a sheltered fork or tree hollow. During the day, between bouts of foraging, birds may rest, perched side by side, sometimes preening each other.

During the breeding season, pairs within a group remain close. The male and female share in nest building, incubation and care of young. Some communal breeding has been recorded, during which unrelated group members or young from the previous year help the parent birds to feed the chicks. In winter, groups flock together and cluster communally at night. Flocks of as many as 200 – 300 birds cram into hollows or simply cling to the trunk of a tree, possibly helping to conserve body heat on cold days.

The Dusky Woodswallow is common throughout its range, preferring open woodland, rural and semi-rural areas. In 1865, John Gould noted its particular fondness for nesting near rural homesteads, and said of the species:

> This Woodswallow must, I think, ever be a general favourite with the Australians, not only from its singular and pleasing actions, but from its often taking up its abode and incubating near the houses, particularly such that are surrounded by paddocks and open pasture-lands skirted by large trees.

In 1909, A.J. North recorded a small flock feeding on boiled sausage on a roadside in the Sydney suburb of Burwood. The species is not uncommon in newly developed urban areas. I have seen them several times around new housing estates in north-western Sydney. Here, it is still surrounded by semi-rural areas but, as the urban sprawl expands, its presence is increasingly limited. Unlike its Masked and Black-faced counterparts, the Dusky Woodswallow will enter fairly dense eucalypt forest, but it prefers to frequent watercourses and natural clearings.

T. Ivison

Dusky Woodswallow

Artamus cyanopterus

Adults at nest; Hattah, Vic; October 1991 (T. & P. Gardner)

OTHER NAMES
Bee Bird, Bluey, Jacky Martin, Skimmer, Sordid Woodswallow, Summerbird, Woodmartin

MEASUREMENTS
LENGTH: 170 – 180 mm
WING: 121 – 133 mm
TAIL: 66 – 78 mm
BILL: 15 mm
TARSUS: 19 mm
WEIGHT: 31 – 46 g

DESCRIPTION
ADULT
Sexes similar. Upperparts deep brown-grey; rump dusky; wings dark blue-grey, outer primaries edged white; underparts deep brown-grey; tail dusky with terminal white bar on outer tail feathers. Eye dark brown. Bill pale blue-brown, tipped black. Black stripe on lower bill. Legs dusky black

IMMATURE
As adult

JUVENAL
Body brown speckled with buff to off-white. Wings as adult. Tail as adult but terminal white tip greatly reduced

IDENTIFICATION
Smoky brown woodswallow with diagnostic white patch on outer wing

VARIATION
Two separate populations recognised as subspecies: A. c. cyanopterus *in the east and* A. c. perthi *in the west*

VOICE
Brassy chirrups and chirps given in flight and when perching; also harsh chattering. Some mimicry

HABITAT
Lightly timbered, grassy areas from woodlands and scrubs to golf-courses and roadsides

FOOD
Insects, mainly taken on the wing, also from ground or gleaned form tree foliage; nectar from flowering plants

HABITS
Nomadic within range; south-eastern population migratory with northern movement in autumn. Moves in flocks of 10 – 30, which nest together in "neighbourhoods". Flocks cluster together to roost day or night

BREEDING
August – January; 1, sometimes 2 clutches per season. Both sexes contribute to nest building, incubation and care of young

NEST: Loose bowl made from twigs, grasses and rootlets, lined with fine grasses; placed in a tree fork, behind bark, stump hollow or fence post; usually 1 – 10 m above ground

EGGS: 3 – 4; dull creamy-white with dusky brown and dark brown blotches; tapered oval; 23 x 17 mm

INCUBATION: 16 days; fledging period 16 – 20 days

DISTRIBUTION
Two separate populations: eastern population from Atherton Tableland, Queensland, south to Tasmania and west to Eyre Peninsula in South Australia; the other in south-west Western Australia

STATUS
Common throughout range

T. Ivison

LITTLE WOODSWALLOW

Artamus minor

This is the smallest of the woodswallows, readily identified by its small size and demure, deep dusky colouring. It resembles a small Dusky Woodswallow but is darker and lacks a white edge to its wings. Confined to the Australian continent, it is sedentary and migratory in different parts of its range, inhabiting dry spinifex plains and areas of acacia scrub in northern and north-western Australia where there are rocky gorges and outcrops. Here the birds are sedentary, preferring to stay close to permanent water sources. In the east, the range encompasses grasslands, where the Little Woodswallow is very nomadic and roams widely.

It usually lives in family groups of seven to 20 birds, but as many as 200 have been recorded at a communal roost. Groups glide gracefully, feed together on abundant food sources, rest together in the heat of the day, and cluster together to roost at night in rock crevices or tree hollows. It is largely insectivorous. Groups perch conspicuously on dead trees and individuals make brief sorties to catch insects on the wing. Other invertebrates, such as ants, are gleaned from trees and rock surfaces. Some nectar is also taken.

In the inland and northern parts of its range, this species utilises the cracks and crevices of rocky outcrops and cliff faces to nest and to roost. A.J. North kept records sent to him by F. Lawson Whitlock, who had the opportunity to observe and record many nests of the Little Woodswallow while collecting from H.L. White in north-western Australia. Whitlock stated:

> I first met with Artamus minor, at Gorge Creek where it haunted little rocky hillsides, and afterwards secured three specimens on the Upper Coongan, a still more rocky locality. This woodswallow never seems to be away from rocks. It is by no means timid and will fly right up and perch within a few feet of an intruder examining its nest. It was not until September 1908, that I found its nest. I had followed a pair of thrushes up a very rocky side creek of the main river, when I observed a Little Woodswallow skim right up the precipitous face of a cliff and back again, and I thought I could distinguish something like a nest in a very small cleft, I climbed up to the spot, and there in a cavity hardly large enough to admit my hand, was a loosely constructed nest, formed of bits of Spinifex pressed into the cleft by the weight of the bird into a shallow cup-shaped depression. Still further down the rocky sides of this creek where it had developed into a massive precipice, I found another nest in a perpendicular joint of the rock, in a small cavernous hollow and about two feet from the floor. A fourth nest I found was by the main Coongan River, and was built in an old mud nest of a Fairy Martin (Petrochelidon ariel) from which the spout had been broken away.

Whether located in a tree or a rock crevice, the nest of the Little Woodswallow is a scant bowl-shaped structure, made from fine twigs and grasses, sparsely lined with small rootlets or leaves. Group members nest in close proximity and both parents contribute to all aspects of incubation and care of the young, often assisted in this activity by unpartnered members of the group.

T. Ivison

LITTLE WOODSWALLOW

Artamus minor

Adults; Sundown National Park, Qld; October 1993 (R. Brown)

OTHER NAMES
None

MEASUREMENTS
LENGTH: 120 – 130 mm
WING: 103 mm
TAIL: 58 mm
BILL: 11 mm
TARSUS: 14 mm
WEIGHT: 13 – 16 g

DESCRIPTION
ADULT
Sexes similar. Head and upperparts dusky chocolate brown; rump dark blue-grey; tail dark blue-grey, outer feathers tipped white; wings mid blue-grey; underparts dusky chocolate brown. Eye dark brown. Bill pale blue-grey, tipped black with black central stripe on lower bill. Legs dusky black

IMMATURE
As adult

JUVENAL
Body brown, speckled buff off-white; wings as adult; tail as adult but white terminal tip greatly reduced

IDENTIFICATION
Dusky brown woodswallow with pale underwing; smallest of Australian woodswallows; no white on flight feathers

VARIATION
None

VOICE
Brisk, chirping "peet, peet"; soft twittering

HABITAT
Open woodland, arid scrubs, rocky gorges and outcrops

FOOD
Insects, mainly taken on the wing, occasionally from the ground or gleaned from foliage; nectar from flowering plants

HABITS
Sedentary in north, nomadic in south. Usually seen in small flocks of 6 – 20 birds; roosts and nests in rock crevices. Flocks cluster together to roost day or night. Often perches on dead timber

BREEDING
August – January, after rains in arid regions; 1 or 2 clutches per year; both sexes contribute to nest building, incubation and care of young

NEST: Scanty, bowl-shaped; made of fine twigs, grasses and rootlets, lined with fine rootlets or small leaves; placed in rock crevice or stump hollow

EGGS: 2 – 3; off-white, spotted and blotched brown and grey-brown, concentrated at larger end; tapered oval; 20 x 14 mm

INCUBATION: Unrecorded; fledging period unrecorded

DISTRIBUTION
Rocky outcrops in northern and inland Australia

STATUS
Widespread but patchy and irregular. Uncommon

T. Ivison

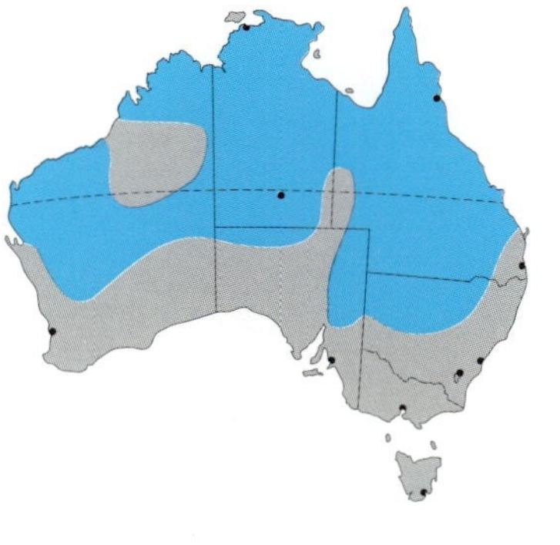

Sponsors: Shanny & Michael Crouch

Magpies, Currawongs and Butcherbirds

The family Cracticidae consists of some very familiar Australian birds: the magpies, currawongs and butcherbirds. They are powerful, medium to large, predatory songbirds with mainly black or strongly pied plumage, which illustrate well the ecological principles of adaptive radiation within a continent, and convergence with outwardly similar species on other continents.

The tree-dwelling currawongs have large bills for tearing and prising; short but strong legs and feet for clambering in trees; and broad, rounded wings and long tails for agile flight within forests. Butcherbirds are perch-and-pounce hunters within forest and woodland; they have rather small and weak feet, but large, strongly hooked bills for seizing and dismembering prey. They share the currawongs' wing and tail proportions but, being smaller, they are faster and more agile in flight among the trees. The Australian Magpie is a large, terrestrial butcherbird with a pointed bill for probing in soil, long legs for walking, and long, pointed wings and short tail for rapid flight in open country. Between them, members of the family occupy all terrestrial habitats on the continent and all vertical strata from the ground to the treetops. The family is restricted to Australia and New Guinea.

Early ornithologists thought that these birds belonged to familiar European bird families. The currawongs were assigned to the crow family Corvidae, reflected in such colloquial names as "Black Jay" and "Jackdaw". Indeed, the Pied Currawong's specific name *graculina* means "like a Jackdaw". The Australian Magpie was named for its superficial resemblance to the European Magpie, a pied member of the Corvidae, but it was long thought that our Magpie and the butcherbirds belong to the shrike family, Laniidae. The shrikes, not represented in Australia, are also hook-billed predatory songbirds which "larder" their prey. Hence they were known colloquially in Europe as "butcherbird", and this common name was applied in Australia. Later ornithological opinion recognised that our birds all belong in a unique Australian family, the Cracticidae. Now the wheel has come full circle, with recent DNA comparisons by Charles Sibley and colleagues revealing that the Cracticidae are related to the Corvidae. Such work has identified the woodswallows, Artamidae, and the strikingly black and red New Guinean *Peltops* "flycatchers" as the closest living relatives of the Cracticidae.

Cracticids share many characters with corvids, particularly in respect of size, loud voice, predatory and scavenging habits, intelligence, gregariousness and breeding behaviour. The main physical difference is in the lack of a dense mat of bristles over the nostrils in the Cracticidae, reflected in the generic name of the Australian Magpie, *Gymnorhina*, meaning "bare nose." Australian corvids are all crows of the genus *Corvus*, entirely black birds with brown, hazel or white eyes and a harsh "cawing" voice. Even the blackest kinds of currawong are readily distinguished from crows by their bright yellow eyes, traces of white in the spread wings or tail, and their more musical ringing calls. The other cracticids are capable of sustained melodious songs.

Cracticids are omnivorous, but mainly insectivorous and carnivorous. They catch a variety of invertebrates, reptiles, birds and small mammals, and eat fresh carrion. Magpies sometimes eat grain and some vegetable matter, and currawongs consume large quantities of berries and other fruits. Butcherbirds are particularly rapacious, pursuing and killing birds, sometimes up to their own size; magpies and currawongs also kill small birds. Currawongs, in particular, rob nests. All cracticids cache excess food for later consumption. Butcherbirds hang or wedge prey in tree crotches for storage or as an aid to dismembering it. They use the hook-tipped bill to strip pieces off the firmly anchored item, much as a hawk uses the opposing pull of claws and bill. Currawongs also do this, but magpies seem not to have the habit. Instead, they hold down the prey with one foot while dismembering it, and cache items in hiding places on or near the ground. Cracticids sometimes carry large prey items in their feet.

Currawongs nest in pairs, but outside the breeding season form large, cohesive flocks which perform seasonal movements. Butcherbirds are sedentary, and some breeding pairs have extra "helpers" at the nest. Magpies have a complex social organisation. They breed in groups, often of more than one adult of each sex, although each female has her own nest, and they defend the communal territory against neighbouring groups or intruders. All cracticids build a bowl-shaped nest of sticks, lined with softer material, in the fork of a tree, usually among foliage. The eggs are pale blue-green to olive-brown, streaked and spotted red-brown and grey. Incubation takes three weeks and the young fledge in four weeks. The female builds the nest and incubates the eggs, but both sexes feed the young. The insistent mewing or whining calls of begging young can be heard for some weeks or months after they fledge. Cracticids are bold and aggressive, vigorously defending their nests and territories by mobbing and sometimes striking humans and other potential predators.

S. J. S. Debus

Black Butcherbird

Cracticus quoyi

The large Black Butcherbird inhabits dense, wet forests in the tropics, such as the rainforest, monsoon forest, mangroves and paperbark forests of northern Australia and New Guinea. Over the rest of Australia, one finds that an intermediate-sized butcherbird co-exists with a smaller one in the eucalypt forests and woodlands: the Pied Butcherbird in the more open woodlands, and the Grey Butcherbird in the denser coastal forests and woodlands and the inland mallee and mulga thickets. In the woodlands of Cape York Peninsula, where the Grey Butcherbird is absent, the small and primarily New Guinean Black-backed Butcherbird has established a foothold.

The Black Butcherbird caused early ornithologists much confusion. As related by A.J. North nearly a century ago, there was much debate about whether the rufous immature birds represented a separate species or a colour form of the Black Butcherbird. The problem was that rufous immatures occur only in a part of the bird's range and, even there, only some immatures are rufous whereas others begin life in black plumage. Furthermore, as in other butcherbirds, some individuals pair while still in immature plumage. The question was resolved by accumulating observations of rufous and black birds paired, rufous young in the nests of black parents, and rufous and black young in the same

Adult feeding on Emerald Dove; Kuranda, Qld; June 1990 (D. Whitford)

Sponsors: Brian & Philippa France

brood. The old belief that rufous birds are adults is explained by the fact that butcherbirds do not attain adult plumage until their second annual moult. Rufous birds seen feeding young are probably yearlings that have stayed in their natal territory, helping to feed the next season's brood.

The Black Butcherbirds of the Northern Territory have caused confusion for a different reason. Michael Sharland related how he was puzzled around Darwin by a call like that of the Pied Currawong, and indeed the locals called it the "currawong". It was not, as anticipated, a range extension, but one of the calls in the local Black Butcherbirds' dialect. Sharland noted that he had:

> ... never heard two birds give such similar calls, except the true mockers, and this one wasn't a mocker. It was just another instance of parallel development in bird voices, although none the less remarkable in its synonymy.

Conversely, the calls of the Black Butcherbird differ considerably from one region to another. Members of a pair sing in duet, each bird alternately contributing a phrase.

The sombre plumage and dense habitat of the Black Butcherbird render it less conspicuous than other butcherbirds. Nevertheless, its rich, loud yodelling and "clonking" calls advertise its presence. Its other habits are typical of butcherbirds. It is secretive and wary, seeking prey from a series of lookout perches, between which it flies swiftly and directly. On sighting prey, it swoops quickly to the ground to seize it. Sometimes it forages on the ground by hopping along and searching. Large prey is carried in the bill to a crotch or crevice in a tree, wedged firmly, and pulled apart by the powerful, hook-tipped bill. Its food includes invertebrates, frogs, reptiles, small birds and eggs, and some fruit.

A pair occupies a territory of perhaps 10 hectares. The bowl-shaped nest of sticks and twigs is built in the fork of a tree. Both sexes are said to build the nest, incubate the eggs and feed the young, but the species has not been studied in depth and there is no information on the duration of the incubation and nestling periods. Adults are bold and aggressive in defence of their young. The Trumpet Manucode often nests in association with the Black Butcherbird, a habit used by the early ornithologists to locate Manucode nests. As related by W. Macgillivray in 1914:

> Towards one another the attitude of these species seems to be one of armed neutrality, the Manucode finding a certain amount of protection from being in the neighbourhood of so warlike a bird, but at the same time keeping a watchful eye on its own eggs or callow young when the ally is near.

The various regional populations of Black Butcherbird have differentiated physically from each other to become subspecies or geographical variants. This may be attributed to the patchy distribution of its habitat in northern Australia, and the reluctance of this sedentary bird to cross expanses of water or unfavourable habitat. Julian Ford analysed this variation, and concluded that "Nothing is achieved by expressing this subspecifically". Nevertheless, it is customary to recognise a massive-billed subspecies in New Guinea, *C. q. quoyi*; a large subspecies in the Northern Territory, Cape York Peninsula, Aru Islands and the southern coast of New Guinea, *C. q. spaldingi*; and a small subspecies with rufous juveniles in north-east Queensland, *C. q. rufescens*. Members of an isolated population on the central Queensland coast are small, like *rufescens*, but lack rufous juveniles; the population has not been formally named.

S. J. S. Debus

Cracticus quoyi

Immature; Cairns, Qld; September 1975 (T. & P. Gardner)

OTHER NAMES
Currawong (in Northern Territory, in error)

MEASUREMENTS
LENGTH: *340 – 440 mm*
WING: *156 – 205 mm*
TAIL: *124 – 163 mm*
BILL: *Not recorded*
TARSUS: *Not recorded*
WEIGHT: *159 – 216 g*
Male slightly larger than female

DESCRIPTION
ADULT
Entirely glossy black with brown eyes; pale blue-grey base to bill; black feet

IMMATURE
Either sooty black, sometimes with slight brown tinge, or rufous marked black (latter only in north-east Queensland, sometimes in same brood as black juveniles); upperparts dusky brown with heavy rufous streaks on body; underparts rufous with black scallops on breast. Bill grey or dark blue-grey

IDENTIFICATION
Black individuals distinguished from other large, glossy black songbirds by large, pale bill. Brown immatures larger and richer rufous than other immature butcherbirds

VARIATION
Two subspecies in Australia: larger C.q. spaldingi *in the Northern territory and on Cape York Peninsula; smaller* C. q. rufescens *in north-east Queensland, in which some immatures are rufous; a third in New Guinea*

VOICE
Advertising call a quick cionking (Queensland) or "currawong" phrase (Northern Territory); harsh, slurred caws in alarm; other mellow caws and ringing notes in contact; a long, yodelling song. Also a soft subsong

HABITAT
Rainforest, mangroves, paperbark forests and fringes

FOOD
Large insects, crustaceans, frogs, reptiles, small birds and eggs, sometimes fruit

HABITS
Solitary or in pairs or family groups. Sedentary. Arboreal, but swoops to ground to seize prey and sometimes forages by hopping along ground, searching

BREEDING
September – February. Both sexes build nest and feed young

NEST: Bowl of sticks, lined with fine twigs in tree fork 2 – 10 m above ground

EGGS: 2 – 3, sometimes 4; grey-green spotted and blotched red-brown, lavender and dark brown; oval-shaped; 35 x 25 mm

INCUBATION: Unrecorded; fledging period unrecorded

DISTRIBUTION
Top End of Northern Territory; Cape York Peninsula and coastal north-east Queensland; also New Guinea and islands

STATUS
Moderately common

S. J. S. Debus

Grey Butcherbird

Cracticus torquatus

Craticus, which means "loud-voiced", is a particularly apt generic name for the Grey Butcherbird, whose penetrating and strident calls are the familiar butcherbird sounds of the heavily settled southern coast. Probably more than any other butcherbird, this species has earned the group an undeserved reputation for cruelty, reflected in one of its early specific synonyms, *destructor*. Its proximity to cities and towns has brought it into conflict with poultry keepers and aviculturists, as well as bird lovers who see small birds killed and hung on a "gibbet".

Butcherbirds sometimes kill chickens and domestic birds, but this can be avoided by adequate protection. In the case of caged birds or poultry, this means a sufficiently large aviary or pen for the inmates to retreat and hide in an enclosed section, and a double wall of mesh, pegged apart by several centimetres, on the wire section. Free-ranging chicks and birds in small cages hung outdoors or on verandas are especially vulnerable and provide an open invitation to a hungry butcherbird.

Butcherbirds regularly prey on small birds and their nestlings, but predator and prey have co-evolved. Under natural circumstances, the prey species' populations and reproductive output can cope with the loss of individuals and a diverse and multi-layered tree and shrub canopy provides many safe hiding places for nests. The predator's impact becomes severe where human activities remove the protective understorey of native shrubs or where increasingly small remnants of forest or woodland are subject to "dieback" (rural tree decline) and invasion by Noisy Miners.

Pair at nest; Goomboorian, Qld; October 1985 (C. Webster)

A butcherbird lacks the strong grasping feet and claws of predatory hawks and eagles. It hangs freshly killed prey in tree crotches so that it can be torn into manageable pieces with the bill. Butcherbirds also use these sites as larders in which to store surplus food, behaviour that represents no more than innate resourcefulness; human equivalents are the gamekeeper's gibbet, the butcher's shop and the suburban refrigerator. In one case observed by Dariel Larkins, a butcherbird brood died in the nest, the bodies were then removed and hung in the parents' larder and later eaten. The energy and resources already expended by the parents on their offspring were thus not wasted, but efficiently recycled.

The Grey Butcherbird is a vigorous perch-and-pounce hunter, waiting motionless on a vantage point then darting to the ground to seize and carry off its prey to another perch. Sometimes it may dash among small birds, seizing one by the neck or killing it with pecks to the head. Birds up to the size of the Painted Button-quail – about the butcherbird's weight – have been recorded as prey. More usually, large insects are taken, also reptiles, mice and some fruits and seeds.

The Grey Butcherbird's voice adds to its sinister image. Many of its calls are harsh and menacing, or rollicking notes like cackling laughter, but it also utters mellow, ringing notes in song, and a soft whisper song that includes mimicry of other species. Members of a pair sing in antiphonal duet, each alternately uttering its individual phrase. All its calls are deeper, quicker and livelier than the measured, pure tones of the Pied Butcherbird.

Dariel Larkins recognised three types of song. The breeding song is a sustained, solo dawn performance by the male on a special singing tree at the start of the breeding cycle. The advertising song is an antiphonal duet by the pair. Quiet singing includes the whisper song uttered by adults in the non-breeding season, and the subsong of practising immatures. Larkins also recognised many types of calls, including a specific or identity call uttered from the roost as the first and last call of the day. Members of a pair use elements of the antiphonal duetting song to keep in contact while they forage independently in their joint territory.

Aspects of the Grey Butcherbird's breeding cycle have been described by K.A. Hindwood, John Disney and Dariel Larkins. All three observers noted that the male feeds the female in courtship or on the nest while she is incubating. Larkins also described how a female butcherbird shaded her nestlings in hot weather, by standing on the nest with spread wings, casting her shadow on the brood.

The territory of one pair of Grey Butcherbirds in Sydney suburban bushland covered about five hectares. The bowl-shaped nest of sticks and twigs, lined with fine grass and other soft fibres, takes about a month to build. Only the female incubates, for 24 to 26 days, but both parents feed the young which fledge in about a month. Immature birds may remain in the natal territory and help to feed and defend their younger siblings in the following season.

S. J. S. Debus

Grey Butcherbird

Cracticus torquatus

Fledglings; Middle Harbour, Sydney; November 1959 (N. Chaffer)

OTHER NAMES
Silver-backed Butcherbird (northern race)

MEASUREMENTS
LENGTH: *260 – 300 mm*
WING: *131 – 160 mm*
TAIL: *102 – 128 mm*
BILL: *Not recorded*
TARSUS: *Not recorded*
WEIGHT: *74 – 110 g*
Male slightly larger than female

DESCRIPTION
ADULT
Head black, with variable white spot in front of each eye; partial white collar; white throat; back grey; rump white; wings dusky brown, with variable white markings; tail black, tipped white; underparts light grey to white. Bill blue-grey, tipped black. Eyes dark brown. Feet dark grey

IMMATURE
As adult but dark areas brown, pale areas washed fawn. Bill dark grey

IDENTIFICATION
Smaller and greyer than Pied Butcherbird, with white (not black) throat; calls differ. In northern Queensland, Black-backed Butcherbird is cleaner black and white with more white in wings, no white loral spots

VARIATION
Four or five subspecies. Tasmanian C. t. cinereus *larger and darker than south-east mainland* C. t. torquatus. *Latter grades into somewhat darker form found in Western Australia, sometimes recognised as* C. t. leucopterus, *which has smaller white loral spots, trace of freckled black "bib" and more white in wings. "Silver-backed"* C. t. argenteus *in Top End of Northern Territory is smaller, cleaner grey and white with more white in wings; lacks white loral spots. Kimberley population,* C. t. latens, *similar to* C. t. argenteus *but wings more pied; broken black "necklace" around throat*

VOICE
Strident series of rollicking notes; upslurred harsh caws; song of mellow, ringing notes uttered rapidly. Soft whisper song includes mimicry of other birds

HABITAT
Eucalypt forest and woodland, mallee and mulga scrub, paperbark and monsoon forest, native fig thickets; in inland, restricted to denser types

FOOD
Solitary or in pairs or family groups. Sedentary. Arboreal, but catches most food by swooping to ground

BREEDING
July – January. Both sexes feed young

NEST: Bowl of sticks and twigs; lined with grass and other fibres, 3 – 10 m above ground, often in sapling or small tree

EGGS: 3 – 5, brownish-green spotted red to brown; oval-shaped; 31 x 23 mm

INCUBATION: 24 – 26 days; fledging period 28 days

DISTRIBUTION
Kimberley of Western Australia and Top End of Northern Territory; southern and eastern Australia, including Tasmania

STATUS
Common

S. J. S. Debus

Black-backed Butcherbird

Cracticus mentalis

The Black-backed Butcherbird is a little-known species of the remote Cape York Peninsula and southern New Guinea. Inhabiting eucalypt savanna woodland, it resembles other butcherbirds in its habits. It is common in its favoured habitat, and around houses and settlements, where it is confiding towards humans. The nest and eggs were discovered in 1910 by H. G. Barnard.

The Black-backed Butcherbird is slightly better known in New Guinea where it has been observed by Harry Bell and Brian Coates. It is primarily a perch-and-pounce hunter, swooping down to snatch prey from the ground or foliage but also searching along branches and in shrubs, investigating potential hiding places of prey. It catches large insects, lizards and small birds up to the Sacred Kingfisher in size, and eats some seeds. It wedges prey into crevices in branches to assist in tearing it apart, or as storehouses for surplus food which is later eaten. Like many other predatory birds, it regurgitates pellets which contain the indigestible remains of its food, such as chitinous insect fragments, bones, feathers and fur.

The male sometimes feeds the female while she is building the nest, perhaps in courtship, although food provision at this stage may also be important for egg formation (as in the birds of prey). The male also feeds the female on the nest while she is incubating the eggs. Both parents feed the young and defend them aggressively. Harry Bell has noted a communal display after the breeding season: a group of four to six territorial neighbours flying up to 50 metres or so in the air, wheeling at each other and calling constantly for about five minutes, after which they suddenly glide down separately to the trees.

Some authors regard the Black-backed Butcherbird as a subspecies of the Grey Butcherbird but although the two are closely related, there are constant differences. The Black-backed form is more dapper in its clean black and white plumage, with a greater amount of white in the wings, and it has a stouter bill and lives in more open woodland. Perhaps most significantly, its voice, by which it signals its identity, is recognisably different. Its calls have the rollicking and bubbling Grey Butcherbird quality but are uttered more slowly.

Male and female sing in duet, sometimes alternately with each bird contributing separate phrases. Song is delivered vigorously, with posturing: head stretched up and bill pointing skywards, body plumage fluffed, while "pumping" or quivering the folded wings. The Black-backed Butcherbird is an accomplished mimic of the calls of other birds.

S. J. S. Debus

Black-backed Butcherbird

Cracticus mentalis

Adult; Cape York Peninsula, Qld (C. & D. Frith/Frithfoto)

OTHER NAMES
White-throated Butcherbird

MEASUREMENTS
LENGTH: *250 – 270 mm*
WING: *137 – 149 mm*
TAIL: *103 – 112 mm*
BILL: *Not recorded*
TARSUS: *Not recorded*
WEIGHT: *Not recorded*

DESCRIPTION

ADULT
Boldly pied. Head black, with partial white collar and white throat; upperparts patterned in black and white, with much white in wings; underparts white. Bill blue-grey, tipped black. Eyes dark brown. Feet dark grey

IMMATURE
Patterned as adult, but dark areas brown (not black) and mottled paler; white areas faintly washed brown. Bill grey

IDENTIFICATION
Small, boldly pied butcherbird. Similar to Grey Butcherbird but clean black and white, no white loral spots, more white in wings. Lacks black throat of larger Pied Butcherbird

VARIATION
None in Australia. Australian species, C. m kempi, *barely distinct from nominate subspecies,* C. m. mentalis, *of New Guinea*

VOICE
Similar in general character to that of Grey Butcherbird, but distinct – variously said to be softer and less musical than Grey's, or like Grey's gurgle but slow and languid. Song an unhurried phrase of long and short, low and high, mellow ringing notes interspersed with disyllabic yodels and croaks. Also a rapid rollicking and yodelling, explosive loud calls and a low muttering gurgle. Harsh, slurred caws in alarm or aggression. Mimics other birds

FOOD
Large insects, lizards, small birds, some seeds

HABITS
Solitary, in pairs or family groups. Sedentary. Arboreal, catching prey by swooping from lookout perch to ground or foliage, or by searching along branches and amongst foliage.

BREEDING
June – November. Female builds nest, both sexes feed young

NEST: Bowl of sticks and twigs lined with grass and other fine fibres, up to 20 m above ground in tree fork

EGGS: 3 – 4; grey-green, sparsely spotted red-brown and slate; oval-shaped; 30 x 27 mm

INCUBATION: Unrecorded; fledging period unrecorded

DISTRIBUTION
Cape York Peninsula; also savanna woodlands of southern New Guinea

STATUS
Common

S. J. S. Debus

Pied Butcherbird

Cracticus nigrogularis

The Pied Butcherbird, the most common and familiar butcherbird of the open woodlands over much of Australia, is best known for its ethereal voice. It also has the conspicuous habit of perching on roadside powerlines and poles in farmland. It is larger and more strikingly black and white (with a black "bib") than the duller Grey Butcherbird, which has a white throat. Its calls also differ. The Grey has a series of rapid rollicking and harsh notes, the Pied a slower series of pure, whistled tones. The Pied Butcherbird is also superficially similar to the Magpie-lark but has a large bill, no white facial markings, a more upright perching stance with drooped tail, and more rapid, undulating flight. On the ground its short legs permit a few shuffling hops, unlike the Magpie-lark's long-legged, elegant strutting.

It preys on a variety of small animals, ranging from large insects to frogs, lizards, small snakes, birds and nestlings, and rodents; it occasionally eats fruit. It can kill mammals up to the size of the Black Rat, but the birds that it takes are usually much less than its own size. It hunts in open situations, watching from a high vantage point then swooping down to the ground. Prey is usually seized quickly and taken to a perch. The typical butcherbird larder, a crevice or fork in a tree, is used for dismembering or storing prey.

A commensal feeding association has developed between the Pied Butcherbird and the Australian Hobby, a small falcon. Falcon and butcherbird sometimes take advantage of each other in an attempt to capture a small bird. A bird being pursued by the falcon may seek shelter in a bush, only to be attacked by a butcherbird while it is reluctant to break cover. Alternatively, a butcherbird's attack may flush a bird out for a watchful falcon. Some episodes rapidly alternate between aerial assault by the falcon and shrubbery assault by the butcherbird, so that it is not clear how the incident started. Sometimes the falcon benefits, sometimes the butcherbird, and sometimes the victim confounds both by escaping despite their combined onslaught. Captures must be frequent enough for the association to persist. Sometimes two butcherbirds work together in pursuit and capture of small birds.

The Pied Butcherbird is more communal than other butcherbirds. It lives in groups of up to six, consisting of two or three adults and several immatures. Brown birds, offspring of previous seasons, stay in their natal territory and assist in the care of the current brood. Lloyd Nielsen observed a group of four adults, in which two females laid in the one nest and shared incubation in shifts. All four defended the nest. At another nest observed by H. Thomas in 1944, an adult and an immature fed the brooding adult and the small nestlings. Sometimes the brooding bird passed food on to the young. The immature also tried, unsuccessfully, to brood beside the adult.

There is little information on territory size. Groups are said to occupy territories of about two square kilometres but the active nests of two neighbouring groups have been seen 75 metres apart. At a nest observed by John Courtney, building took about a month. Eggs were laid on alternate days and took about 20 days to hatch. Hatching was spread over four days, suggesting that incubation started with the first of the four eggs. Only a single young survived to leave the nest at about 30 days old.

This illustrates that butcherbird nestlings, like those of other birds, are vulnerable to predators, the weather or food shortage.

The Pied Butcherbird has a special dawn song which it utters at the start of the breeding season from a song perch. This sustained song surpasses its other calls. Nevertheless, to quote A.J. North, the Pied Butcherbird's other songs and contact calls are also:

> ... clear and prolonged notes ... among the most musical of those of any of our Australian birds ... If the bird were clad in sombre hue, instead of a conspicuous and strikingly contrasted plumage, its "native wood notes wild" would render it one, if not the most attractive, of all our arboreal species.

The advertising song of a pair consists of animated, bubbling phrases of ringing whistles. The two sing in antiphonal duet, alternately bowing and stretching as they contribute their individual phrases of low and high notes, with body feathers fluffed and slightly opened wings vibrating with the effort. The Pied Butcherbird sometimes sings on moonlit nights.

S. J. J. Debus

Adult; Kakadu National Park, N.T.; July 1980 (K. Griffiths)

Sponsor: Douglas B. Ash

Pied Butcherbird

Cracticus nigrogularis

Adult and immature helper at nest; Goomboorian, Qld; September 1970 (C. Webster)

OTHER NAMES
Black-throated Butcherbird

MEASUREMENTS
LENGTH: *320 – 360 mm*
WING: *171 – 190 mm*
TAIL: *118 – 152 mm*
BILL: *Not recorded*
TARSUS: *Not recorded*
WEIGHT: *83 – 155 g*
Male slightly larger than female

DESCRIPTION

ADULT
Upperparts boldly patterned in black and white, with full black hood extending to breast; broad and complete white collar; remainder of underparts white. Bill blue-grey, tipped black. Eyes dark brown. Feet black

IMMATURE
Duller and paler than adult. Dark areas brown (not black) with indistinct "bib"; white areas faintly washed brown; pale eyebrows. Youngest birds lack dark "bib".

IDENTIFICATION
Larger than Black-backed and Grey Butcherbirds; cleaner black and white than Grey Butcherbird, and distinguished from both by solid black throat and breast. Immature larger and paler than immature Grey Butcherbird, and acquires faint brown "bib" which also separates it from immature Black-headed Butcherbird

VARIATION
Negligible. Slight clinal increase in size from north to south

VOICE
Clear, high-pitched lilting whistles, pure in tone, delivered slowly. Also more animated, bubbling phrases of ringing whistles. Strident shriek in alarm or aggression, higher pitched than Grey Butcherbird's cawing. Soft whisper song includes mimicry of other birds

HABITAT
Eucalypt and other dry woodlands; farmland

FOOD
Large insects, frogs, lizards, small snakes, small birds and nestlings, and rodents

HABITS
Solitary, in pairs or family groups; more communal than other butcherbirds. Sedentary. Arboreal, though captures prey by swooping to ground

BREEDING
August – November in south; May – November in north. Group members feed brooding female and young

NEST: Bowl of sticks, lined with grass and other soft fibres, in upright tree fork 5 m or more above ground

EGGS: 3 – 5; dull green, spotted brown or black; oval-shaped; 33 x 24 mm

INCUBATION: about 20 days, by female; fledging period about 30 days

DISTRIBUTION
Australian mainland, except southern arid areas and south-east coast

STATUS
Common

S. J. S. Debus

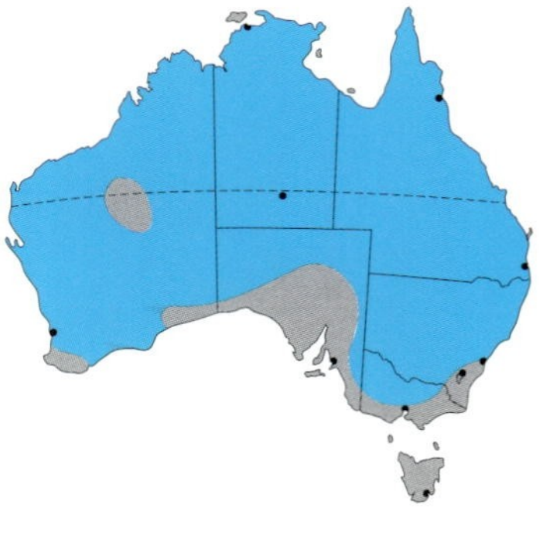

Australian Magpie

Gymnorhina tibicen

The Magpie is one of Australia's most conspicuous, successful and ubiquitous birds, one of the few species to benefit greatly from human activities. Its readiness to forage on farmland, playing fields or suburban lawns, coupled with its loud carolling and group displays of territoriality, render it obvious to the most casual observer. Its behaviour in the breeding season commands the attention even of those who are normally oblivious of birds. Less obvious is the fact that the Magpie is a large butcherbird, adapted for life on the ground. It shares the butcherbirds' boldly pied plumage and blue-grey, black-tipped bill. It also shares their melodious songs, antiphonal duetting and singing posture with uplifted bill, body plumage fluffed and wings slightly open. It differs in its pointed, rather than hook-tipped, bill, its longer and stronger legs, and its longer and more pointed wings. It lacks the butcherbirds' white tail-tip, having a broad black tip to a shorter tail.

These physical divergences from the tree-dwelling butcherbirds reflect the Magpie's way of life. It is much more a ground-dweller, foraging by walking about and probing in the soil in open or even treeless areas. It requires trees only for roosting, nesting or shelter, to which it commutes by rapid, direct flight.

Nevertheless, the Magpie's relationship to the butcherbirds is such that some authorities have even placed it in the genus *Cracticus*. This close relationship is revealed by one authenticated case of hybridisation in the wild with the Pied Butcherbird. Dr Jane Hughes confirmed by genetic comparison that the individual concerned was indeed a hybrid, as its appearance suggested – from different angles it looked like either parent species, so that observers were initially undecided whether it was butcherbird or Magpie. It associated with Magpies, but its parentage was unknown. Other such supposed hybrids have usually been cases of aberrant Magpie plumage, in which rare individuals "throw back" to a Pied Butcherbird type of pattern with white belly, white in the secondaries and a black tail. A more frequent aberration is a white belly starting just anterior to the legs, coupled with a white back, in a population of normally black-backed birds. Albino and wholly black individuals occur. Even in normal plumage, regional populations of Magpies differ in the dorsal pattern of black and white to such an extent that they were regarded as separate species. Now they are treated as geographical subspecies that interbreed and intergrade where they come into contact.

The Magpie departs somewhat from butcherbirds in its feeding habits. It consumes a wide variety of ground invertebrates, by searching then suddenly jabbing its bill into the soil or snapping up items. It sometimes shuffles one foot in the grass to flush out prey. In an ingenious series of experiments, Robert Floyd determined that it detects prey such as scarab grubs moving under the soil surface by listening. It also catches lizards and mice, and pursues small birds relentlessly. These it sometimes catches and eats, but many chases seem to be in aggression or defence of territory; occasionally it will kill a bird yet not eat it. It does not share the butcherbirds' habit of robbing nests in trees, but opportunistically seizes the young of ground-nesting birds when these are encountered.

A pioneering study by Angus Robinson and later detailed research by Robert Carrick revealed the Magpie's complex social system, these studies having been elegantly summarised by Ian Rowley. The strongest and most aggressive adults defend the best territories with sufficient resources like food and nest sites for successful breeding. The breeding unit often consists of more than a pair of adults: frequently a dominant male with two or more females, and sometimes one or more subordinate males. The dominant male achieves most of the matings, and one of the females is also dominant over the others. She receives some assistance from the dominant male in nest defence and feeding of young; the nests of subordinate females often fail because they are preoccupied with defence. Territorial groups contain birds of all ages, but immatures are evicted when they reach the age of independence. These non-breeding birds form large feeding flocks which wander locally over treeless areas unsuitable for breeding.

Gradually, individuals form social groups and manage to defend specific feeding areas, at first without suitable trees for nesting, eventually fighting their way up the hierarchy to establish themselves in a breeding territory when a vacancy occurs. The benefit of such a system seems to be security for the breeding female, without which she cannot raise offspring. By supporting the system with group defence, the subordinate individuals eventually attain breeding status by replacing dominant individuals that die. For some it may be a long wait, some breeding birds having lived for up to 25 years.

The nesting habits of the Magpie frequently bring it into conflict with humans. Magpie nests of wire, built on utility poles, may interfere with powerlines by causing short circuits, leading to blackouts or grass fires. Humans must bear some responsibility for this situation, because groups of Magpies living in marginal territories are forced to accept substandard nest sites. Since Magpies prefer to build in leafy trees, the solution lies in restoring some of the natural tree cover and in cleaning up sources of wire scraps.

Another source of conflict is attacks by Magpies on humans in the breeding season. Immature Magpies learn to attack humans by observing and imitating the behaviour of their parents, who pass on to their offspring the "knowledge" that humans are enemies. Magpie aggression increases under hormonal influence in the breeding season, and humans inadvertently approaching a nest too closely may trigger a defensive response. Strikes can be avoided by carrying a walking stick, leafy branch or umbrella and raising it when under attack, or by wearing a robust, broad-brimmed hat. Since Magpies attack stealthily from the rear, and are reluctant to strike while being watched, strikes may be prevented by turning to face the attacking bird, or by having large, false "eyes" on the rear of one's hat. Cyclists find that a tall flag at the rear of the cycle is helpful. Humans can also take some initiative, by avoiding the vicinity of active Magpie nests.

At the start of the breeding season, during or after nest building, females solicit mating by crouching with the bill uplifted, closed wings slightly raised and quivered, and the white tail flicking up and down as an obvious signal. The male arrives with a show of aggression, fluffing his plumage and leaving his wings slightly open to increase his apparent size, before mounting. Mating appears to be rather violent, and subordinate males sometimes steal matings by force.

Adult male; White-backed race (T. Howard)

Sponsor: Jim & Chris Hamann for Lameroo North

In contrast, the post-fledging behaviour of young Magpies and even their parents is often gentle and playful, although the aggressive, competitive undercurrent is always present. Young Magpies chase and play-fight like puppies, grappling and rolling over on the ground, sometimes joined by their parents. Young birds also play with objects by pouncing on them, grappling them with their feet and rolling about. They may jump up to seize branches, leaves or even garments by swinging upside-down on clothes lines. Young Magpies avoid parental aggression by submissive displays of posturing like a soliciting female and "whimpering", squatting hunched on the ground, or even rolling over on the side or back. Sometimes Magpies sunbathe on warm days by lying flat on the ground, with plumage ruffled, wings stretched out and tail spread, often on damp soil or in humid conditions. Their eyes are closed, seemingly in enjoyment, but they are alert and quickly regain their composure if approached.

The play of young Magpies serves as practice for later life. At first they are fed by their parents, each of which often takes charge of a specific fledgling in broods of two. Foraging adults accept human food scraps for themselves but selectively feed natural food such as insects to their young. Adults eventually reduce their feeding of the young, which start to feed themselves. They beg at food placed in front of them, then learn to pick it up. They progress to capturing small items, and gradually acquire their parents' foraging skills. Magpies vigorously rub tough or potentially dangerous food, such as cooked meat or stinging insects, on the ground to soften it up or remove offensive parts. They hide surplus food in soil and in ground vegetation.

Young Magpies practise their parents' and group's song phrases, which they learn by imitation, in quiet subsong of extended warbling. They start to counter-sing loudly in aggressive encounters, and join in the group's antiphonal choruses. Like butcherbirds, adult males deliver a melodious nocturnal or dawn song from a song perch at the start of the breeding season. The songs of Magpie groups sound similar within a region, but vary geographically with distinct regional dialects. Within and between groups, individuals have recognisable phrases; some of these are characteristic of each sex. In quiet whisper song, Magpies imitate the calls of other bird species. Tame Magpies also learn to imitate human speech and to whistle tunes. There is no truth to the persistent myth that in order to "talk" a tame Magpie needs to have its tongue split. The birds are quite capable of doing so without such cruel or drastic action, and their vocalisations are controlled by the syrinx, not the tongue.

S. J. S. Debus

AUSTRALIAN MAGPIE

Gymnorhina tibicen

Immatures; Warrumbungle National Park, N.S.W. (K. Griffiths)

OTHER NAMES
None in general use, other than Black-backed, White-backed and Western Magpie for various forms

MEASUREMENTS
LENGTH: 380 – 440 mm
WING: 228 – 300 mm
TAIL: 121 – 168 mm
BILL: Not recorded
TARSUS: Not recorded
WEIGHT: 206 – 387 g
Male slightly larger than female

DESCRIPTION

ADULT
Black with large white patch on upper and under surface of innerwing; undertail coverts white; tail white with broad black tip; dorsal surface variable, either white from nape to rump or interrupted by black saddle of variable width and completeness. Female has white areas from nape to lower back mottled grey; in some populations black saddle also mottled. Bill pale blue-grey with black tip. Eyes red-brown. Feet black

IMMATURE
Head and body mottled brown where adults black; white areas from nape to lower back mottled grey. Bill grey. Eyes brown

IDENTIFICATION
Unmistakable. Easily distinguished from currawongs by pale bill, dark eyes, different distribution of white in plumage, long, pointed wings and short tail, fast flight, voice

VARIATION
Complex. About six subspecies which vary in size and dorsal plumage pattern. Smallest birds in New Guinea and Tasmania; clinal increase in size from north to south on Australian mainland. Tropical G. t. papuana *has most massive bill; males white-backed or with narrow black saddle but females and young have broadest black saddles of all, with heavily mottled nape.* G. t. tibicen *of eastern Australia is clean black and white with broad black saddle.* G. t. leuconota *of south-eastern Australia has white-backed males.* G. t. hypoleuca *of Tasmania is similar to* leuconota *but smaller.* G. t. dorsalis *of south-western Australia has white-backed males and black-backed females with mottled saddles.* G. t. longirostris *of northern and central Australia has long bill; it resembles* tibicen *but males may have narrow saddle or white back (latter in central Australia, sometimes regarded as* G. t. finki*)*

VOICE
Strident rallying shrieks and cries in aggression and alarm; loud carolling song with antiphonal duetting; softer melodious warbling, often with mimicry of other bird species in whisper song. Also harsh squawk in alarm; mournful descending whistle. Juveniles beg insistently with whining cries

HABITAT
Open woodland, farmland, grassland and urban areas

FOOD
Worms, snails, crustaceans, arthropods, frogs, reptiles, small birds, mice, human food scraps, carrion; some plant matter, grain, seeds and fruits

HABITS
Sedentary. Breeding groups, often consisting of more than two adults, defend permanent territories. Average group size varies from 3 – 4 adults in southern and eastern Australia to 8 in south-western Australia. Non-breeding birds form a graduated series of social groups from large, locally nomadic flocks to resident groups in marginal territories, fighting their way up the hierarchy until they obtain a place in a dominant group with a viable territory. Forages on the ground by walking and searching. Roosts and nests in trees

BREEDING
June – December. Female builds nest; both sexes feed young

NEST: A bowl of sticks, lined with grass, feathers, wool, fur and other soft material; placed in the fork of a live or dead tree or on an artificial structure such as a telegraph pole; usually 5 – 20 m above ground, occasionally down to 1 m from ground in shrub in treeless country. Some nests made partly or entirely of lengths of wire, sometimes barbed, and lined with plastic strips, cloth, string and other material of human origin

EGGS: 1 – 6, usually 3 – 5; pale blue or green, spotted and streaked red-brown; oval-shaped; 38 x 27 mm

INCUBATION: 20 days, by female; fledging period 28 days

DISTRIBUTION
Most of mainland Australia and some large islands; Tasmania; also savanna woodlands of southern New Guinea. Introduced to New Zealand

STATUS
Abundant

S. J. S. Debus

Pied Currawong

Strepera graculina

As a group, the currawongs are to some extent related to the butcherbirds, which also include the Australian Magpie. Currawongs are larger and more crow-like, with stout black bills and generally more sombre plumage, relieved by smaller amounts of white. They also lack the butcherbirds' melodious songs and instead utter loud, raucous or ringing cries.

The Pied Currawong is one of the most successful and adaptable members of the family, and certainly the most successful currawong. Bird lore has commonly claimed that it breeds in mountain forests and winters in large flocks on the coastal plain such as around Sydney. It became a permanent resident in parts of Sydney around 1960 and by 1970 it was a resident breeder in parts of the lower North Shore of that city. It now breeds in parks and gardens as well as urban bushland, and recently some birds have taken to nesting on artificial structures such as sheltered scaffolding in the suburbs. Although large flocks move in autumn and winter to milder coastal and inland plains, many areas now have more or less resident or locally nomadic flocks. Some areas visited by winter flocks, such as Armidale in northern New South Wales, also have resident breeding pairs in bush and city. This transition to urban existence has also occurred in Melbourne, Canberra and cities in Queensland.

The Pied Currawong's population explosion is attributable to human activities. Being an omnivore and scavenger, the currawong has benefited from an increased food supply, particularly in the form of introduced berry-bearing trees and shrubs. It has taken to Privet, Firethorn, Wild Olive and Camphor Laurel with great gusto, and even the fruits of the Canary Island Palm. It digests the pulp and later regurgitates the seeds in pellets. It is responsible for Ochna and palm seedlings sprouting in Sydney gardens, and for Privet, Firethorn and Camphor Laurel invading bushland, as feeding birds disperse the seeds up to five kilometres. In northern New South Wales, Hawthorn is spreading at an annual rate of 120 metres, now turning some paddocks into dense Hawthorn thickets. On the western slopes and plains, the currawong's pellets contain the fruits of the Peppercorn Tree. This artificial abundance of food has enabled high winter survival of immatures, which swell the next season's breeding population and the cycle continues as the currawongs propagate their new food sources.

The Pied Currawong also feeds on road-killed animals, human food scraps, pet food, offal, orchard fruits and introduced birds and rodents, all available in abundance. A capable predator, it searches guttering for nestling Common Starlings and House Sparrows, seizes adult Common Mynas at their communal evening roosts in Canary Island Palms, and drags young Ringtail Possums from their dreys. Pairs of currawongs cooperate to catch large prey such as rats. Prey too large to swallow is carried, sometimes in the feet, to a jagged stump or other crotch where it is wedged firmly and pulled apart. Birds return to these larders until the prey is consumed.

In suburban gardens and in rural bushland remnants, these predators of nestlings and eggs may be causing a decline in small, native bush birds. The increased currawong population, combined with

Adult (M. Seyfort)

Sponsor: Mrs John A. Minter

a reduced understorey of protective native shrubs, may be lowering the breeding success of the smaller species which many people value as insectivorous songbirds. Selective culling of the currawong population has been proposed as an interim measure but past culling of thousands has had no impact, despite the incentive of bounties. In the longer term the solution lies in an ecological approach. Native trees and shrubs should replace exotic berry-bearing species as ornamental plants in parks and gardens, these problem plants should be controlled or eradicated in bushland and farmland, harbours for introduced birds should be removed, and the native tree and shrub cover should be rehabilitated on degraded farmland. In parts of Canberra where there is a dense and varied native tree and shrub layer, small birds persist despite a large currawong population.

Paradoxically, the Pied Currawong may be threatened on Lord Howe Island where the endemic subspecies numbers about 40 breeding pairs. The small size of the island, and consequent small area of habitat, limits the maximum population size. Its depredations on domestic chickens and on the endangered Woodhen mean that, as elsewhere, it is prone to destruction by humans.

On the mainland, the Pied Currawong does not have everything its own way. Adults and nestlings fall prey to large reptiles such as the Lace Monitor and Diamond Python, and to large predatory birds such as hawks and owls. Its nests are commonly parasitised by the Channel-billed Cuckoo, which lays its eggs in currawong nests and leaves the hosts to rear one or two young cuckoos. The currawongs' eggs are not removed; the young cuckoos simply out-compete and out-grow the currawong hatchlings, which starve. Furthermore, flocks of currawongs eat the nestlings of other nest predators, including ravens and even their own species. Nestlings are vulnerable until they grow feathers and start to look like currawongs, after which cannibalism is inhibited.

Harry Recher's study, published in 1976, remains the only detailed account of the Pied Currawong's breeding behaviour. Both sexes gather sticks from a dead shrub but the female builds the nest, which takes about a month from the commencement of construction to the laying of eggs. The female solicits mating by raising her tail, spreading and quivering her wings and giving begging calls. The male rears up with wings and tail fully spread, quivering his wings to accentuate the white flashes, before mounting. The pair defends a territory of almost a hectare. The male feeds the female during nest building and incubation, and also during the nestlings' first week when the brooding female passes food to them. Thereafter, both parents feed the young, which remain in the nest for a month and in the territory for a further two months. Adults defend their young against humans by swooping, clapping the bill and passing overhead with "woofing" wing beats. Pied Currawongs live up to 14 years in the wild.

S. J. S. Debus

Pied Currawong

Strepera graculina

Adult in flight; Aireys Inlet, Vic.; August 1987 (M. Seyfort)

OTHER NAMES
Black Magpie, Mountain Magpie, Scrub Magpie, Chillawong, Tallawong

MEASUREMENTS
LENGTH: 440 – 480 mm
WING: 225 – 260 mm
TAIL: 200 – 235 mm
BILL: Not recorded
TARSUS: Not recorded
WEIGHT: 222 – 470 g
Male slightly larger than female

DESCRIPTION
ADULT
Black with large white patch in wing (revealed in flight); white undertail coverts and tail tip; white bases to tail feathers usually conspicuous dorsally though not visible in some southern birds; female sometimes greyer ventrally than male

IMMATURE
Duller and browner than adult; gape yellow

IDENTIFICATION
Unmistakable over most of range, where Grey Currawong is much paler and lacks white tail base, although white edges of undertail coverts may be visible dorsally. In western Victoria, Pied Currawong may lack visible white tail base and could be confused with dark Murray Mallee form of Grey Currawong, which lacks white in wings

VARIATION
Three similar subspecies. Isolated S. g. magnirostris *of Cape York Peninsula has large bill. In* S. g. graculina *there is a clinal increase in size from north to south. There is also a decrease in amount of white in plumage, with north Queensland birds having most white, southern having progressively less white, to south-western Victoria where some birds lack visible white on base of tail feathers.* S. g. crissalis *inhabits Lord Howe Island*

VOICE
Various renditions of "currawong" call: a rollicking or wailing croak with a ringing quality, repeated. Also long "wolf whistle", soft whistles and croaks

HABITAT
Most terrestrial types within its range: forests, woodlands, farmland with trees, urban parks and gardens

FOOD
Earthworms, snails, crustaceans, arthropods, lizards, birds and their eggs and young, small mammals, carrion, offal, refuse, flowers, fruits, seeds, leaves, nectar and lerps

HABITS
Regionally variable, and changing as birds adapt to living in cities. In pairs, family groups and large flocks. Tropical birds sedentary. Southern birds sedentary in breeding season, with breeding pairs and non-breeding flocks permanently resident in mild areas. Some altitudinal migration, with flocks leaving highlands and wintering in lower tableland, coastal and inland areas. Arboreal and terrestrial, foraging by clambering about in trees, poking and prising, or walking on ground. Also chases small birds, robs other predators

BREEDING
August – January; July in north. Both sexes build nest and feed young

NEST: Bowl of sticks, lined with roots, grass and bark, 7 – 25 m above ground in fork of tree canopy

EGGS: Usually 3, light brown spotted and blotched dark brown; oval-shaped; 42 x 30 mm

INCUBATION: About 21 days, by female; fledging period about 30 days

DISTRIBUTION
Eastern Australia, Lord Howe Island

STATUS
Abundant; increasing. Lord Howe Island population small

S. J. S. Debus

Black Currawong

Strepera fuliginosa

In Latin,"*strepere*" means "to make a noise", a particularly apt label for the Pied Currawong and its close relative, the Black Currawong of Tasmania. The latter has variously been considered an early colonist of Grey Currawong stock, or only a subspecies of the Pied Currawong. It is clearly of the same stock as the Pied Currawong, as revealed by its massive, slightly hooked bill, black plumage and the nature of its voice. It has been isolated for sufficiently long to develop distinct characteristics which justify its recognition as a separate species. Its bill is even more bulbous than that of the Pied Currawong, the distribution of white in its plumage is different, and it has a slightly shorter tail. Its calls, although of Pied Currawong quality, are recognisably different. Its habits are similar to those of the Pied Currawong, it lives in the same habitats, and forages in the same way.

Early ornithologists such as A.J. North thought that the Black Currawong also occurred in south-eastern Australia but this is attributable to confusion with other species. Southern Pied Currawongs have less white in their plumage than northern birds, and may lack visible white bases to their tail feathers; the mallee form of the Grey Currawong is dusky with no white in the wings. The Tasmanian form is almost black but is readily distinguished by its white wing patches, white undertail coverts, and slender, pointed (rather than hook-tipped) bill, and different calls.

The Black Currawong is poorly known, the only details of its habits those quoted by North in 1918 and J.A. Fletcher in 1922. It is omnivorous, consuming a variety of small animals, carrion, seeds and fruits. Prey includes invertebrates, small and nestling birds, and mice. It is bold, taking human food scraps, unguarded chickens and ducklings, and orchard fruit. In turn, its flesh is palatable to humans and many Black Currawongs were eaten by the early settlers.

The Black Currawong forages in trees, prising amongst the foliage, branches and loose bark. It also forages on the ground or on beaches, and around shallow water where it sometimes wades. Large prey items are wedged into crevices and pulled apart with the bill, the bird returning repeatedly to its larder to feed again. Competition with the Grey Currawong is avoided by differences in habitat preferences and, to some extent, in distribution. In Tasmania the Grey Currawong is mainly confined to lowland dry eucalypt forest and woodland. The Black Currawong lives in highland and other wet forests; in winter it moves to farmland and other open country in the lowlands.

Little is known about the breeding of the Black Currawong, beyond the fact that it builds the usual bowl-shaped nest of sticks high in a forest tree and that adults defend the young against humans. J.A. Fletcher observed that a parent Black Currawong with a beakful of insects for its young sometimes paused to accept human food scraps. It ate these, then picked up the insects again to take to its young. Fletcher also observed that in a pair of currawongs with two fledglings, each parent had charge of a specific young bird. As the young grew, food was simply placed in front of them. These aspects of behaviour are similar to those of the Australian Magpie.

S. J. S. Debus

Black Currawong

Strepera fuliginosa

Adult; Cradle Mountain, Tas.; December 1988 (T. & P. Gardner)

OTHER NAMES
Black Magpie, Black Jay, Mountain Magpie, Muttonbird

MEASUREMENTS
LENGTH: *470 – 490 mm*
WING: *237 – 289 mm*
TAIL: *155 – 215 mm*
BILL: *Not recorded*
TARSUS: *Not recorded*
WEIGHT: *290 – 560 g*

DESCRIPTION
ADULT
Sooty black with white tip to tail; trace of white wing bar; white tips to flight feathers. Bill black. Eyes yellow. Feet black

IMMATURE
Duller than adult; gape yellow

IDENTIFICATION
Black Currawong with traces of white only in wing tips and tail; massive bill. Dark Tasmanian form of Grey Currawong has larger white wing bar; white undertail coverts; slender, more pointed bill; different calls

VARIATION
None

VOICE
Long, wailing or rollicking cry, described as chattering scream or rending, rolling croak. Flocks produce a babble of sound. Also long whistle, soft whistles and croaks

HABITAT
Highland wet forests; farmland and open country in lowlands; beaches; parks and gardens

FOOD
Large insects, small birds and nestlings, mice, carrion, seeds, berries and other fruits

HABITS
Resident and partial altitudinal migrant. Breeding pairs sedentary. In non-breeding season, forms large flocks which move to lowlands. Arboreal and terrestrial

BREEDING
August – December

NEST: Bowl of sticks built high in tree fork

EGGS: 2 – 4; grey-buff, blotched red-brown and purple; oval-shaped; 40 x 29 mm

INCUBATION: Not recorded; fledging period not recorded

DISTRIBUTION
Tasmania, Bass Strait islands including Flinders Island, King Island

STATUS
Common

S. J. S. Debus

Grey Currawong

Strepera versicolor

The subspecific name of the Grey Currawong's Tasmanian subspecies, *S. v. arguta*, means "clear-voiced". As remarked by John Gould, it is appropriate for this species whose high-pitched, shrill cries at once separate it from the Pied and Black Currawongs. This is a useful field character because, among the Grey Currawong's many subspecies, are forms which resemble the Pied and Black Currawongs. In south-western Victoria the Pied Currawong has the least amount of white in its plumage, and may lack visible white bases to the tail feathers. It is precisely in this area, and in adjoining South Australia, that the darkest mainland Grey Currawongs occur; they lack white in the wings. Similarly, the Grey Currawong is very dark in Tasmania, where it occurs with the blackest of all the currawongs.

Another useful field character is the shape of the bill. That of the Grey Currawong is rather slender and pointed, lacking the slightly hooked tip of the more massive and arched bills of the Pied and Black Currawong. This difference is related to differences in feeding. Over much of its range, the Grey Currawong avoids competition by living in drier and more open forests and woodlands, and partly in areas not occupied by the other species. Nevertheless, the two species overlap widely in south-eastern Australia and Tasmania. There, the Grey Currawong is a more specialised bark-forager, clinging to tree trunks and using its finer bill to lever off bark and poke into cracks. As observed by Tess Kloot, it also inserts its bill into the curled ribbons of eucalypt bark, opening its mandibles as it slides the bill along to unravel the ribbon. On rocky ground it slides its bill sideways under stones and, rotating the bill, opens its mandibles to lift the stone and expose hidden prey. The Grey Currawong is omnivorous, eating arthropods, frogs, lizards, birds and their eggs and young, rodents, seeds and fruits.

Variation among the regional populations of the Grey Currawong is complex. Of the four forms currently recognised, early ornithologists such as A.J. North thought that all were separate species. They also recognised the "Brown Currawong", called *Strepera intermedia*, as a distinct species, a situation that is partly explained by an anomalous population of exceptionally dark birds in the Murray Mallee region. South-eastern and south-western Grey Currawongs resemble each other, although the latter are darker. In between is the mallee "Black-winged Currawong" which is darker than either. The "Brown Currawong" was apparently based on immature birds of south-western or mallee populations near their zone of contact in South Australia.

Pairs of Grey Currawongs are sedentary and territorial. Flocks of up to a dozen also occur on the mainland. In Tasmania, loose groups of up to 30 – 40 birds wander, sometimes into open country, in search of food in winter. Other aspects of the Grey Currawong's biology, particularly its breeding behaviour, are poorly known. The typical currawong nest, a bowl of sticks, is built in a fork of a leafy tree canopy. The female alone incubates the two or three buff, spotted eggs, and both sexes feed the young.

Like the Pied and Black Currawongs, Grey Currawongs apply beakfuls of wet clay to their feathers. Pied Currawongs also bathe in water, then apply dry earth to their feathers, or bathe in water after applying clay. This behaviour may have a function in the control of feather parasites or in plumage maintenance.

S. J. S. Debus

Adult at nest; Sleaford Bay, S.A.; August 1977 (C. Gill)

Sponsors: E.W. & M.J. Abrahams

Grey Currawong

Strepera versicolor

Immature (T. Howard)

OTHER NAMES
Grey Jay, Grey Magpie, Mountain Magpie, Squeaker, Rainbird; subspecies variously Black-winged Currawong (south-eastern mallee), Clinking Currawong (Tasmania)

MEASUREMENTS
LENGTH: 450 – 510 mm
WING: 260 – 321 mm
TAIL: 200 – 270 mm
BILL: Not recorded
TARSUS: Not recorded
WEIGHT: 311 – 561 g

DESCRIPTION
ADULT
Subspecies variably mid-grey to dusky, with white tail tip and white undertail coverts which may be visible dorsally as white edges to base of tail. Most forms have white wing patches (revealed in flight) and white edges and tips to flight feathers. Darkest mainland form, of south-eastern mallee, lacks white in wings. Bill black. Eyes yellow. Feet black

IMMATURE
Similar to adult but duller and browner, with yellow gape

IDENTIFICATION
Grey mainland forms unmistakable. All forms have slender, pointed (not hooked) bills and shrill calls. Dark south-eastern mallee form similar to southern Pied Currawong, which may lack visible white on base of tail but has white wing patches. Dark Tasmanian form similar to Black Currawong but has larger white wing patches, white undertail coverts, slender bill

VARIATION
Complex. Four subspecies. Nominate S. v. versicolor *of south-east mainland palest.* S. v. plumbea *of Eyre Peninsula, Central and Western Australia similar but darker. Tasmanian* S. v. arguta *largest and darkest. South-eastern mallee* S. v. melanoptera *(separating* versicolor *and* plumbea*) is darkest mainland form and lacks white in wings*

VOICE
High-pitched, shrill, ringing or metallic note, repeated two or three times. Also soft mewing and squeaks

HABITAT
Eucalypt open forest and woodland, mallee, dense mulga

FOOD
Arthropods, lizards, small birds, eggs and nestlings, rodents, fruits, seeds and nectar

HABITS
Solitary, in pairs or family parties. Sedentary. In Tasmania, some flocking and local movements in winter

BREEDING
July – November. Both sexes feed young

NEST: Bowl of sticks, lined with grass and other fibres; in tree fork among foliage; 3 – 14 m above ground

EGGS: 2 – 3; buff, spotted and blotched brown and grey; oval-shaped; 2 x 30 mm

INCUBATION: By female; incubation and fledging periods not recorded

DISTRIBUTION
Southern Australia, Tasmania

STATUS
Moderately common in coastal and near-coastal areas, rare inland

S. J. S. Debus

Crows and Ravens

In the Australasian region, the family Corvidae is represented by crows and ravens of the almost cosmopolitan genus *Corvus* – big black birds with loud, cawing voices. Elsewhere, the family is well represented by a variety of jays, magpies and choughs, many of which are pied or brightly patterned in blues and greens. They are not closely related to the White-winged Chough or Australian Magpie, which were named for their superficial resemblance to Old World corvids.

Corvids are large passerines. They share glossy plumage, strong legs and feet, powerful flight, harsh voices, and stout bills with a dense mat of bristles over the nostrils. Noisy and gregarious, they are often regarded as the peak of avian evolution because of their intelligence, complex social organisation, adaptability and success, however this is an anthropocentric view. Among their nearest relatives are the shrikes, family Laniidae, not represented in Australia.

According to Charles Sibley and colleagues, many of the endemic Australian passerine families can be grouped together with the Corvidae and relatives to form the "corvine assemblage" which originated on the drifting fragment of Gondwana that became Australia. Early members of the assemblage evolved in isolation to become the Australian robins, shrike-thrushes and flycatchers that convergently resemble their Old World, unrelated namesakes. The ancestors of the crows and jays must have dispersed early and evolved elsewhere, the genus *Corvus* perhaps colonising Australia in recent times when that continent drifted close enough to Asia. Alternatively, early dispersal of crows via a southern (Gondwanan) route may have given rise to the Australasian members of the genus *Corvus* which, according to Jollie, are an ancient stock.

The crows (about 40 species in the genus *Corvus*) range in size from the Jackdaw, *Corvus monedula*, to the Northern Raven *Corvus corax*. Most are black but some are pied or have brown or grey in their plumage. The Australian species, of intermediate size and entirely glossy black, are divided into crows and ravens on the basis of their down colour and throat hackles. Crows have white bases to the head and neck feathers and short throat hackles, while ravens are larger, with grey feather bases and long throat hackles. This division is based on the characters of European crows and ravens but in fact all five Australian species are closely related to each other and are derived from common stock. All have white eyes in adulthood, an unusual feature among crows but one that is shared with the Moluccan Long-billed Crow, *Corvus validus*, and pale-eyed species in New Guinea and nearby islands. As a group, the Australasian crows seem most closely related to the Jungle Crow, *Corvus macrorhynchos*, of south-east Asia.

Typical crows are omnivorous and opportunistic, being both scavengers and predators. They hold food down under one or both feet while tearing it with the bill and transport food in a pouch under the bill; excess food is cached for later consumption. They walk and fly strongly, sometimes perform aerobatics, and frequently mob larger predators. Non-breeding individuals roost communally.

Crows are monogamous, and form permanent pair bonds. The five Australian species build large, bowl-shaped nests of sticks in trees and sometimes on artificial structures such as telegraph poles and windmill platforms. Both parents share in the building and lining of the nest with bark, grass, wool and feathers. Construction takes two to three weeks. The usual clutch of four or five eggs is incubated for about 19 days by the female and she is fed on the nest by the male. The male also supplies all the

Adult Australian Raven; Yanchep, W.A.; April 1972 (G. Chapman)

Sponsors: Mason & Katharine Frecker

food to the chicks via the brooding female, who also starts gathering food for the young when they begin to grow feathers. Young remain in the nest for 30 to 37 days (small species) or about 43 days (large species). After leaving the nest, the young are dependent on the parents for a further month or two, but forage for themselves within three months. Independent young wander off to join nomadic flocks of immature birds until they find a vacant territory.

In places where they are persecuted – particularly in agricultural areas – crows are wary of humans. However, in recent years they have become bold and tame around our capital cities. They now breed freely in suburban streets, parks and gardens, and forage brazenly in backyards and with urban scavengers such as pigeons and gulls.

Ian Rowley has shown that the five Australian crows provide an excellent example of speciation. During past climatic fluctuations, the ancestral species was separated into isolated refuges. When conditions allowed the populations to spread and come into contact, they had diverged sufficiently to prevent interbreeding and had also developed sufficient ecological differences for them to coexist.

Over much of Australia two species occur together: the Australian Raven and Little Raven; Australian Raven and Little Crow; Torresian Crow and Little Crow. In each case, one species is a large resident, each pair occupying a large permanent territory and nesting solitarily in tall trees; the other is a small nomad, nesting semi-colonially in small trees and with greater flexibility of breeding time in relation to seasonal conditions. Unlike the large species, the territory of which supplies all requirements, the small species have shorter nesting cycles and occupy territory only long enough to rear their brood. Their small territories are used only for nesting, foraging being conducted outside in areas unsuitable for breeding.

Although all species are omnivorous, there are some differences in the feeding habits and diet of the large and small species: the Australian Raven prefers carrion and the Little Raven specialises in insects. Where two rather similar species occur together, as in the Australian and Little Ravens or Torresian and Little Crows, natural selection has emphasised the differences between them so that their identity is maintained and competition is minimised. An exception occurs in the New England region of New South Wales where the ranges of the three large species converge. The Australian Raven, Forest Raven and Torresian Crow are very similar in ecology and social organisation, and interspecific competition for breeding territories occurs in the overlap zone. The Australian Raven appears to be winning the contest, displacing the Forest Raven and perhaps also the Torresian Crow.

The occurrence of five closely related and superficially similar species, with three or even four in one area, presents identification problems. Apart from the colour of feather bases on the head and neck and relative hackle length, they are best identified in the field by their calls, displays, characteristic behaviour patterns and differences in social organisation and ecology. Distribution is also a useful guide.

S. J. S. Debus

AUSTRALIAN RAVEN

Corvus coronoides

The Australian Raven is the most familiar corvid in the settled parts of southern Australia, particularly the sheep country. It occurs in the south-west corner, across the Great Australian Bight in a thin coastal strip, thence over most of the eastern half of the country except for some tropical and subtropical coastal areas. It is the common species in Perth and Sydney. It has long throat hackles which are fanned out when calling from a prominent perch. Its calls are very distinctive, high-pitched, slow and wailing, and dying away as if the bird had been strangled.

It is an omnivorous scavenger and predator. However, studies by Ian Rowley and colleagues in the CSIRO found that it is not the threat to the sheep industry that it was once thought to be. Only the Australian Raven and Forest Raven are capable of feeding efficiently on a lamb or sheep carcass, and even these species cannot kill a healthy lamb. Lambs die mainly from bad weather, poor feeding or abandonment by the ewe, and virtually all lambs fed upon by ravens are starving or already dead. Its true role in lambing paddocks is that of scavenger, cleaning up dead lambs or afterbirth and sometimes hastening the inevitable end of sick or weak lambs and sheep. It also eats many insects,

Immature; Kosciusko National Park, N.S.W.; March 1985 (C. Andrew Henley/Larus)

Sponsor: Norman Chaffer

including grasshoppers and locusts, flies and maggots, and it will remove blowfly maggots from living sheep. Because it has difficulty in penetrating the hide of a sheep, a raven must start at soft spots on the carcass. It is for this reason that it often removes the eyes first.

Breeding pairs are permanently resident in large territories of about a square kilometre. The territory provides all requirements throughout the year, including a tall nest tree and other high look-out perches, and is defended against intruders. Territory owners call loudly from high perches, or from a display flight in which the downstroke is slightly laboured and the upstroke has a pronounced flicking action. During such flights, the Australian Raven has a particularly rakish outline with pointed, back-swept wings and conspicuously long throat hackles. In aggressive encounters with other ravens, the hackles are fanned and the feathers above and behind the eyes are erected into small "ear" tufts.

Members of a mated pair spend all year in each other's company, and apparently have little need for courtship displays in the mating season. They occasionally perform chasing or following flights, and often preen each other's head and neck feathers. If one is absent for a time, it signals its return with fluttering flight and a tremulous, descending wail.

The nest is a large bowl of sticks, lined with bark, grass and wool, built by both members of a pair. It is placed in the fork of a tall tree or on an artificial structure with a commanding view, but neighbouring pairs cannot see each other on their respective nests, almost a kilometre apart. The clutch of four or five pale green, spotted eggs is laid in late winter and incubated by the female for about 20 days. The male feeds the female on the nest during this period.

Young are naked and blind at hatching and, at first, are brooded constantly by the female. The male supplies all food via the female until the young are feathered, at about six weeks of age when the female can leave them unattended while she shares in the foraging. Fledglings flutter clumsily to nearby trees, where they are fed by their parents over the ensuing days until their wings and tail are more adult in proportion, and they begin to follow their parents to the ground.

S. J. S. Debus

AUSTRALIAN RAVEN

Corvus coronoides

Group feeding on dingo; Birdsville, Qld; July 1991 (R. Brown)

OTHER NAMES
Crow

MEASUREMENTS
LENGTH: 480 – 540 mm
WING: 337 – 397 mm
TAIL: 193 – 238 mm
BILL: 50 – 66 mm
TARSUS: 59 – 69 mm
WEIGHT: 540 – 820 g
Female slightly smaller than male

DESCRIPTION
ADULT
Entirely glossy black, with long, floppy throat hackles, grey feather bases. Bare grey skin on sides of chin. Bill and legs black. Eyes white with blue inner ring

JUVENAL
Duller than adult, with shorter throat hackles. Fledglings have pink gape and skin around bill. Blue-grey eyes which turn brown during first year, then hazel

IDENTIFICATION
Large black bird with high-pitched cawing and wailing voice. Distinguished from other crows by voice and long throat hackles. Looks large and rakish in flight, with tapered wings, slender tail; long throat hackles often visible. Glides on level, "fingered" wings

VARIATION
No subspecies, but a slight clinal increase in size from west to east, also from south to north in eastern Australia

VOICE
High-pitched, powerful, slow "aah...aah...aaah" of three or four notes, the last drawn-out and descending in a "dying" finish, uttered from a horizontal posture on a prominent perch with the throat hackles conspicuously fanned. Other high, child-like wailing calls; also soft tenor caws and low rattle

HABITAT
Most terrestrial types except dense forest, particularly sheep paddocks, country roads, rubbish dumps

FOOD
Grain, fruit, invertebrates, small vertebrates, eggs, refuse, carrion. Mainly carnivorous

HABITS
Breeding pairs solitary, sedentary and permanently territorial. Immatures wander in nomadic flocks. Arboreal and terrestrial, gleaning prey from ground and trees. Also searches aerially for carrion by patrolling roads

BREEDING
Eggs laid July – September, one brood per season. Both sexes share nest building and feeding of young

NEST: Large bowl of sticks, lined with bark and wool, placed more than 10 m above ground in a tree or artificial structure

EGGS: 4 – 5; pale green, spotted and blotched brown; long oval; 45 x 30 mm

INCUBATION: 20 days, by female; fledging period 43 days

DISTRIBUTION
Eastern half of Australia except Cape York Peninsula; south-west Western Australia

STATUS
Endemic species. Abundant

S. J. S. Debus

Forest Raven

Corvus tasmanicus

The Forest Raven's bass cawing is heard in wild, cold, windswept areas. It is the only corvid occurring in Tasmania and is also found in two small areas on the mainland: in forests from Wilsons Promontory around coastal Victoria to the Coorong in South Australia; and on the New England Tablelands and adjacent coast in New South Wales. It is a large, relatively long-legged corvid with slow, very deep calls. Its hackles are not as long as those of the Australian Raven, giving it a neater appearance than that species. Southern birds have a conspicuously short tail.

The Forest Raven was only recently recognised as a distinct species, following an investigation by Ian Rowley and colleagues in the CSIRO into the impact of crows on lambs. It had long been thought that there was only one species of raven in Australia but, after recognition of the Little Raven as a distinct species in the 1960s, further work demonstrated that ravens in Tasmania and isolated pockets

Adult with Willie Wagtail (T. Howard)

Sponsor: R. Arnheim (Parkes)

on the mainland were not a subspecies of the Little Raven, but were specifically distinct. Although similar to the Little Raven, in having medium-length throat hackles and a well-feathered chin, the Forest Raven is larger, with a more massive bill, and is the largest of the Australian corvids. Its social organisation, displays and behaviour resemble those of the Australian Raven rather than those of the Little Raven.

The two populations of the Forest Raven have been separated long enough for subspecific differences to develop between them. McAllan and Bruce have recently suggested that the northern subspecies deserves elevation to full species rank as *Corvus boreus* but this view is not taken here. Physical differences between the subspecies are slight: shorter wings and tail in southern birds, giving a different wing to tail ratio and a rounder-winged, shorter-tailed silhouette in flight. The voice of the northern subspecies seems not quite as deep as that of southern birds, and some of the phrases uttered are different. An unresolved question is whether the two populations have diverged sufficiently to prevent interbreeding if they came into contact again.

The Forest Raven, particularly the northern subspecies, has not been studied intensively. From the few details known of its breeding biology and behaviour it resembles the Australian Raven. Pairs appear to be permanently resident in large territories. On the New England Tablelands, two pairs were observed nesting in diagonally opposite corners of a block of forest two square kilometres in area, and foraging into adjacent pasture.

Territory is advertised and defended by loud calls from a high perch, in a special posture with the throat hackles erected and the tail depressed, or from a display flight. The Forest Raven's aerial wing-flicking display is similar to that of the Australian Raven, except that the downstroke is more laboured and sweeping. Southern birds are distinctly blunt-winged. Pairs of the northern subspecies sometimes soar in spirals, and a single adult has been seen flying in a wide circuit high over its territory and calling with its hackles erect. Forest Ravens perform a fluttering "returning home" display flight but the tremulous, descending call is deeper and harsher than that of the Australian Raven.

Few nests of the Forest Raven have been examined. Nine nests in Tasmania were, on average, 20 metres above the ground, and several on the New England Tablelands were 10 to 20 metres above the ground. A clutch of four eggs is laid in spring in Tasmania. No clutches have been examined on the mainland, but known-age young indicate laying in late winter and spring in New South Wales, where broods of three fledglings are common. In both subspecies, only females have brood patches, indicating that as in other corvids, only the female incubates. Both parents feed and defend their fledglings.

S. J. S. Debus

FOREST RAVEN

Corvus tasmanicus

Immature; Wynyard, Tas.; August 1992 (F. Dowling)

OTHER NAMES
Crow, New England Raven, Relict Raven

MEASUREMENTS
LENGTH: 520 – 540 mm
C. t. tasmanicus
WING: 333 – 377 mm
TAIL: 180 – 211 mm
BILL: 57 – 63 mm
TARSUS: 61 – 69 mm
WEIGHT: 500 – 800 g
C. t. boreus
WING: 373 – 394) mm
TAIL: 211 – 227 mm
BILL: 57 – 59 mm
TARSUS: 66 – 70 mm
WEIGHT: 645 – 740 g
Females of both subspecies slightly smaller than males

DESCRIPTION
ADULT
Entirely glossy black, with medium-length throat hackles, grey feather bases. Bill and legs black. Eyes white with blue inner ring

JUVENAL
Duller than adult, with short throat hackles. Fledglings have pink gape and slight pink edges to underside of bill. Blue-grey eyes which turn brown during first year, then hazel

IDENTIFICATION
Large black bird with deep, harsh cawing voice. Distinguished from other crows by voice. Looks large and ponderous in flight, with rounded, broad-based wings, broad tail; glides lazily on drooped, "fingered" wings. Southern subspecies has conspicuously shorter tail than other ravens

VARIATION
Two subspecies: C. t. tasmanicus *in Tasmania, Bass Strait islands and nearby mainland coasts has short tail;* C. t. boreus *on New England Tablelands and adjacent coast in northern New South Wales has longer tail, proportioned more like other ravens*

VOICE
Rich, far-carrying bass "karr...karr...karr", slow and measured, with "gravelly" quality, and final note sometimes drawn-out and descending. Delivered from a high perch with the throat hackles erected, and tail exaggeratedly depressed. Other soft, guttural calls

HABITAT
In Tasmania, most terrestrial types. On mainland, mainly eucalypt forests but extends into neighbouring open areas

FOOD
Grain, fruit, invertebrates, small vertebrates, eggs, refuse, carrion. Mainly carnivorous

HABITS
Breeding pairs solitary, sedentary and permanently territorial; immatures wander in nomadic flocks. Arboreal and terrestrial, gleaning prey from ground and trees. Also searches aerially for carrion by patrolling roads. Pairs sometimes soar in spirals

BREEDING
Eggs laid August – November, one brood per season. Both sexes share feeding and defence of young

NEST: Large bowl of sticks, lined with bark and wool, placed more than 10 m above ground in a tree

EGGS: 4; pale green spotted and blotched brown; long oval; 46 x 31 mm

INCUBATION: Probably 20 days, by female; fledging period probably 42 days

DISTRIBUTION
Tasmania, Bass Strait islands, southern coasts of Victoria and coastal south-eastern South Australia. Also Barrington Tops, Northern Tablelands and mid-north Coast of New South Wales

STATUS
Endemic species. Abundant in Tasmania, locally common on mainland

S. J. J. Debus

Little Raven

Corvus mellori

Crows and ravens have traditionally been regarded as enemies of graziers, for which reason Ian Rowley and his colleagues in the CSIRO investigated the impact of the Australian Raven on the sheep industry in the 1960s. This study involved the collection of many specimens of crows and ravens in order to analyse their stomach contents. Because crows are wary and notoriously difficult to shoot, advantage was taken of their very aggressive response to imitations of their territorial call. However, on one occasion a target pair did not respond to imitations of the Australian Raven's call. One of these birds was eventually collected and found to be slightly smaller than the Australian Ravens that also bred in the area. It also had shorter throat hackles and more complete feathering under the beak. Subsequent study revealed that the "new" form differed in voice, behaviour and ecology and did not interbreed with the Australian Raven. Because it had grey feather bases, it was put in the raven group and called the Little Raven, *Corvus mellori*. This discovery prompted a nationwide study into the characters and distribution of all the Australian corvids, and resulted in the recognition of a fifth species, the Forest Raven, *Corvus tasmanicus*, which differs from the other two and does not interbreed with them. What early ornithologists called the "Australian Raven" thus consists of three species.

The Little Raven occurs only in south-eastern Australia, in temperate farmland and alpine areas, and is the common corvid in Adelaide, Melbourne and Canberra. It often occurs in large flocks, and its calls are a series of rapid, guttural "barks", sharply cut off. When calling from a perch it often flicks its wings above its back. It lacks the long, shaggy throat hackles of the Australian Raven.

Pairs of Little Ravens are resident and territorial only during the breeding season. Their territories of two or three hectares are much smaller than those of the Australian Raven, and are used only for nesting. Although it breeds semi-colonially and nests are less than 200 metres apart, neighbours achieve visual seclusion by building in thick cover.

With such a small territory to defend, the Little Raven has no need for long-range signals and does not engage in the conspicuous high-perching or aerial advertisement displays of the Australian Raven. Instead, it performs a stationary wing-flipping display from a low perch or on the ground. In courtship, the male performs a "promenade" on the ground, strutting past the female with his wings drooped and tail cocked. Neither display requires much space. Foraging occurs away from the nesting territory and a returning bird performs a fluttering "returning home" display flight like that of the Australian Raven, but with a deeper, creaky call.

Both members of a pair cooperate in building a nest of sticks, similar to that of the Australian Raven. However, the nest of the Little Raven is placed below 10 metres in small trees – even occasionally on or near the ground. It can therefore breed in areas unsuited to the Australian Raven and its clusters of territories enable some communal defence against the larger species.

The clutch of four or five pale green, spotted eggs is incubated by the female. Eggs are laid in spring in lowland areas and in summer in alpine areas, but this may be delayed by drought or occur in autumn after the breaking of a drought. They are incubated by the female for 19 days, during which time she is fed on the nest by the male. Young fledge in about 37 days, a week less than the time taken by the Australian Raven. Fledglings leave with their parents to join foraging flocks which vacate the breeding areas.

Over its entire range, the Little Raven coexists with the Australian Raven. The Little Raven avoids competition by breeding later in sites unacceptable to Australian Ravens, and by completing its breeding cycle in as short a time as possible, before departing for areas where the Australian Raven does not breed. The Little Raven has a slender bill and is more insectivorous. In the non-breeding season, nomadic foraging flocks exploit insect outbreaks in forests or on treeless plains.

Little Ravens are sociable. They soar together and large flocks perform spectacular massed flights on their way to communal roosts.

S. J. S. Debus

Adult; Mt Kosciusko, N.S.W.; December 1968 (G. Chapman)

Sponsor: Sir Ron Brierley

LITTLE RAVEN

Corvus mellori

Flock; Kosciusko National Park, N.S.W.; February 1989 (T. Howard)

OTHER NAMES
Crow

MEASUREMENTS
LENGTH: 480 – 500 mm
WING: 303 – 370 mm
TAIL: 183 – 216 mm
BILL: 48 – 56 mm
TARSUS: 56 – 66 mm
WEIGHT: 407 – 660 g
Female slightly smaller than male

DESCRIPTION

ADULT
Entirely glossy black, with medium-length throat hackles, grey feather bases. Bill and legs black. Eyes white with blue inner ring

JUVENAL
Duller than adult, with shorter throat hackles. Fledglings have pink gape, blue-grey eyes. Eyes turn brown during first year, then hazel

IDENTIFICATION
Large black bird with guttural barking or cawing voice. Distinguished from other crows by voice and wing-flipping display. Compact and agile in flight, with tapered wings. Often in large, tight flocks

VARIATION
Negligible

VOICE
Guttural baritone, rapid and clipped "ark-ark-ark-ark". When perched, each "bark" often accompanied by upwards flick of closed wings. Other soft baritone caws, occasionally with "dying" finish like Australian Raven, and a long, creaky call

HABITAT
Most terrestrial types within range

FOOD
Grain, fruit, invertebrates, small vertebrates, eggs, refuse, carrion. Mainly insectivorous

HABITS
Adults resident and territorial only when breeding, otherwise nomadic in flocks. Some regular summer migration from inland areas towards coast. Arboreal and terrestrial, gleaning prey from ground and trees. Pairs breed semi-colonially in low trees, forage in flocks away from small nesting territories. Flocks soar and perform coordinated mass aerobatics

BREEDING
Eggs laid May – December, depending on region and seasonal conditions; one brood per season. Both sexes share in nest-building and feeding of young

NEST: Large bowl of sticks lined with bark, wool and Emu feathers, placed low (often below 10 m) in trees or on artificial structures in treeless areas, occasionally on cliffs

EGGS: 4 – 5; pale green spotted and blotched brown; long oval; 44 x 30 mm

INCUBATION: 19 – 20 days, by female; fledging period 35 – 41 days

DISTRIBUTION
South-eastern Australia from northern New South Wales to Eyre Peninsula and Victorian coasts; King Island

STATUS
Endemic species. Abundant

S. J. S. Debus

Little Crow

Corvus bennetti

The Little Crow is a short-legged, magpie-sized desert species that is quite tame around outback towns from western Queensland through Alice Springs to the west coast. Its very nasal calls are somewhat deeper and more prolonged than those of the Torresian Crow and lack their staccato quality. The Little Crow is much more gregarious than the Torresian Crow in the inland, and is unique among the Australian corvids in placing mud in its nest.

In social behaviour and ecology, the Little Crow is the arid-zone equivalent of the Little Raven. Pairs are resident and territorial only when breeding, and forage away from their nesting areas. They nest semi-colonially in small territories of less than a hectare each, below 10 metres in small, dense trees with as little as 15 metres between neighbouring nests. Solitary pairs also nest on telegraph poles in treeless areas, at sufficient distances for visual seclusion from their neighbours. In courtship, a male and female preen each other's head and neck feathers.

The Little Crow's calls are short-range signals from low perches or on the ground, with no special posture, although birds on perches often depress the tail with each syllable. Breeding pairs sometimes perform a dipping "currawong flight" or cuckoo-like undulation around the nest, with a two-note "quark-quark" call. They appear not to perform long-range aerial displays like the large species,

Group of immatures; Pilbara, W.A.; circa 1980 (M. & I. Morcombe)

although one bird has been seen flying with laboured, sweeping downstrokes and loud calling. Breeding birds rarely perform the fluttering "returning home" display flight, with a creaky call, like the Little Raven, but individuals often perform it without calling when joining perched members of their flock.

Nests of the Little Crow are similar to those of the Little Raven, but have a layer of mud or clay between the stick bowl and the lining. The mud is presumably applied wet, confining Little Crows which are breeding to the vicinity of surface water. The mud layer may have a role in regulation of humidity or temperature in the nest, perhaps a necessary strategy in a hot, dry climate.

The clutch of three to six eggs (usually four) is normally laid in late winter or spring, but also in summer and autumn after rain. The Little Crow is the most flexible of the Australian corvids in the timing of its breeding and is thus able to take advantage of good conditions in the desert. Its incubation and nestling periods, at about 17 and 30 days respectively, are the shortest among the Australian corvids and enable it to complete its breeding cycle rapidly when conditions permit.

The Little Crow is able to coexist with the Australian Raven or Torresian Crow for the same reasons that the Little Raven and Australian Raven are able to do so. The Little Crow avoids competition by nesting in sites unacceptable to the larger species, by completing its breeding cycle in as short a time as possible, and then vacating the breeding area. It is a versatile feeder, agile enough to catch insects in flight, and its fine bill enables it to pick scraps of carrion left by larger scavenging birds.

Little Crows are sociable, foraging and travelling in large, tight flocks. They soar in spirals together and perform coordinated diving aerobatics.

S. J. S. Debus

Little Crow

Corvus bennetti

Adult; Moomba, S.A.; October 1987 (G. Chapman)

OTHER NAMES
Crow

MEASUREMENTS
LENGTH: 450 – 480 mm
WING: 310 – 346 mm
TAIL: 163 – 200 mm
BILL: 42 – 52 mm
TARSUS: 52 – 61 mm
WEIGHT: 345 – 500 g
Female slightly smaller than male

DESCRIPTION
ADULT
Entirely glossy black, with short throat hackles, white feather bases on head and neck. Bill and legs black. Eyes white with blue inner ring

JUVENAL
Duller than adult; feather bases may be washed grey. Fledglings have pink gape and blue-grey eyes which turn brown during first year, then hazel

IDENTIFICATION
Large black bird with nasal cawing voice. Distinguished from other crows by voice and behaviour. Small, compact and blunt-winged in flight, with slender tail. Flight agile, often in large, tight flocks. Looks small and short-legged compared to other Australian crows, with small bill, high forehead

VARIATION
None

VOICE
Baritone, nasal, rapid "nark-nark-nark-nark" with each note slightly prolonged; hoarse and monotonous. Also a rapid, high-pitched, three-note "kup-kup-kup". Other soft baritone caws, a warbling or bubbling call, and a long creaky call

HABITAT
Mostly terrestrial types in arid and semi-arid zones

FOOD
Grain, fruit, invertebrates, small invertebrates, eggs, refuse and carrion. Less dependent on large carrion than other species

HABITS
Adults resident and territorial only when breeding, otherwise nomadic in flocks. Some regular summer migration to coast in south-western Australia. Arboreal and terrestrial, gleaning prey from ground and trees. Tame around outback towns. Breeds semi-colonially in low trees, forages in flocks away from small nesting territories. Flocks soar and perform coordinated mass aerobatics. Breeding birds perform an undulating, cuckoo-like display flight

BREEDING
Eggs laid June – September or after rain in south and inland, February – May in north; one brood per season. Both sexes share the feeding of young

NEST: Large bowl of sticks, lined with bark, wool and Emu feathers, often with mud under lining. Placed low (often below 10 m) in tree or on artificial structure in treeless areas

EGGS: 3 – 6, usually 4; pale green, spotted and blotched brown; long oval; 39 x 27 mm

INCUBATION: 17 – 19 days, by female; fledging period 29 – 31 days

DISTRIBUTION
Arid and semi-arid zones of Australia

STATUS
Endemic species. Locally common to abundant

S. J. S. Debus

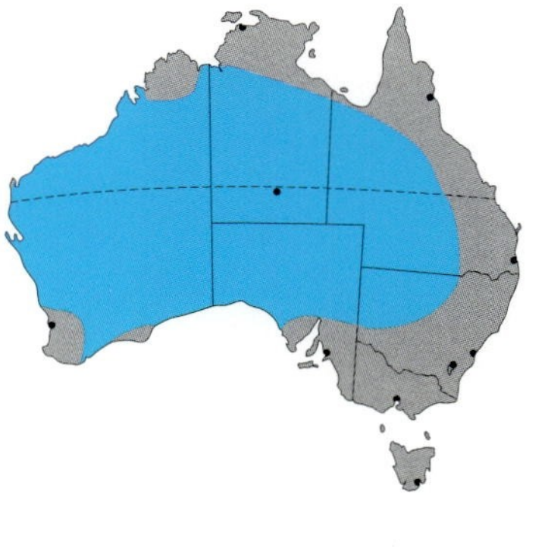

Torresian Crow

Corvus orru

The Torresian Crow is a tropical species, occurring in the northern two-thirds of Australia and most abundant in the better-watered areas where it is characteristic of cultivation paddocks. It is the common species in coastal towns and cities from Forster in New South Wales, through Brisbane and Darwin to Shark Bay in Western Australia. Its calls are usually a rapid, monotonous honking or quacking, but it also produces a harsher, nasal version of the Australian Raven's "strangled" call. It often shuffles its wings several times upon alighting on a perch.

In many aspects of its behaviour and ecology, the Torresian Crow resembles the Australian Raven. Breeding pairs are sedentary in large territories of a square kilometre or more, with neighbouring nests about a kilometre apart.

Territory is advertised by characteristic, clipped "uk-uk-uk" calls from a high perch. Sometimes a bird adopts an exaggerated posture with the neck stretched forward, bill tilted up and tail depressed, as it utters longer, slower notes. During aggressive encounters, individuals utter harsh, snarling calls with the final notes prolonged, often from an aerial wing-flicking display. Compared to the similar displays of the Australian Raven, the Torresian Crow has shorter wings and tail, and a longer pause on the downstroke. The short throat hackles, although erect, serve only to make the head look larger than normal. Torresian Crows travelling in groups, or resident birds traversing their territories, perform a dipping, "currawong flight" or cuckoo-like undulation, with a single-note "owk" call. In the inland, where it coexists with the similar Little Crow, the Torresian Crow gives a very distinctive, high-pitched yodelling caw.

Little is known of its courtship displays, but males and females have been seen preening each other's head and neck feathers. Breeding birds perform the "returning home" fluttering flight like the Australian Raven, but the descending call is more nasal (rather than tremulous) and finishes abruptly.

Both members of a pair build a rather frail, scantily lined nest of sticks, placed within the leafy canopy of a tall tree, perhaps because the birds require shade in a tropical climate. However, they also build on exposed artificial structures such as telegraph poles, and seem less particular about nest sites than the Australian Raven. The breeding season varies with latitude. In the temperate south and inland it breeds mainly in spring, while in tropical coastal areas the season extends through summer and even into autumn and winter. The usual clutch of five eggs is incubated by the female for about 20 days and she is fed on the nest by the male. Young fledge at about 41 days. Nests are often parasitised by the Channel-billed Cuckoo.

In eastern Australia, the Torresian Crow coexists with the Australian Raven. It is slightly smaller, with a less massive bill, and avoids competition to some extent by being more granivorous and less selective about nest sites, and by breeding at a slightly different time of year. Nevertheless, breeding pairs of each species defend territories against one another as well as against members of their own species. In northern New South Wales, interspecific conflict with the Forest Raven also occurs.

S. J. S. Debus

Torresian Crow

Corvus orru

Adults feeding on dead Brolga; East Kimberley, W.A.; June 1992 (J. Lochman/Lochman Transparencies)

OTHER NAMES
Crow, Australian Crow

Measurements
LENGTH: *480 – 540 mm*
WING: *315 – 382 mm*
TAIL: *166 – 216 mm*
BILL: *50 – 60 mm*
TARSUS: *57 – 67 mm*
WEIGHT: *430 – 670 g*
Female slightly smaller than male

DESCRIPTION
ADULT
Entirely glossy black, with short throat hackles, white feather bases on head and neck. Bill and legs black. Eyes white with blue inner ring

JUVENAL
Duller than adult; feather bases may be washed grey. Fledglings have pink gape and blue-grey eyes which turn brown during first year, then hazel

IDENTIFICATION
Large black bird with staccato, nasal cawing voice. Distinguished from other crows by voice and behaviour. Compact and blunt-winged in flight, with short, broad tail; glides lazily on drooped wings. Often shuffles wings exaggeratedly on alighting

VARIATION
Slight variation within Australian population, C. o. cecilae. *Western birds smaller, but with larger bill, than eastern birds. Other subspecies in New Guinea and nearby islands*

VOICE
Tenor, staccato nasal honking: rapid "uk-uk-uk-uk" of more than three notes, sometimes finishing with longer notes. Also longer caws and a harsh, snarling, nasal "arr...arr...arr" finishing with prolonged notes. Western birds also utter a high-pitched, yodelling caw. Long calls given from high perch, in horizontal posture with neck and bill stretched forward and short hackles erected. Also a soft rhythmic warble or rattle, and a long creaky call

HABITAT
Most terrestrial types, especially agricultural areas. In arid zone, restricted to areas with tall trees

FOOD
Grain, fruit, invertebrates, small vertebrates, eggs, refuse, carrion. Somewhat more granivorous than other species

HABITS
Breeding pairs solitary, sedentary and permanently territorial. Immatures wander in nomadic flocks which may be large in coastal areas. Arboreal and terrestrial, gleaning prey from ground and trees. Searches aerially for carrion by patrolling roads. Perches on heads of standing crops. Travelling birds perform an undulating, cuckoo-like display flight

BREEDING
Eggs laid July – December in south, December – August in north; one brood per season. Both sexes share nest building and feeding of young

NEST: A fine bowl of sticks, scantily lined with grass, placed more than 10 m above ground in canopy of tall tree or on tall artificial structure

EGGS: 5; pale green, sparsely or heavily spotted and blotched brown; long oval; 44 x 30 mm

INCUBATION: 19 – 20 days, by female; fledging period 38 – 42 days

DISTRIBUTION
Northern two-thirds of Australia in suitable habitat. Also New Guinea, Moluccas, New Britain, Tanimbar

STATUS
Endemic to Australasian region. Abundant in tropics and on coast, patchy and local in inland Australia

S. J. S. Debus

House Crow

Corvus splendens

The House Crow, common in urban areas from Pakistan to Burma, occasionally makes its way to Australia. It is black, with a broad grey-brown collar extending onto the breast, and brown eyes. It has successfully colonised other countries around the Indian Ocean, and has appeared in Australian ports after travelling aboard ships from south-east Asia. More than 50 birds, all destroyed, have been recorded in Western Australia during the twentieth century, and there have been at least three records (two involving three birds each) from Victoria, the latest in 1991. It is capable of dispersing widely from ports, and has bred once in Western Australia.

It is very social (more so than the Little Raven or Little Crow of Australia) and may associate in large flocks, even during the breeding season. Traditional roosts, used year after year, are made in banyan trees, coconut or forestry plantations and palm groves. Large numbers of birds converge on a roost at sunset, making considerable noise which may be added to by mynas and parakeets that share the roost. The House Crow is monogamous and pairs usually mate for life. In its native and introduced range it is a useful scavenger but also an agricultural pest and a predator of nestling birds and eggs. Its establishment in Australia would therefore be likely to have an adverse economic and ecological impact. Sightings should be reported immediately to agricultural authorities who are keen to prevent its establishment in Australia.

S. J. S. Debus

HOUSE CROW

Corvus splendens

Sri Lanka; October 1980 (C. Andrew Henley/Larus)

OTHER NAMES
Indian Crow, Ceylon Crow, Colombo Crow

MEASUREMENTS
LENGTH: *420 – 440 mm*
WING: *266 – 294 mm*
TAIL: *162 – 175 mm*
BILL: *51 – 56 mm*
TARSUS: *45 – 51 mm*
WEIGHT: *310 – 362 g*
Female slightly smaller than male

DESCRIPTION

ADULT
Glossy black with grey-brown collar, often restricted to nape but variably extending to breast and upper back. Bill and legs black. Eyes brown

JUVENAL
Similar to adult but less glossy; inside of mouth pink for some months

IDENTIFICATION
Large black bird with pale collar, harsh cawing voice. Distinguished from other crows in Australia by grey-brown on neck

VARIATION
Subspecies vary in contrast between grey and black areas

VOICE
High-pitched rasping "ka" or "kow"; low-pitched "kowk"

HABITAT
In Australia, environs of port cities. Elsewhere, urban and agricultural areas

FOOD
Omnivorous

HABITS
In Australia, solitary birds leave incoming ships and frequent port or urban environs and disperse widely. Elsewhere, a gregarious urban commensal of man and a serious agricultural pest. Bold around human habitation, where it enters buildings to seize food. Forms huge communal roosts and breeding colonies. Breeding adults sedentary, with several pairs nesting in one tree. Dispersing birds "hitch-hike" aboard ships. Such birds, and others introduced by human agency, have founded colonies around the Indian Ocean in Africa, the Middle East and non-native parts of south-east Asia

BREEDING
Has bred in Australia once; could become established unless prevented. Elsewhere, breeds almost throughout the year, mainly April to June in India

NEST: Bowl or platform of sticks or occasionally wire, lined with fibres, in a tree

EGGS: 3 – 5, sometimes 6; blue-grey, spotted and blotched brown and grey; 36 x 27 mm

DISTRIBUTION
Native to Asia. Deliberately introduced or ship-assisted birds have established feral colonies in coastal East and South Africa, around the Suez and Persian Gulfs, and in south-east Asia. In Australia, regularly arrives at Perth on ships from Asia and has occurred around Melbourne

STATUS
Abundant, successful commensal of man in other countries. Occasional ship-assisted immigrant to Australia, where its establishment has so far been prevented

S. J. S. Debus

Adult male Goldfinch at nest; Christchurch, New Zealand; November 1984 (D. Hadden)

Sponsor: Dr Norman Wettenhall

Eggs

Buntings

Sparrows

Weavers

Grass-finches

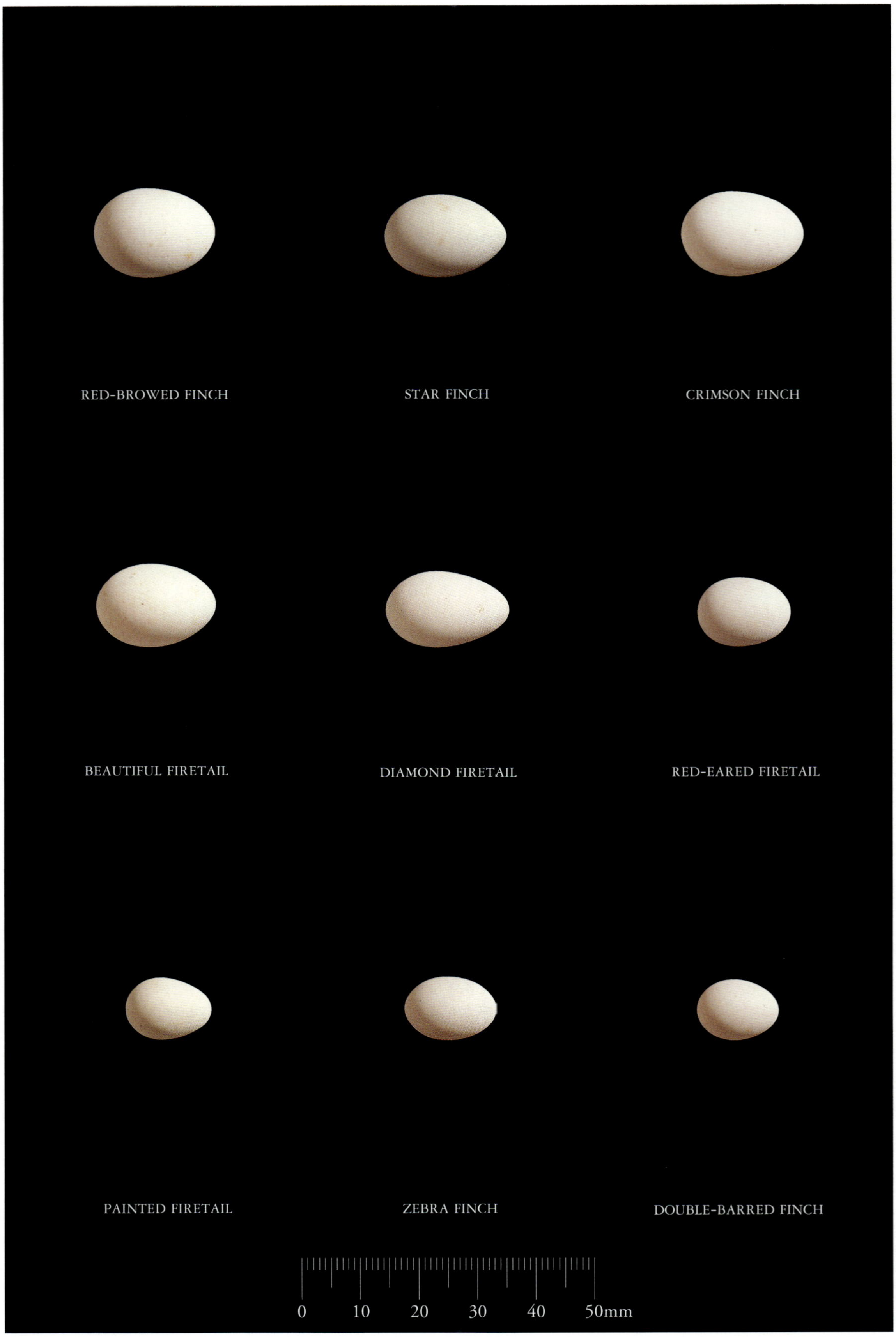

GRASS-FINCHES

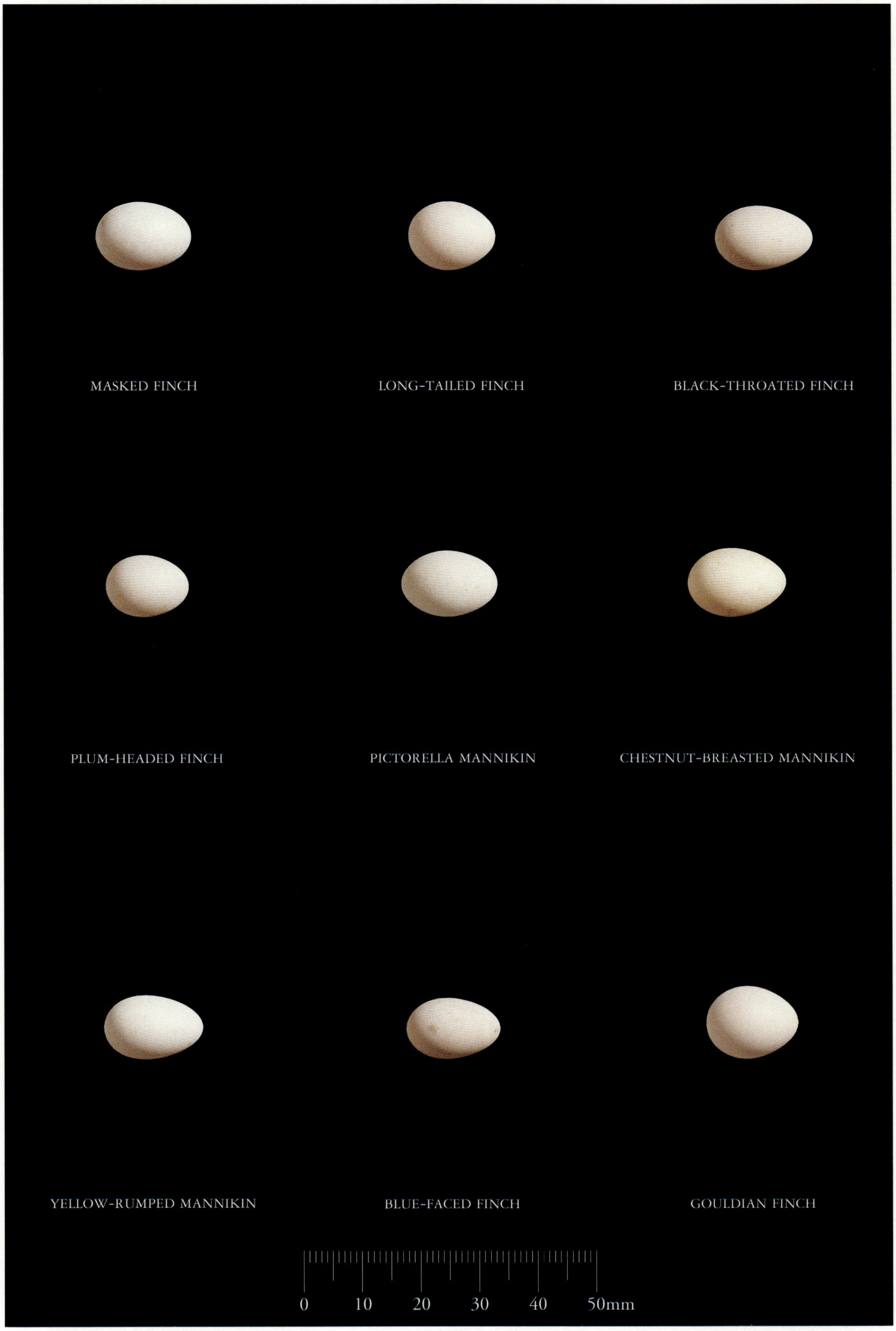

Finches

Orioles and Figbirds

Drongos

Starlings and Mynas

Magpie-larks

Birds of Paradise

Mud-nest Builders

Woodswallows

WHITE-BREASTED WOODSWALLOW

MASKED WOODSWALLOW

WHITE-BROWED WOODSWALLOW

BLACK-FACED WOODSWALLOW

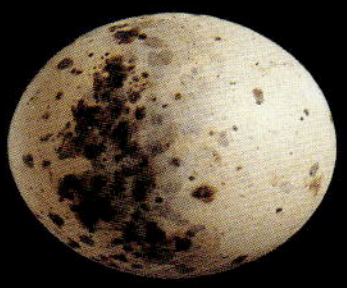

DUSKY WOODSWALLOW

LITTLE WOODSWALLOW

BOWERBIRDS

GOLDEN BOWERBIRD
SATIN BOWERBIRD
REGENT BOWERBIRD
SPOTTED BOWERBIRD
GREAT BOWERBIRD
FAWN-BREASTED BOWERBIRD
TOOTH-BILLED BOWERBIRD
SPOTTED CATBIRD
GREEN CATBIRD
0
10
20
30
40
50mm

MAGPIES, CURRAWONGS AND BUTCHERBIRDS

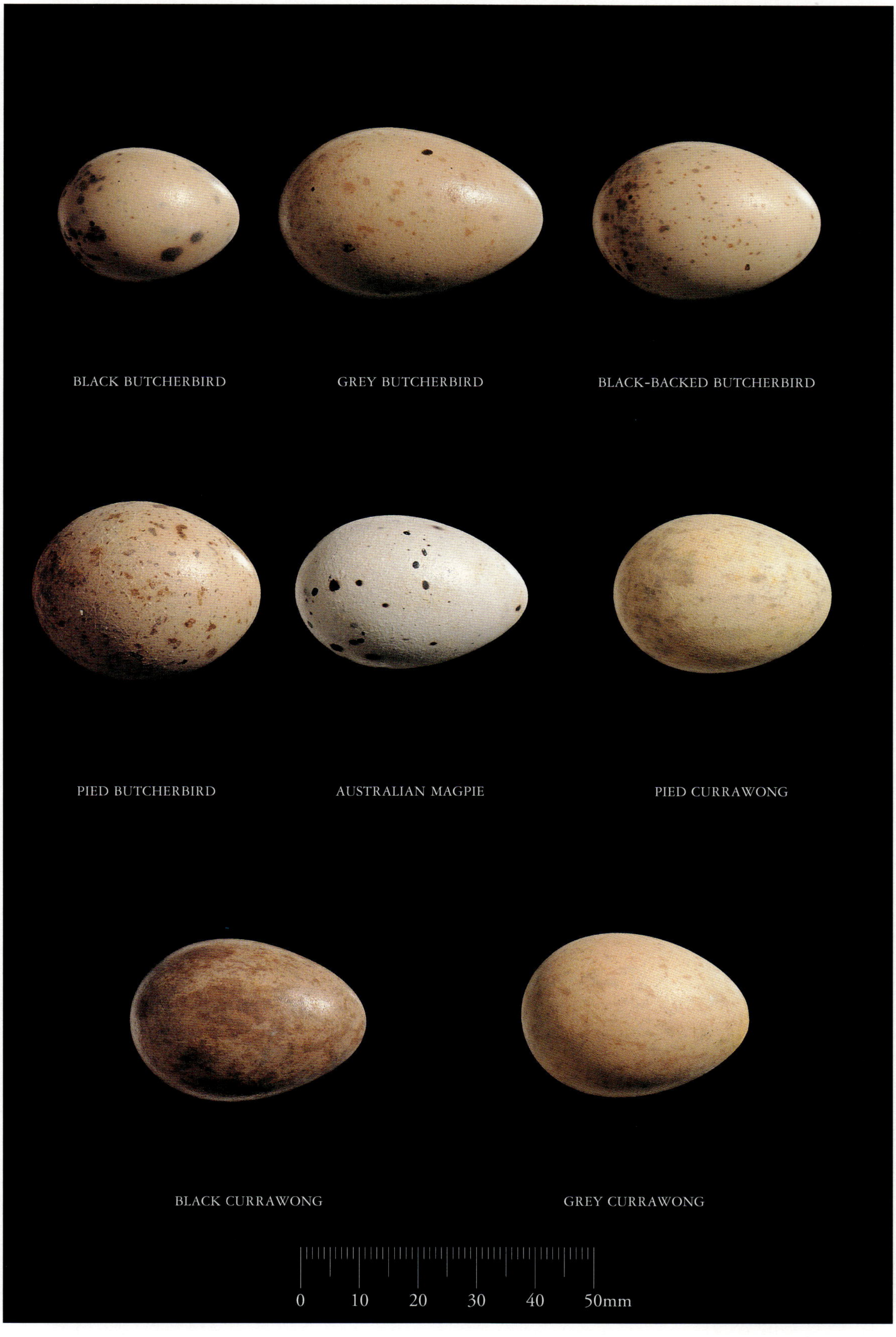

CROWS AND RAVENS

Adult female Redpoll at nest; Christchurch, New Zealand; February 1987 (D. Hadden)

BIBLIOGRAPHY

Ali, S. and Dillon Ripley, S. 1972, *Handbook of the Birds of India and Pakistan, Vol. 5*, Oxford University Press, Bombay.

Allison, B. 1993, From pet food to persimmons, *Bird Observer* 727: 10–14.

Amadon, D. 1943, The genera of starlings and their relationships, *Amer. Mus. Novit.* 1247: 1–16.

Amadon, D. 1951, Taxonomic notes on the Australian butcher-birds, *Amer. Mus. Novit.* 1504: 1–33.

Amadon, D. 1944, The genera of Corvidae and their relationships, *Amer. Mus. Novit.* 1251: 1–201.

Ames, P.L. 1975, The application of syringeal morphology to the classification of the Old World Insect Eaters (Muscicapidae), *Bonn. Zool. Beitr.* 26: 107–134.

Anon. 1889, *Aust. Mus. Memoirs, No. 2. Lord Howe Island, its Zoology, Geology, and Physical Characteristics*, Government Printer, Sydney

Ash, J. 1949, Roosting behaviour of chaffinches, *Brit. Birds* 42: 289.

Aston, H.I. 1988, Communal Breeding by the Australian Magpie-lark, *Emu* 88: 112–114.

Baldwin, M. 1971, Group Movements in the White-winged Chough, *Aust. Bird Watcher* 4: 69–77.

Baldwin, M. 1974, Studies of the Apostlebird at Inverell. Part 1: General Behaviour, *Sunbird* 5: 77–88.

Baldwin, M. 1975, Studies of the Apostlebird at Inverell. Part 2: Breeding behaviour, *Sunbird* 6: 1–7.

Baldwin, M. 1979, Nest of the Australian Magpie, *Aust. Birds* 14: 40–42.

Banfield, E.J. 1908, The Spangled Drongo-shrike, *Emu* 7: 178–181.

Bannerman, D.A. 1953, *The Birds of the British Isles*, Oliver & Boyd, London.

Barker, R.D. and Vestjens, W.J.M. 1990, *The Food of Australian Birds 2: Passerines*, CSIRO, Melbourne.

Barrett, C. 1920, Wood-swallows and Ants, *Emu* 19: 256.

Barrett, C.L. 1945, *Australian Bird Life*, Oxford University Press, Melbourne.

Bass, D.A. 1989, Seasonal changes in the behaviour and abundance of Pied Currawongs *Strepera graculina* and the consequences for seed dispersal, *Aust. Bird Watcher* 13: 78–80.

Bass, D.A. 1990, Pied Currawongs and seed dispersal, *Corella* 14: 24–27.

Beecher, W.J. 1978, Feeding adaptations and evolution in the starlings, *Bull. Chicago Acad. Sci.* 11: 269–298.

Beehler, B. 1989, The birds of paradise, *Scientific American* 261: 116–123.

Beland, P. 1977, Mimicry in Orioles of south-eastern Queensland, *Emu* 77: 215–218.

Bell, H.L. 1970, Distribution of the Blue-faced Parrot-finch, *Emu* 70: 32.

Bell, H.L. 1961, The Introduced Spice Finch in North-eastern Queensland, *Emu* 61: 94–96.

Bentz, G.D. 1979, The appendicular myology and phylogenetic relationships of the Ploceidae and Estrildidae (Aves: Passeriformes), *Bull. Carnegie Mus. Nat. Hist.* 15: 1–25.

Beruldsen, G.R. 1980, *A Field Guide to Nests and Eggs of Australian Birds,* Rigby, Adelaide.

Black, A. 1986, The taxonomic affinity of the New Guinean Magpie *Gymnorhina tibicen papuana, Emu* 86: 65–70.

Blakers, M., Davies, S.J.S.F. and Reilly, P.N. 1984, *The Atlas of Australian Birds*, Melbourne University Press, Melbourne.

Bock, W.J., and Morony, Jr., 1978, Relationships of the Passerine Finches (Passeriformes: Passeridae), *Bonn Zool. Beitr.* 29: 122–147.

Boehm, E.F. 1949, The Australian Raven in relation to other species of birds, *S. Austr. Orn.* 19: 26.

Boles, W.E. and Edden, R. 1986, *Birds of the Australian Rainforests,* Currawong Press, Sydney.

Boles, W.E. 1987, Bird in the hand: Grey Butcherbird *Cracticus torquatus, Corella* 11: 124–126.

Boles, W.E. 1987, Comments on the Subspecies of Native and Introduced Australian Finches, *Emu* 87: 20–24.

Borgia, G. 1985, Bower quality, number of decorations and mating success of male Satin Bowerbirds *(Ptilonorhynchus violaceus):* an experimental analysis, *Animal Behav.* 33: 266–271.

Bourke, P.A. 1957, Introduced birds, *Emu* 57: 263–264.

Breeden, S. and Cooper, W.T. 1992, *Visions of a Rainforest: A Year in Australia's Tropical Rainforest,* Simon and Schuster, Sydney.

Brooks-King, M. 1940, A note on the song of Chaffinches and Yellow Buntings, *Brit. Birds* 34: 159.

Brown, E.D., Farabaugh, S.M. and Veltman, C.J. 1988, Song sharing in a group-living songbird, the Australian Magpie *Gymnorhina tibicen*. Part I. Vocal sharing within and among social groups, *Behaviour* 104: 1–28.

Brown, E.D. and Farabaugh, S.M. 1991, Song sharing in a group-living songbird, the Australian Magpie *Gymnorhina tibicen*. Part III. Sex specificity and individual specificity of vocal parts in communal chorus and duet songs, *Behaviour* 118: 244–274.

Brown, E.D. and Veltman, C.J. 1987, Ethogram of the Australian Magpie, *Gymnorhina tibicen,* in comparison with other Cracticidae and Corvus species, *Ethology* 76: 309–333.

Buchanan, R. 1978, Pied Currawongs: their diet and weed dispersal in some urban bushland, *Sydney, Bull. of the Ecological Society of Aust.* 8: 6.

Burton, R. 1985, *Bird Behaviour,* Granada, London.

Burton, T.C. and Martin, A.A. 1976, Analysis of hybridization between Black-backed and White-backed Magpies in south-eastern Australia, *Emu* 76: 30–36.

Cade, T.J., Tobin C.A. and Gold, A. 1965, Water economy and metabolism of two estrildine finches, *Physiol. Zool.* 38: 9–33.

Cameron, A.D. and Harrison, C.J.O. 1978, *Bird Families of the World,* Elsevier Publishing, Oxford.

Campbell, A.J. 1900, *Nests and Eggs of Australian Birds,* Pawson & Brailsford, Sheffield.

Campbell, B. and Lack, E. 1985, *A Dictionary of Birds*, T. & A.D. Poyser, Staffordshire.

Carnaby, I.C. 1965, The White-browed Wood-swallow in Western Australia, *Emu* 65: 74.

Cayley, N.W. 1990, *What Bird Is That?* Angus & Robertson, Sydney.

Cayley, N.W. 1992, *Australian Finches in Bush and Aviary*, Angus & Robertson, Sydney.

Chaffer, N. 1949, Notes on two New Guinea bowerbirds, *Emu* 49: 19–25.

Charteris, Hon. G. 1948, The proportions of sexes in roosting Chaffinches, *Brit. Birds* 30: 117–119.

Chisholm, A.H. 1948, *Bird Wonders of Australia (3rd ed.)*, Angus & Robertson, Sydney.

Chisholm, A.H. 1929, *Birds and Green Places*, J.M. Dent & Sons, London.

Chisholm, A.H. 1971, Various Bird Notes, *Aust. Bird Watcher* 4: 41–43.

Chisholm, A.H. and Chaffer, N. 1956, Observations on the Golden Bower-bird, *Emu* 56: 1–39.

Christidis, L. and Boles, W.E. 1994, *The Taxonomy and Species of Birds of Australia and its Territories, The Royal Australasian Ornithologists Union Monograph 2*, R.A.O.U., Melbourne.

Christidis, L. 1987, Phylogeny and Systematics of Estrilildine Finches and Their Relationships to Other Seed-eating Passerines, *Emu* 87: 119–123.

Clancy, G.P. 1982, Some Records of the Olive-backed Oriole and Spangled Drongo Mimicking Other Birds Calls, *Aust. Birds* 16: 62–63.

Clancy, G.P. 1982, Chestnut-breasted Mannikin and Forest Kingfisher Food Records, *Corella* 16: 65.

Clarkson, K, and Zann, R. 1988, Extra-pair courtship and mate-guarding in wild zebra finches, *Taeniopygia guttata, Anim. Behav.* 26: 1853–1856.

Clayton, N.S., D. Hodson, and Zann, R.A. 1991, Geographic Variation in Zebra Finch Subspecies, *Emu* 91: 2–11.

Coates, B.J. 1990, *The Birds of Papua New Guinea, Volume 2, Passerines*, Dove, Brisbane.

Coleman, E. 1948, Swarming of the Dusky Wood-swallows, *Vic. Nat.* 64: 230.

Condon, H.T. 1949, The introduced Grenadier Weaver or Red Bishop-bird, *S. Austr. Ornithol.* 19: 1.

Cooper, C. and Cooper, R. 1981, Some observations on the food sources utilised by Pied Currawongs, *Aust. Birds* 15: 50–52.

Cooper, R.P. 1972 Daylight Clustering of the Dusky Woodswallow, *Aust. Bird Watcher* 4: 112–116.

Cooper, R.P. 1968, The Little Woodswallow, *Aust. Bird Watcher* 3: 77–81.

Cooper, W.T. and Forshaw, J.M. 1977, *The Birds of Paradise and Bower Birds*, Collins, Sydney.

Counsilman, J.J. 1974, Waking and Roosting of the Indian Myna, *Emu* 74: 135–148.

Counsilman, J.J. 1977, Visual displays of the Indian Myna during pairing and breeding, *Babbler* 1: 1–13.

Courtney, J. and Marchant, S. 1971, Breeding details of some common birds in south-eastern Australia, *Emu* 71: 121–133.

Craig, J.F.K. 1982, Mate attraction and breeding success in the Red Bishop, *Ostrich* 53: 246–248.

Craig, J.F.K. 1980, Behaviour and Evolution in the Genus Euplectes, *J. Ornith.* 121: 144–161.

Crouch, H.W. 1970, The Olive-backed Oriole in South Australia, *South Aust. Orn.* 25: 195–200.

Crouther, M.M. and Crouther, M.J. 1983, Observations on the Breeding of Figbirds and Common Koels, *Corella* 7: 89–92.

D'Ombrain, A.F. 1934, The White-browed Woodswallow, *Emu* 33: 292–297.

Day, D. 1981, *The Encyclopedia of Vanished Species*, McLaren Publishing, Hong Kong.

Diamond, J.M. 1982, Rediscovery of the Yellow-fronted gardener Bowerbird, *Science* 216: 431–434.

Diamond, J.M. 1986, Biology of birds of paradise and bowerbirds, *Annual Review of Ecology and Systematics* 17: 17–37.

Disney, H. J. de S. 1965, Desert Birds, *Austr. Nat. Hist.* 15: 124–127.

Disney, H.J. de S., et al 1974, *Bird in the Hand*, The Bird Banders', Association of Australia, Sydney.

Disney, H.J. de S. 1974, Nesting of Grey Butcherbirds, *Aust. Bird Bander* 12: 59–60.

Disney, H.J. de S. 1966, House and Tree Sparrows, Passer spp. (in) S.G. Lane (ed.) *Bird in the Hand*, Bird Banders Association of Australia, Sydney

Dobben, W.H. van 1949, Nest-building techniques of the Icterine Warbler and Chaffinch, *Ardea* 37: 89–97.

Donaghey, R.J. 1981, *The ecology and evolution of bowerbird mating systems*, PhD Thesis, Monash University, Melbourne.

Dow, D.D. 1980, Communally Breeding Australian Birds with an Analysis of Distribution and Environmental Factors, *Emu* 80: 121–140.

Dunn, A.M. 1991, Unusual Behaviour of a Female Zebra Finch: Possible Egg Dumping, *Corella* 15: 150–152.

Eales, L.A. 1985, Song learning in zebra finches: some effects of song model availability on what is learnt and when, *Anim. Behav.* 33: 1267–1268.

Edden, R. and Boles, W. 1986, *Birds of the Australian Rainforest*, Reed, Sydney.

Edwards, E.O. 1946, Mobbing Behaviour of Dusky Woodswallows, *Aust. Bird Watcher* 6: 86.

Farabaugh, S.M., Brown E.D. and Veltman, C.J. 1988, Song sharing in a group-living songbird, the Australian magpie *Gymnorhina tibicen.* Part II. Vocal sharing between territorial neighbours, within and between geographical regions, and between sexes, *Behaviour* 104: 105–125.

Farner, D.S. and Serventy, D.L. 1960, The timing of reproduction in birds in the arid regions of Australia, *Anat. Rec.* 137, 354.

Feare, C. 1984, *The Starling, Oxford University Press, Oxford.*

Felix, J. and Hisek, K. 1978, *The Illustrated Book of Birds*, Octopus, London.

Fisher, C.D., Lindgren E. and Dawson, W.R. 1972, Drinking patterns and behaviour of Australian desert birds in relation to their ecology and abundance, *Condor* 74: 111–136.

Fleming, A. 1974, Mobbing Behaviour of Dusky Woodswallows, *Aust. Bird Watcher* 6: 86.

Fletcher, J.A. 1922, Field notes on the Black Bell-Magpie *Strepera fuliginosa, Emu* 22:60–63.

Fletcher, J.A. 1918, The Black Bell-Magpie *(Strepera fuliginosa), Emu* 17: 227–228.

Floyd, R.B. and Woodland, D.J. 1981, Localization of soil dwelling scarab larvae by the Black-backed Magpie, *Gymnorhina tibicen* (Latham), *Anim. Behav.* 29: 510–517.

Ford, J. and Sedgwick, E.H. 1967, Bird distribution in the Nullarbor Plain and Great Victoria Desert region, Western Australia, *Emu* 67, 99–124.

Ford, J. 1981, Hybrid Phenotypes in Male Figbirds *Specotheres viridis* in Queensland, *Emu* 82: 126–130.

Ford, J. 1983, Taxonomic notes on some mangrove-inhabiting birds in Australasia, *Rec. W.A. Mus.* 10: 381–415.

Ford, J. 1975, Systematics and Hybridization of Figbirds, Spechotheres, *Emu* 75: 163–171.

Ford, J. 1978, Hybridization Between the White-vented and Black-vented Forms of the Black-faced Woodswallow, *Emu* 78: 105–114.

Ford, J. 1979, A new subspecies of Grey Butcherbird from the Kimberley, Western Australia, *Emu* 79: 191–194.

Forshaw, J. (ed.) 1975, *Birds; An Illustrated Survey of Bird Families of the World,* Landsdowne, Melbourne.

Forshaw, J. (ed.) 1991, *Encyclopedia of Birds,* Smithmark, New York.

Frith, C.B. 1989, A construction worker in the rainforest, *Birds International* 1: 29–39.

Frith, C.B. and Frith, D.W. 1990, Notes on the nesting biology of the Great Bowerbird *Chalmydera nuchalis* (Ptilonorhynchidae), *Australian Bird Watcher* 13: 137–148.

Frith, C.B. and Frith, D.W. 1993, Courtship display of the Tooth-billed Bowerbird *Scenopoeetes dentirostris* and its behavioural and systematic significance, *Emu* 93: 129–136.

Frith, H.J. 1969, *Birds in the Australian High Country,* A.H. & A.W. Reed, Sydney.

Frith, H.J. and Tilt, R.A. 1959, Breeding of the zebra finch in the Murrumbidgee irrigation area, New South Wales, *Emu* 59: 289–295.

Frith, H.J. (ed.) 1976, *Reader's Digest Complete Book of Australian Birds,* Reader's Digest, Sydney.

Gibson, J.D. 1954, Dust-bathing or anting? *Emu* 54: 279.

Gibson, J.D. 1961, Colombo Crows in Victoria, *Emu* 61; 244–245.

Gill, H.B. 1970, Birds of the Innisfail and Hinterland, *Emu* 70: 105–116.

Gilliard, E.T. 1969, *Birds of Paradise and Bower Birds,* Weidenfeld & Nicholson, London.

Gogerley, J.F.H. 1922, Change in Colour of Bill and Iris of the Oriole, *Emu* 21: 315.

Goodwin, D. 1982, *Estrildid Finches of the World,* Cornell, Ithaca.

Gotch, A.F. 1981, *Birds Their Latin Names Explained,* Blanford Press, Pool Dorset.

Gottsch, M.D. 1959, Notes on the White-winged Chough, *Aust. Bird Watcher* 1: 8–10.

Gould, J. 1865, *Handbook of the Birds of Australia,* Gould, London.

Greenway, J.C., Jr. 1962, Family Oriolidae, (in) Mayr, E. and J.C. Greenway, Jr. (eds.), *Checklist of Birds of the World* 15: 122–137.

Hall, B.P. (ed.) 1974, *Birds of the Harold Hall Australian Expeditions, 1962–1970,* British Museum (Natural History), London.

Hall, M.F. 1962, Evolutionary aspects of estrildid song, *Synops. Zool. Soc. Lond.* 8: 37–55.

Harrison, C.J.O. 1974, Song and Display of the Plum-headed Finch, *Emu* 74: 254–255.

Hastwell, K. 1991, Use of mud during preening by currawongs Strepera spp., *Aust. Bird Watcher* 14: 73.

Heinson, R.G., Cockburn A. and Cunningham, A. 1987, Foraging, Delayed Maturation, and Advantages of Cooperative Breeding in White-winged Choughs, *Corcorax melanorhamphos, Ethology* 77: 177–186.

Hermes, N. 1981, Mixed species flocks in a Dry Sclerophyll Forest in Autumn and Winter, *Corella* 5: 41–45.

Hinde, K. 1958, The courtship of the Greenfinch *Chloris chloris, Behav.* 7: 207–232.

Hinde, K. 1953, The conflict between drives in the courtship and copulation of the Chaffinch, *Behav.* 5: 1–31.

Hinde, K. 1952, The Behaviour of the Great Tit *(Parus major)* and other related species, *Behaviour,* Suppl. 2: 1–201.

Hinde, R.A. 1955, A comparative study of the courtship of certain finches (Fringillidae), *Ibis* 97: 706–745.

Hinde, R.A. 1954, The courtship and copulation of the greenfinch *(Chloris chloris), Behaviour* 6: 207–232.

Hindwood, K.A. and McGill, A.R. 1958, *The Birds of Sydney (County of Cumberland), New South Wales,* Roy. Zool. Soc. NSW, Sydney.

Hindwood, K.A. 1967, Notes on the Grey Butcher-bird, *Aust. Bird Watcher* 3: 40–42.

Hindwood, K.A. 1948, A communal roost of the Indian Myna, *Emu* 47: 315–317.

Hindwood, K.A., 1940. The Birds of Lord Howe Island. *Emu* 40: 1–75.

Hindwood, K.A. 1967, The White-winged Widow-bird, *Proc. Roy. Zool. Soc. N.S.W. 1966–1967*: 27–29.

Hobbs, J.N. 1980, Australian Magpie-lark Taking Bat, *Aust. Birds* 14: 56.

Hobbs, J.N. 1971, Use of Tools by the White-winged Chough, *Emu* 71: 84–85.

Hobbs, J.N. 1979, Black-faced Woodswallow Attempting to Feed on a Skink, *Aust. Bird Watcher* 2: 154.

Holland, L. 1965, Vocal Mimicry by the White-breasted Wood-swallow, *Aust. Bird Watcher* 2: 154.

Holland, L.M. 1962, Vocal Mimicry by the Southern Figbird, *Emu* 62: 205.

Hone, J. 1978, Introduction and Spread of the Common Myna in New South Wales, *Emu* 78: 227–230.

Hudson, W.H. 1951, *Adventures Among Birds,* J.M. Dent and Sons, London.

Hornby, H.E. 1967, The Breeding cycle of the White-winged Widow Bird *Euplectes albonotatus, Ostrich* 38: 5–10.

Hughes, J.M. 1982, An explanation for the asymmetrical "hybrid" zone between Black-backed and White-backed Magpies, *Emu* 82: 50–53.

Hughes, J.M., Pearce B.J. and Vockenson, K. 1983, Territories of the Australian Magpie *Gymnorhina tibicen* in south-east Queensland, *Emu* 83: 108–111.

Hutton, I. 1990, *Birds of Lord Howe Island Past and Present,* Ian Hutton, Coffs Harbour.

Immelmann, K. 1965, *Australian Finches,* Angus & Robertson, Sydney.

Ingraham, G. 1972, Notes on the Feeding of White-browed Wood-swallow *Artamus superciliosus, Sunbird* 3: 64–65.

Ingram, C.J. and Roberts G.J. 1983, Redpolls *Carduelis flammea* of the Lord Howe Island Group, *Sunbird* 13: 70–71.

Iredale, T. 1950, *Birds of Paradise and Bower Birds,* Georgian House, Melbourne.

Jackson, S.W. 1909, In the Barron River Valley, North Queensland, *Emu* 8: 233–283.

Jackson, S.W. 1919, Haunts of the Rufous Scrub-bird (*Atrichornis rufescens*, Ramsay) – Discovery of the female on the Macpherson Range, south-east Queensland, *Emu* 19: 258–272.

Jakob, R.M. 1977, Observations on a Captive Group of

Apostlebirds, *Avicultural Magazine* 83: 77–85.
Jasper, T. 1962, Nesting Notes on the Red-browed Finch, *Emu* 62: 177–180.
Jerrard, C.H.H. 1931, Vocal Powers of the Yellow Oriole, *Emu* 31: 42–43.
Jollie, M. 1978, Phylogeny of the species of Corvus, *The Biologist* 60: 73–108.
Jones, L, Darrock, R.F., Gilding J. and Bennett, D.H. 1980, A review of seasonal and ecological factors in Australian Magpie *Gymnorhina tibicen* attacks on people, *Bird Behav.* 2: 113–116.
Joseph, L. and Drummond, R. 1982, Food item of the Black Butcherbird, *Sunbird* 12: 49–50.
Kear, J. 1962, Food selection in finches with special reference to interspecific differences, *Proc. Zool. Soc. Lond.* 138: 163–204.
Keast, A. 1958, Infraspecific Variation in Australian Finches, *Emu* 58: 219–246.
Keast, J.A. 1958, Seasonal Movements and Geographic Variation in the Australian Woodswallows (Artamidae), *Collected Papers 1943–1961.*
Keast, J.A. 1945–1955, Variation in the Australian Oriolidae, *Collected Papers 1943–1961.*
Keast, J.A. (ed.) 1981, Ecological Biogeography of Australia, *Monogr. Biol.* 41. W. Junk, The Hague.
Kendeigh, S.C. 1952, Parental care and its evolution in birds, *Illinois Biol. Monogr.* 22: 1–356.
Kershaw, J.A 1918, Australian Green-backed Finch *Erythrura trichroa macgillivrayi, Emu* 18: 1.
King, B.F. and Dickinson, E.C. 1975, *A Field Guide to the Birds of South-east Asia,* Collins, London.
Lack, D. 1940, Variation in the introduced English Sparrow, *Condor* 42: 239–241.
Lack, D. 1941, Notes on territory, fighting and display in the Chaffinch, *Brit. Birds* 24: 216–219.
Larkins, D. 1979, Shading and sunning in the Grey Butcherbird, *Aust. Birds* 13: 43–46.
Larkins, D. 1980, Shading and evidence for other involuntary behaviour in Grey Butcherbirds, *Aust. Birds* 14: 72–75.
Larkins, D. 1982, Vocal behaviour in the Grey Butcherbird at Turramurra, New South Wales, *Aust. Birds* 16: 53–59; 17: 32.
Lenz, M. 1990, The Pied Currawong in urban Canberra: friend or foe? *Canberra Bird Notes* 15: 2–9.
Liddy, J. 1968, The Olive-backed Oriole: an occasional Disseminator of Mistletoe, *Corella* 6: 93.
Lloyd, J.L. 1981, *Introduced Birds of the World,* Reed, Sydney.
Long, J.L. 1981, *Introduced Birds of the World,* Reed, Wellington.
Longmore, N.W. and Scouller, R.M. 1989, A display of the Spangled Drongo *Dicrurus hottentottus, Corella* 13: 160–161.
Lord, E.A.R. 1956, Birds of Murphy's Creek District Southern Queensland, *Emu* 56: 100–128.
MacGillivray, W. 1910, Along the Great Barrier Reef, *Emu* 10: 216–233.
MacGillivray, W. 1914, Notes on some north Queensland birds, *Emu* 13: 132–186.
MacGillivray, W. 1917, Ornithologists in North Queensland, *Emu* 17: 72–73; 208–209.
Mack, D.B. 1966, Mimicry by Thornbills and Woodswallows, *South Aust. Orn.* 24: 112.
Macworth-Praed, C.W. and Grant, C.H.B. 1960, *African Handbook of Birds, Series One (2nd ed.),* Longman, London.
Marchant, S. 1992, A Bird Observatory at Moruya, NSW 1975–1984, *Occasional Publication No. 1,* Eurobodalla Natural History Society, Moruya.
Marler, P. 1956, Behaviour of the Chaffinch, *Behaviour, Suppl.* 1–184, R.J. Brill, Leiden.
Marshall, A.J. 1948, The Breeding and Distribution of *Erythrura trichroa* in Australia, *Emu* 47: 305–310.
Marshall, A.J. 1954, *Bower-birds. Their Displays and Breeding Cycles: a preliminary statement,* Clarendon Press, Oxford.
Matthews, G.M. 1925–1927, *The Birds of Australia Vol. 12,* Witherby, London.
Matthews, G.M. 1928, *The Birds of Norfolk and Lord Howe Islands and the Australian South Polar Quadrant with Additions to "The Birds of Australia",* Witherby, London.
McCulloch, E.M. 1970, Feeding of Dusky Wood-swallows, *Emu* 70: 200.
McEvey, A. 1976, Osteological notes on Grallinidae, Cracticidae and Artamidae, *Proc. XVI Int. Orn. Congr* 1794: 150–160.
McFarland, D.C. 1994, Notes on the Lord Howe Currawong *Strepera graculina crissalis, Aust. Bird Watcher* 15: 310-313.
McGill, A.R. 1949, Australian status of the Colombo Crow, *Emu* 49: 43–44, 83.
McGill, A.R. 1944, Hybridism in Woodswallows, *Emu* 44: 153–154.
McKean, J.L. 1963, Nocturnal Movement of the Dusky Wood-swallow, *Aust. Bird Watcher* 2: 60.
McLean, G.L. (ed.) 1969, *Proceedings of the Third Pan-African Ornithological Congress, The Ostrich, Supplement No. 8,* South African Ornithological Society.
Mitchell, I.G. 1966, Courtship patterns in some species of Collius passer, *Ostrich* 37: 47–53.
Nichols, O.G., Watkins D. and Kabay, E.D. 1982, The distribution of the Red-eared Firetail *Emblema occulata* in relation to Bauxite Mining in the Northern Jarrah Forest, *Emu* 82: 169–172.
Monk, J.F. 1954, The breeding biology of the Greenfinch, *Bird Study* 1: 2–14.
Morris, D. 1958, The comparative ethology of grassfinches (Erythrurae) and mannikins (Amandinae), *Proc. Zool. Soc. Lond.* 131: 389–439.
Morris, D. 1954, The reproductive behaviour of the Zebra Finch *(Poephila guttata)* with special reference to pseudo-female behaviour, *Behav.* 6: 271–322.
Mulvaney, M. 1986, Birds, berries and bad bushes, *Canberra Bird Notes* 11: 94–99.
Newton, I. 1964, The breeding biology of the chaffinch, *Bird Study* 11: 47–68.
Paynter, R.A., Jr. (ed.) 1968, *Checklist of Birds of the World* 14: 306–390, Harvard University Press, Cambridge.
Nichols, D.G. and Woinarski, J.C.Z. 1988, Longevity of Pied Currawongs at Timbertop, Victoria, *Corella* 12: 43–47.
Nielsen, L. 1965, Double clutch of Pied Butcher-bird, *Emu* 64: 113.
North, A.J. 1909, Nests and Eggs of Birds Found Breeding in Australia and Tasmania, *Vol. 2, Part 3, Aust. Mus. Sydney Special Catalogue, No. 1.*
North, A.J. 1901–1904, *Nests and Eggs of Birds Found Breeding in Australia and Tasmania, Vols 1–2,* Australian Museum, Sydney.
Noske, R.A. 1980, Diurnal clustering of White-breasted

Woodswallows, *Aust. Bird Watcher* 8: 202–203.
O'Grady, D. 1964, Southern Figbird as "usurper" and mimic, *Emu* 63: 338.
O'Grady, G.Y. and Lindsey, T. 1979, *Australian Birds and Their Young,* Cassell, Australia.
O'Neill, M.G. and Taylor, R.J. 1984, Co-operative hunting by Pied Currawongs *Strepera graculina, Corella* 8: 95–96.
Paters, E.S. 1937, The White-winged Chough, *South Aust. Orn.* 14: 67.
Paton, D.C. and Ford, H.A. 1977, Pollination by Birds of Native Plants in South Australia, *Corella* 1: 68–69.
Pearson, T.G. (ed.) 1936, *Birds of America,* Doubleday, United States.
Pellis, S.M. 1981, A description of social play by the Australian Magpie *Gymnorhina tibicen* based on Eshkol-Wachman notation, *Bird Behaviour* 3: 61–79.
Pellis, S.M. 1981, Exploration and play in the behavioural development of the Australian magpie *Gymnorhina tibicen, Bird Behaviour* 3: 37–50.
Peters, J.L. 1931–1979, *Checklist of the Birds of the World, Vol 15,* Museum of Comparative Zoology, Cambridge.
Pizzey, G. 1980, *A Field Guide to the Birds of Australia,* Collins, Sydney.
Pratt, E.K. 1968, Notes on the Southern Figbird, *Aust. Bird Bander* 6: 56–57.
Prawiradilaga, D.M. 1994, Caching behaviour of breeding Pied Currawongs *Strepera graculina, Aust. Bird Watcher* 15: 275-276.
Pruett-Jones, M.A. and S.G. 1985, Food caching in the tropical frugivore, Macgregor's Bowerbird *(Amblyornis macgregoriae), Auk* 102: 334–341.
Ramsay, E.P. 1863, Notes on the Birds Breeding in the Neighbourhood of Sydney, New South Wales, *Ibis* 5: 177–180.
Ramsay, J. 1919, Notes on birds observed in the upper Clarence River district, New South Wales, September-December, 1918, *Emu* 19: 2–9.
Readshaw, J.L. 1968, The distribution, abundance, and seasonal movements of the Pied Currawong, *Strepera graculina* (Shaw), an important predator of Phasmatidae in eastern Australia, *Aust. J. Zool.* 16: 37–47.
Recher, H.F. 1976, Reproductive behaviour of a pair of Pied Currawongs, *Emu* 76: 224–226.
Recher, H.F. and Schultz, M. 1983, Observations on the breeding of White-browed Woodswallows, *Corella* 7: 1–6.
Restall, R.L. 1975, *Finches and Other Seed-eating Birds,* Faber & Faber, London.
Ripley, S.D. 1964, A systematic and ecological study of birds of New Guinea, *Bull. Peabody Mus. Nat. Hist.* 19: 1–85.
Rix, C.E. 1970, Birds of the Northern Territory, *South Aust. Orn.* 25: 147–190.
Roberts, N.L. 1963, Some aspects of post-fledging behaviour in the Black-backed magpie, *Emu* 63: 234–241.
Roberts, N.L. 1976, Reproductive behaviour of a pair of Pied Currawongs, *Emu* 76: 224–226.
Robinson, A. 1956, The annual reproductory cycle of the Magpie, *Gymnorhina dorsalis* Campbell, in South-western Australia, *Emu* 56: 233–336.
Rodgers, K., Rodgers, A. and Rodgers, D. 1986, *Banders Aid: A Guide to Aging and Sexing Bush Birds,* A. Rodgers, Victoria.
Rowley, I. 1965, White-winged Choughs, *Aust. Nat. Hist.* 15: 81–85.
Rowley, I. 1975, *Bird life*, Collins, Sydney.
Rowley, I. 1978, Apostlebirds gathering material for their mud-nest, *Aust. Bird Watcher* 7: 166–169.
Rowley, I. 1978, Communal Activities Among White-winged Choughs, *Corcorax melanorhamphos, Ibis* 120: 178–197.
Rowley, I. 1970, The genus Corvus (Aves: Corvidae) in Australia, *CSIRO Wildl. Res.* 15: 27–71.
Sage, B.L. 1957, Remarks on the taxonomy, history and distribution of the House Sparrow introduced into Australia, *Emu* 57: 349–352.
Salim, A. and Ripley, S.D. 1974, *Handbook of the Birds of India and Pakistan, Vol. 10, Flowerpeckers to Buntings,* Oxford Univ. Press, Bombay.
Sanderson, K. and Crouch, H. 1993, Vocal repertoire of the Australian magpie *Gymnorhina tibicen* in South Australia, *Aust. Bird Watcher* 15: 162-163.
Schodde, R. and Tidemann, S.C. (eds) 1988, *Reader's Digest Complete Book of Australian Birds,* Reader's Digest Services, Sydney.
Reader's Digest, 1989, *Reader's Digest Complete Book of Australian Birds (2nd ed.)*, Reader's Digest Services, Sydney.
Schodde, R., Fullager, P. and Hermes, N. 1983, A review of Norfolk Island birds: past and present, *Aust. Nat. Parks and Wildl. Serv. Spec. Publ.* 8: I-VIII.
Sedgwick, E.H. 1946, Notes on the Great Bowerbird in the Northern Territory, *Emu* 25: 300.
Seebohm, H. 1885, *History of British Birds,* Porter, London.
Serventy, D.L. and Whittell, H.M. 1976, *Birds of Western Australia, (5th ed.),* University of Western Australia Press, Perth.
Sharland, M. 1964, *A Territory of Birds,* Rigby, Adelaide.
Sibley, C.G. 1976, Protein evidence of relationships of some Australian passerine birds, *Proc. XVI Int. Orn. Congr.* 1974: 557–570.
Sibley, C.G. and Ahlquist, J.E. 1985, The phylogeny and classification of the Australo-Papuan passerines, *Emu* 85: 1–14.
Sibson, R.B. 1983, Winter Flocking of Chaffinches in Northern New Zealand, *Notornis* 30: 73–80.
Sides, J.A.L. 1971, Apostle-bird assists peewees with nesting, *Sunbird* 2: 80–81.
Simpson, K. and Day, N. 1984, *The Birds of Australia,* Lloyd O'Neil, South Yarra.
Skead, C.J. 1956, A study of the Red Bishop, *Ostrich* 27: 112–126.
Slater, P., Slater, P. and Slater, R. 1986, *The Slater Field Guide to Australian Birds,* Landsdowne-Rigby, Sydney.
Stokes, T. and Hermes, N. 1979, Cluster Roosting in the Black-faced Woodswallow, *Emu* 79: 84–86.
Storr, G.M. 1986, Birds of the Southeastern interior of Western Australia, *Records of the Western Australian Museum Supplement No. 26,* Western Australian Museum, Perth.
Storr, G.M. 1984, The Distribution of estrildine Finches in Queensland, *Sunbird* 14: 1–6.
Storr, G.M. 1984, Birds of the Pilbara Region, Western Australia, *Records of the Western Australian Museum Supplement No. 16,* Western Australian Museum, Perth.
Storr, G.M. 1985, Birds of the Mid-eastern interior of Western Australia, *Records of the Western Australian*

Museum Supplement No. 22, Western Australian Museum, Perth.

Storr, G.M. and Johnstone, R.E. 1979, *Field Guide to The Birds of Western Australia,* Western Australian Museum, Perth.

Strong, M. and Cuffe, E. 1985, Drongos Pursue Microchiropters, *Sunbird* 15: 42–43.

Tarr, J.N.S. 1981, An apparent nesting by a hybrid Woodswallow, *Aust. Birds* 15: 76.

Taylor, I.M. 1988, Nest height and tree selection: a comparison of the Australian Magpie-lark and Australian Magpie, *Canberra Bird Notes* 13: 36–38.

Taylor, M. and Canberra Ornithologists Group 1992, *Birds of the Australian Capital Territory: An Atlas,* Canberra Ornithologists Group and National Capital Planning Authority, Canberra.

Thomas, D.F. 1935, *Birds of Cape York Peninsula,* Government Printer, Melbourne.

Thomas, H. 1951, Notes on the Pied Butcher-bird, *Emu* 51: 165–168.

Tideman, S.C., et al 1992, Predation on Gouldian Finch *Erythrura gouldiae* by reptiles, *Corella* 16: 105–106.

Van Der Liet, J. and Grindle, J. 1978, *Bird Families of the World,* Elsevier, Oxford.

Van Dobben, W.H. 1949, Nest-building techniques of the Icterine Warbler and Chaffinch, *Ardea* 37: 89–97.

Vestjens, W.J.M. and Carrick, R. 1974, Food of the Black-backed Magpie, *Gymnorhina t. tibicen,* at Canberra, *Aust. Wildl. Res.* 1: 71–83.

Vincent D. 1975, Clustering of the Dusky Woodswallow, *South Aust. Orn.* 27: 22–23.

Wheeler, R. 1947, Wood-swallows and north winds, *Emu* 47: 55.

White, H.L. 1922, A collecting trip to Cape York Peninsula, *Emu* 22: 99–116.

Wilson, H. 1946, The life history of the Western Magpie, *Emu* 45: 233–244; 271–286.

Wimbush, D.J. 1969, Studies on the Pied Currawong *Strepera graculina* in the Snowy Mountains, *Emu* 69: 72–80.

Witherby, H.F., et al 1940, *The Handbook of British Birds, Vol. 1*, Witherby, London.

Woinarski, J.C.Z. and Tidemann, S. 1992, Survivorship and some population parameters for the endangered Gouldian Finch *Erythrura gouldiae* and two other Finch species at two sites in tropical northern Australia, *Emu* 92: 33–38.

Woodall, P.F. 1984, Communal Breeding in the Figbird, *Sunbird* 14: 40–41.

Woodall, P.F. 1990, Double-barred Finch on North Stradbroke Island, *Sunbird* 20: 1617.

Woodall, P.F. 1987, Sparrows and Finches (Family: Fringillidae, Passeridae and Ploceidae) in the Brisbane region, 1972–1983. Relative numbers, distribution and species interaction, *Sunbird* 17: 37–51.

Yates, J.N. 1982, Unusual Bathing Behaviour in the Figbird, *Sunbird* 12: 55.

Zann, R. 1976, Distribution, status and breeding of Black-throated *Poephila cincta* in northern Queensland, *Emu* 76: 201–206.

Zann, R. 1977, Pair-bond and bonding behaviour in three species of Grassfinches of the genus *Poephila* (Gould), *Emu* 77: 97–106.

Adult male Star Finch; Pilbara, W.A.; 1980 (M. & I. Morcombe)

INDEX

Acridotheres tristis 118
Aidemosyne modesta 82
Ailuroedus crassirostris 185
Ailuroedus melanotis 180
Aplonis fusca 112
Aplonis metallica 110
Apostlebird 217
Artamidae 221
Artamus cinereus 230
Artamus cyanopterus 232
Artamus leucorhynchus 223
Artamus minor 234
Artamus personatus 225
Artamus superciliosus 227

Birds of Paradise 189
Bishop, Red 36
Bowerbird, Fawn-breasted 171
Bowerbird, Golden 142
Bowerbird, Great 165
Bowerbird, Regent 151
Bowerbird, Satin 146
Bowerbird, Spotted 157
Bowerbird, Western 161
Bowerbird, Tooth-billed 175
Bowerbirds 138
Buntings 1
Butcherbird, Black 238
Butcherbird, Black-backed 244
Butcherbird, Grey 241
Butcherbird, Pied 246
Butcherbirds 236

Carduelis carduelis 10
Carduelis chloris 13
Carduelis flammea 19
Catbird, Green 185
Catbird, Spotted 180
Chaffinch 16
Chlamydera cerviniventris 171
Chlamydera guttata 161
Chlamydera maculata 157
Chlamydera nuchalis 165
Chough, White-winged 213
Corcoracidae 211
Corcorax melanorhamphos 213
Corvidae 263
Corvus bennetti 275
Corvus coronoides 266
Corvus mellori 272
Corvus orru 278
Corvus splendens 280
Corvus tasmanicus 269
Cracticidae 236
Cracticus mentalis 244
Cracticus nigrogularis 246
Cracticus quoyi 238
Cracticus torquatus 241
Crow, House 280
Crow, Little 275
Crow, Torresian 278
Crows 263
Currawong, Black 258
Currawong, Grey 260
Currawong, Pied 254
Currawongs 236

Dicruridae 132
Dicrurus hottentotus 135
Drongo, Spangled 135
Drongos 132

Emberiza citrinella 3
Emberizidae 1
Emblema pictum 62
Erythrura gouldiae 103
Erythrura trichroa 100
Estrildidae 39
Euplectes albonotatus 33
Euplectes orix 36

Figbird 129
Figbirds 121
Finch, Black-throated 79
Finch, Blue-faced 100
Finch, Crimson 49
Finch, Double-barred 69
Finch, Gouldian 103
Finch, Long-tailed 76
Finch, Masked 73
Finch, Plum-headed 82
Finch, Red-browed 42
Finch, Star 45
Finch, Zebra 65
Finches 7
Firetail, Beautiful 52
Firetail, Diamond 55
Firetail, Painted 62
Firetail, Red-eared 59
Fringilla coelebs 16
Fringillidae 7

Goldfinch 10
Grallina cyanoleuca 208
Grallinidae 206
Grass-finches 39
Greenfinch 13
Gymnorhina tibicen 249

Heteromunia pectoralis 85

Lonchura castaneothorax 88
Lonchura flaviprymna 91
Lonchura malacca 94
Lonchura punctulata 97

Magpie, Australian 249
Magpie-lark, Australian 208
Magpie-larks 206
Magpies 236
Mannikin, Black-headed 94
Mannikin, Chestnut-breasted 88
Mannikin, Nutmeg 97
Mannikin, Pictorella 85
Mannikin, Yellow-rumped 91
Manucode, Trumpet 203
Manucodia keraudrenii 203
Mud-nest builders 211
Myna, Common 118
Mynas 107

Neochmia phaeton 49
Neochmia ruficauda 45
Neochmia temporalis 42

Oriole, Olive-backed 126
Oriole, Yellow 123
Orioles 121
Oriolidae 121
Oriolus flavocinctus 123
Oriolus sagittatus 126

Paradiseidae 189
Passer domesticus 24
Passeridae 22
Ploceidae 31
Poephila acuticauda 76
Poephila cincta 79
Poephila personata 73
Prionodura newtoniana 142
Ptilonorhynchidae 138
Ptilonorhynchus violaceus 146
Ptiloris magnificus 199
Ptiloris paradiseus 191
Ptiloris victoriae 195

Raven, Australian 266
Raven, Forest 269
Raven, Little 272
Ravens 263
Redpoll 19
Riflebird, Magnificent 199
Riflebird, Paradise 191
Riflebird, Victoria's 195

Sericulus chrysocephalus 151
Sparrow, House 24
Sparrow, Tree 27
Sparrows 22
Sphecotheres viridis 129
Stagonopleura bella 52
Stagonopleura guttatum 55
Stagonopleura oculatum 59
Starling, Common 115
Starling, Metallic 110
Starling, Tasman 112
Starlings 107
Strepera fuliginosa 258
Strepera graculina 254
Strepera versicolor 260
Struthidea cinerea 217
Sturnidae 107
Sturnus vulgaris 115

Taeniopygia bichenovii 69
Taeniopygia guttata 65

Weavers 31
Woodswallow, Black-faced 230
Woodswallow, Dusky 232
Woodswallow, Little 234
Woodswallow, Masked 225
Woodswallow, White-breasted 223
Woodswallow, White-browed 227
Woodswallow 221
Wydah, White-winged 33

Yellowhammer 3